Springer Undergraduate Mathematics Series

T0222678

For further volumes:
www.springer.com/series/3423

Nigel J. Cutland · Alet Roux

Derivative Pricing
in Discrete Time

 Springer

Nigel J. Cutland
Department of Mathematics
University of York
York, UK

Alet Roux
Department of Mathematics
University of York
York, UK

ISSN 1615-2085   Springer Undergraduate Mathematics Series
ISBN 978-1-4471-4407-6   ISBN 978-1-4471-4408-3 (eBook)
DOI 10.1007/978-1-4471-4408-3
Springer London Heidelberg New York Dordrecht

Library of Congress Control Number: 2012946458

Mathematics Subject Classification: 91G20

Printed on acid-free paper

Springer is part of Springer Science+Business Media (www.springer.com)

Preface

Mathematical finance—the mathematical theory of financial markets—not only underpins modern financial practice, but is a thriving area of mathematical research. This book provides an introduction to the way in which real world financial markets are modelled, and how derivatives can be priced in a rational way. Derivatives are, as the name suggests, financial entities whose value is derived from the value of other more concrete assets such as stocks and commodities. *Options* and *forwards* are two of the better known types of derivative.

We consider only *discrete time models*; that is, models where time proceeds in steps (minutes, hours, days, etc.). The main reason for doing so is to keep the mathematics as simple as possible, while explaining the basic principles, without the need for the sophisticated mathematical tools required by the more complicated *continuous time* theory. All of the basic ideas can be seen in even the simplest of discrete time models, and there is a precise sense in which the continuous theory is simply the limit of the discrete theory. For simplicity we also restrict attention to models with only finitely many scenarios (or price histories), thus avoiding the need for general measure and integration theory.

There are many texts that deal with discrete time mathematical finance. This one provides a treatment that aims to be more rigorous and comprehensive than most, while remaining elementary. For example, since *fair pricing* of derivatives is a fundamental issue, we provide a rigorous definition (often lacking or a little fuzzy) and complete proofs concerning the identification of fair prices (in many texts the fair price is not actually shown to be fair). We also cover comprehensively the fair pricing of derivatives (including American options) in incomplete markets, a topic that is often omitted in introductory courses.

The development proceeds slowly and carefully in the early chapters. The introductory chapter (Chapter 1) gives a little background on derivatives and may be skipped by the student wishing to get straight into the theory. It is surprising that many of the basic ideas can be explained in simple financial models with only two time steps—"today" and "tomorrow"—so we spend considerable time looking at such models (Chapters 2 and 3). Chapters 4 and 5 extend the theory to models with several time steps. The more advanced topics comprise (in Chapter 6) an introduction to the way in which a continuous model can be seen as the limit of discrete models and (in Chapter 7) a complete and rigorous treatment of American options in a discrete setting. Finally, in Chapter 8, we give a brief introduction to a number of topics that would form natural further study.

The mathematical prerequisites are no more than what should have been encountered in the first two years of an undergraduate mathematics programme: mainly finite dimensional linear algebra and elementary probability theory, combined with the ability to follow a line of mathematical reasoning—since we place strong emphasis on complete proofs. To that extent the treatment is more "pure" than "applied"—although the subject matter is by definition relevant to the real world. Related to this is the emphasis in the main on general theory rather than specifics, although plenty of worked examples and exercises are provided. Solutions to a selection of exercises are provided in the appendix. The mark † next to an exercise indicates that we have provided a partial solution (a hint or numerical answer without the working) and ‡ indicates an exercise (typically one that is less routine or more theoretical) for which we have provided a full solution. Instructors can obtain a complete set of solutions from the publisher.

A selection of the material (omitting Sections 3.1.4, 3.6, 5.6, 6.4 and 7.3 onwards) would be suitable for an undergraduate course. The material in the sections mentioned requires a little more mathematical maturity but is still elementary in the sense that it does not require any additional mathematical prerequisites. It would be suitable for inclusion in a rigorous course at MSc level where, in our experience, students are frequently graduates with a more limited mathematical background coming from disciplines such as economics or physics.

In real-world financial markets the buying and selling of financial instruments is often done by institutions such as companies, investment firms and governments. For convenience of language, we shall use the words "he" and "him" in this book in a non-gender-specific sense when referring to agents in financial models, with the understanding that these agents are as likely to be institutions as individuals.

The book grew out of the introductory module developed by the authors for the MSc in Mathematical Finance at the Universities of Hull and York, and a corresponding module at undergraduate level. It has been taught also by Kasia Grzesiak and Ekkehard Kopp, who thus contributed to its evolution, and to whom thanks are therefore due.

The website http://www.springer.com/978-1-4471-4407-6 may be consulted for the latest corrections to the book. It also explains how instructors may access the complete set of solutions to the exercises.

York, UK   Nigel J. Cutland
                                                                            Alet Roux

Contents

List of Figures and Models

1

Derivative Pricing and Hedging

This book is about the mathematical modelling of financial markets and the pricing of derivatives. This chapter provides a brief introduction to the idea of a derivative and the pricing problem. We start with a brief overview of the ingredients of a financial market.

A *financial market* is an environment in which a range of financial objects or assets (generally called *securities*) is traded. Among the better known types of security are *shares* and *bonds*.

A shareholder in the stock of a company (such as BP or Anglo American) owns a part of the company including its assets and earnings. The value of these is affected by economic conditions, management decisions and other factors, and so the value of a share will vary over time, and in particular cannot be determined in advance. This is described by saying that shares are *risky*; their value may rise or fall in the future. The term BP *share price* or BP *stock price* means the current price at which one share of BP stock may be traded on the stock exchange. The share price of a given stock is modelled mathematically by a random variable that changes over time.

Bonds are issued by companies, governments or banks as a way to borrow money. The owner of a bond is entitled to receive from the issuer a fixed sum, the principal, at the maturity date of the bond; in addition a bond will yield interest over its lifetime. The simplest example is a fixed income bond yielding a fixed rate of interest, but in practice yields (that is, interest rates) vary with time. Bonds (especially government bonds) tend to be less risky than stock,

N.J. Cutland, A. Roux, *Derivative Pricing in Discrete Time*,
Springer Undergraduate Mathematics Series,
DOI 10.1007/978-1-4471-4408-3_1, © Springer-Verlag London 2012

though they are not entirely risk free since a company or a government may *default* (i.e. go bankrupt).

In this book we consider only mathematical models of financial markets having risky assets (which we think of as stocks) and a risk free asset (which we think of as a fixed rate bond) for which the future price is deterministic, and given by a fixed interest rate, making it completely *risk-free*. In practice bond prices in the market are not entirely deterministic, but we make this simplification in order not to obscure the essential ideas. Thus, in the models we consider, investing funds in bonds is equivalent to depositing them in a bank or building society savings account with a fixed interest rate. To reflect this simplification, some textbooks refer to this as this investing in a money market account with a fixed rate of interest. The way the theory extends to bonds or bank accounts with variable yields is discussed briefly in Chapter 8.

A *derivative*, sometimes called a *derivative security*, is a financial asset whose value is derived from one or more other assets, called the *underlying securities*. This rather abstract definition is more easily understood by the example of a *European call option*, which is one of the simplest types of derivative that is traded in today's markets.

Example 1.1 (European call option)

A *European call option* on a stock such as BP is a financial asset, taking the form of a contract or binding agreement that entitles its owner to buy 1 share of BP stock at a fixed *strike price* K (written into the contract) at a fixed time T in the future (also written into the contract)—for example one month from now.

The owner or holder of the option is not *obliged* to buy a share for this price: he may opt not to exercise this right. The choice of taking up or *exercising* the option at time T will depend on the actual market price of BP stock at time T, which is of course unknown at the present time.

At time T the current price S_T of BP stock will be known; what should the owner of the option do?

– If $S_T > K$ then it makes sense to exercise the option, whether the option holder wants a BP share or not. Having bought an asset worth S_T that only cost K, selling it will yield a profit of $S_T - K$.

– If $S_T \leq K$, then it doesn't make sense to exercise the option. If the option holder wishes to own a BP share, then he can do just as well or better by buying one on the open market for S_T.

At the time that a European call option is bought or sold, the future price S_T is unknown so one cannot say which of the two outcomes above will occur.

However, it is clear that the option is something worth owning as it is a contract without obligations for the owner; at worst it is worth nothing one month from now, but could well have positive value at that time. It also is clear that the value of a European call option is derived from that of the actual underlying BP stock, so that it is in fact a *derivative*.

A precise mathematical description of the value of the option at time T is the random quantity

$$\max\{0, S_T - K\} = \begin{cases} S_T - K & \text{if } S_T > K, \\ 0 & \text{if } S_T \leq K \end{cases}$$

which is always nonnegative. A trader should therefore expect to have to pay *something* to acquire such an option; but how much?

This example illustrates the main question studied in this book, namely how much should one pay *now* for a derivative that offers an unknown but well defined payoff in the future. For the seller (or *writer*) of such a derivative the question is how much to sell it for *now*, when he does not know what the future holds.

The chief topic of this book is the answer provided by the remarkable mathematical theory of *derivative pricing*, which shows that in many situations there is a price for a derivative such as a European call option that guarantees that neither the buyer or the seller of a derivative can gain an unfair advantage (known as *arbitrage*), even though they do not know the future. Such a price is called a *fair* or *arbitrage-free* or *correct price*.

The theory developed here deals only with *discrete time*, which means that time is regarded as proceeding in steps (minutes, hours, days, etc.), with events taking place only at one of these time steps. The reason for doing so is again to keep the mathematics as simple as possible, while explaining the basic principles of derivative pricing, thus paving the way for understanding the more complicated *continuous time* theory.

Before developing the theory of derivative pricing in this simple setting, it is necessary to construct a mathematical model of a financial market in discrete time. We will use fixed yield bonds and stock as the underlying securities. Our *mathematical* story begins in Chapter 2, where the first task is to represent bonds and stocks *mathematically*.

The rest of this chapter gives a little more information about types of derivative, and how derivatives are commonly used in financial markets. *The mathematical development does not depend on the material in the remainder of this chapter, so it is not necessary to read or fully to understand it before proceeding to the theory that commences in Chapter 2. Some readers may, however,* be interested to get at least a feel for the real world of finance, and an idea of where the simple theory of this book eventually leads.

1.1 Types of Derivative

As we have seen, a derivative is a financial instrument or asset whose value depends on the value of some underlying assets. An underlying asset could be any financial security, such as a share in a company, or a corporate or government bond. Since many of the ideas in financial markets extend (in theory and in practice) to markets that include commodities (for example agricultural products such as wheat or coffee or minerals such as oil or iron ore), these can also be underlying assets for a derivative.

Mathematically a derivative is defined as *any* instrument or contract that gives to its owner at some time in the future a payoff that depends on some other underlying assets, and the theory developed in the following chapters takes this approach. In practice, however, there is a restricted range of derivatives that are actively traded, and here we mention some of the more common ones, and how they are traded.

The first distinction is in the way in which derivatives are traded. *Over-the-counter derivatives* are contracts that are privately negotiated between interested parties, without using an exchange or other intermediary. The parties to over-the-counter derivatives are often highly sophisticated agents such as banks or large financial institutions who need, or provide, tailor-made financial packages. On the other hand, *standardized derivatives* are increasingly being traded on exchanges around the world in large volumes.

1.1.1 Options

An option, such as a European call option described above, is a financial instrument that gives its holder the right, without any obligation, to make a specific trade with another market participant, the *option writer* or *seller*. The choice of exercising this right lies completely with the *option holder* (the *buyer*): if he decides to exercise the option, then the option writer (the seller) must execute his part of the agreed trade. Thus the positions of the writer (seller) and holder (buyer) of an option are asymmetrical: the holder has a right and the writer has an obligation. The writer of an option is said to have a *short position* in the option, while the holder of an option takes a *long position*.

The best-known and simplest types of option are the *call option* and the *put option*. A *call option* gives its owner the right to buy one unit of the underlying asset at a predetermined *strike price*. By contrast, the holder of a *put option* has the right to *sell* one unit of the underlying asset at the strike price. There are many other types of option, such as the *butterfly spread, straddle, strangle,* and *bull* and *bear spreads*. For each of the these the holder has the right (but not the obligation) to exercise the option and receive a payoff that depends in some way on the underlying assets—in some cases not only their value at the exercise time but the history of their performance to date.

Options can be classified according to when they may be exercised. A *European option* may only be exercised at its *expiry date*. By contrast, an *American option* may be exercised at any trading date (which may be chosen by the holder) from the time it was written, until its expiry date. Other varieties include *Bermudan options*, which allow exercise only at certain pre-specified dates, or *chooser options*, which allow the holder to choose at a fixed time before expiry between a call and a put option.

European and American put and call options, often referred to as *vanilla options*, have been in use for centuries, and are commonly traded on exchanges. Options with non-standard or unusual specifications (sometimes referred to as *exotic options*) and *compound options* (options whose value depends on other derivatives) are usually traded over-the-counter. New types of option are regularly introduced by forward-looking market participants; such options often have geographic names, for example, *Asian, Russian, Israeli* and *Parisian*. The modern approach to the standardization of derivatives is conservative and therefore new options are usually only available over-the-counter.

1.1.2 Forwards and Futures

A *forward contract* (or just a *forward*) is a binding agreement between two parties to buy or sell one unit of some underlying asset on a fixed date in the future, called the *delivery time*, for a price specified in advance, called the *forward price*. The party who agrees to buy the asset is said to be taking a *long position* in the forward contract, while the party who agrees to sell the asset is in a *short position* in the forward contract.

A position in a forward contract is a derivative because its actual value at the delivery time depends on the market price of the asset concerned at delivery time: this value is just the difference between the market price and the agreed forward price. Many everyday contracts are undertaken on a forward basis, often implicitly. For example, it is common practice when buying shares on an exchange to only pay for them once the contract note has been issued (usually a day or so later), by which time the current stock price may have moved.

The differences between a forward and an option are:

– A forward contract is made without cost to either party, and both parties have an obligation.

– An option is bought and sold at the time of the contract, and the seller has an obligation but not the buyer.

In the case of a forward, if the market price of the asset at delivery differs significantly from the pre-agreed forward price, it could happen that one of the parties is not able to fulfil his obligation under the contract. Because of this risk of default (called *counterparty risk*), forwards are traded over-the-counter, between banks and large corporations, and are not usually sold on to third parties.

By contrast, *futures* (or *futures contracts*) are designed to retain the essential properties of a forward, while eliminating counterparty risk. This is achieved by a process called "marking to market" at each day until the delivery date. This is explained in more detail in Chapter 8. The idea is that as with a forward, a futures contract made *today* (without exchange of cash) to buy or sell in the future at a fixed price will be worth something (positive or negative) *tomorrow*, so tomorrow there is an exchange of cash to bring its value back to zero. Due to their popularity, futures are now traded freely on exchanges, with delivery prices quoted alongside the prices of their underlying assets. Thus, for example, one talks in July of the "December oil price".

1.1.3 Swaps

A swap contract is the exchange between two parties of some future cashflows according to pre-agreed criteria that depend on the values of some underlying assets. A typical *interest rate swap* may involve two parties agreeing to pay each other cash flows that are equal to the interest earned on some notional principal, where one party pays according to a pre-agreed fixed rate, and the other at some floating (or current) interest rate: the effect of such a swap is to exchange a fixed rate loan and a floating rate loan. A *currency swap* involves the exchange of fixed interest payments on a loan in one currency for fixed interest payments on an equal loan in another currency. Mathematically one can consider a swap as a portfolio of forward contracts.

In this book, we restrict our attention mainly to derivatives with a single specified payoff date, such as European call and put options. In the penultimate chapter we see how the pricing theory for these needs to be modified to deal with the more sophisticated American options. Here the fact that the holder may choose to exercise the option at any time up to the expiry date makes the theory significantly more complicated.

1.2   Uses of Derivatives

Financial markets provide an environment in which to trade financial assets, and as such are essential to modern economies. They facilitate the transfer of capital in the form of primary financial resources, such as cash, bonds or stock, and the provision of credit in the form of loans and debentures.

Derivatives provide a vehicle for transferring the risk inherent in these primary financial securities—a procedure known as *hedging*. They are also used as sophisticated instruments by speculators who are prepared to take risks to make possible gains, based on their beliefs about future market behaviour. A further important use of derivatives is in providing *liquidity* when they are written on commodities such as property, precious metals or manufactured goods that are both valuable and durable but are not easily traded.

1.2.1   Hedging

The aim of *hedging* is to reduce a market agent's exposure to adverse price movements, essentially by balancing one risk against a similar risk. Since the future price movements of commodities or financial assets are often highly uncertain, it is natural for companies, entrepreneurs or institutional investors to try to safeguard themselves against possible losses by hedging in financial markets; that is, trying to ensure that a possible loss would be balanced by an equivalent possible gain.

Forwards and options provide a natural hedging tool. Consider for example the situation of an investor who wishes to sell some BP shares at some point in the near future. He faces the risk that the stock price may fall before he can sell his shares, due to decreased demand for BP stock, or due to other shareholders wishing to reduce their holdings. This risk may be managed by means of a forward contract to fix a price at which the share can be sold. Although this might not achieve the best price, it would be a way to avoid a very bad price, and moreover would provide some certainty for planning purposes if needed.

In the same situation, another way to manage the risk of a drop in the BP stock price would be to buy a put option on it with suitable exercise time and strike price. If the stock price at the time of exercise is above the strike price of the option, then the investor can sell his shares at the market price. If the stock price is below the strike price, then the investor will sell his shares at the higher strike price. This procedure guarantees a minimum stock price at the time of sale, but there is an outlay of funds at the outset to achieve this (the cost of purchasing the put option).

1.2.2 Speculation and Arbitrage

Whereas hedgers participate in financial markets with the purpose of reducing or balancing risk, *speculators* are market participants who consciously take on risk with the intention of profiting from their investments. Speculators and hedgers use the same market instruments and derivatives to achieve their respective aims. Speculators may trade with other speculators as well as with hedgers.

Suppose, for example, that a speculative trader suspects that the price of a specific stock will rise in the immediate future. Such a trader could purchase a call option on that stock, with a strike price below what he believes the future price to be. If the stock price is above the strike at the time of exercise, then the trader would exercise the option and sell the stock at the market price, thus making a profit (after recouping the initial cost of purchasing the call option). However, if the stock price at the time of exercise is below the strike price, then the trader would be left with nothing, except the loss from the initial purchase of the call option.

The same trader, if he believes that the price of a particular stock will drop in the future, would purchase a put option on that stock, which would give a profit if the stock dipped below the strike price.

Arbitrageurs are market participants who seek to exploit market deficiencies to create riskless profit, called *arbitrage*. They often do this by simultaneously holding positions in many assets to take advantage of possible mispricing between them. An *arbitrage opportunity* is a market condition where the pricing of certain assets allows a risk free profit to be made by a combination of trades involving these assets.

Of course, if a certain position in a market creates an arbitrage opportunity, then any rational investor would seek to acquire unlimited quantities of this position, demand would overwhelm supply, and market forces would act to remove this opportunity. It follows that, if arbitrage opportunities are found in real markets, they must necessarily yield very small profit margins and be available for a very short time before market prices adjust to eliminate them.

1.3 Pricing of Derivatives

The basic problem in the pricing of derivatives is to identify a *fair price* (or a range of fair prices) for any given derivative. This will be defined mathematically as any price for which neither party to a transaction can guarantee a risk-free profit; in other words, any price that prevents arbitrage. So arbitrage (or rather, lack of it) is one of the keys to the theory of pricing derivatives. Fundamental to the mathematical treatment that follows (and especially the definition of a fair price) is the observation above: in practice arbitrage opportunities are either absent or short-lived. In general, even in a simplified mathematical model, there will often be a range of fair prices, perhaps better described as prices that are not unfair. That is to say, there will be a range of prices where neither buyer nor seller can make a risk free profit, and so other factors come into play when determining the price at which to trade a derivative. This will be discussed further when we examine incomplete markets.

2

A Simple Market Model

In the simplest possible market model there are two assets (one stock and one bond), one time step and just two possible future scenarios. Many of the basic ideas of mathematical finance arise even in such a simple model. It also turns out that many more complex models can be viewed as a succession of one-period models such as this.

2.1 Model Assumptions

The aim in this chapter is to build a mathematical model of a simple financial market. To begin, we describe the mathematical objects involved and their properties. In order to keep the model tractable, it is necessary to make some simplifying assumptions. This will aid in understanding the most important basic ideas and principles of mathematical finance, while avoiding unnecessary mathematical difficulty.

The first assumption is that the model involves just two moments in time. It is often helpful to think of the first, earlier, time as "now" or "today", and of the second, later, time as "the future" or "tomorrow".

Assumption 2.1 (Two trading dates)

Trading takes place at time 0 and time 1.

N.J. Cutland, A. Roux, *Derivative Pricing in Discrete Time*,
Springer Undergraduate Mathematics Series,
DOI 10.1007/978-1-4471-4408-3_2, © Springer-Verlag London 2012

Two basic kinds of asset are traded in this model: one risk-free and one risky. These are called the *primary* or *underlying* assets. In later development there will be other types of asset or security, whose value is derived in some way from the two *primary* or *underlying* assets. These are the so-called *derivative securities*, or just *derivatives*, our main subject of study.

We normally think of the *risk-free asset* as a bond, or a unit of money in a bank or savings account, that is earning interest.

Assumption 2.2 (Risk-free asset)

The model contains a *bond* with initial price B_0. The interest rate $r \in \mathbb{R}$ is known in advance, and the value of the bond at time 1 is

$$B_1 = (1 + r)B_0.$$

Remark 2.1

The price movement of the bond is completely determined by any of the pairs (B_0, r), (B_0, B_1) and (r, B_1).

The next assumption encapsulates the idea that at the future time 1 there can be only two possible outcomes or *scenarios*.

Assumption 2.3 (Number of scenarios)

There are two possible future scenarios, denoted u and d; the set of possible scenarios is written Ω so we have $\Omega = \{u, d\}$. Scenario u occurs with probability $p \in (0, 1)$, and scenario d occurs with probability $1 - p \in (0, 1)$.

Any model with just two scenarios is called a *binary* model. The probability assignment $(p, 1 - p)$ is called the *real-world probability* or the *market probability*. It turns out that this probability plays no part in what follows, which is one of the reasons why the mathematical theory of derivative pricing was described as "remarkable" in Chapter 1.

We will think of the *risky asset* as a stock or share, although in principle it could be a unit of any commodity for which the future price is unknown.

Assumption 2.4 (Risky asset)

The model contains a single *stock* whose initial price S_0 is known at time 0. The price S_1 of the stock at time 1 is a *random variable* whose value depends on the scenario. (Recall that the *stock price* or *share price* of a given stock is the market price of one share of that stock.)

Thus, at time 0 we know the initial stock price S_0 and we know that at time 1 it will be *either* $S_1(\mathrm{u})$ *or* $S_1(\mathrm{d})$. Only at time 1 will we know which it actually is. The labels of the scenarios are chosen to suggest either an upwards (u) or downwards (d) movement from the initial price S_0, and for convenience we always assume that

$$S_1(\mathrm{d}) < S_1(\mathrm{u}). \tag{2.1}$$

The final assumption on the prices of the bond and stock for the moment is that they are strictly positive.

Assumption 2.5 (Positivity of prices)

We have $B_t > 0$ (equivalently $B_0 > 0$ and $r > -1$) and $S_t > 0$ for $t = 0, 1$.

It is helpful to represent the stock prices in this model as a tree (on its side), with the branches representing the possible future scenarios, such as in Figure 2.1.

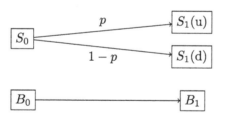

Figure 2.1 One-period single-stock binary model

A trader in this model may form a *portfolio* of assets at time 0 and hold this portfolio, unchanged, until time 1. This is formally stated as follows.

Assumption 2.6 (Divisibility, liquidity and short selling)

For any $x, y \in \mathbb{R}$ an investor may form a *portfolio* consisting of x bonds and y shares. Such a portfolio is denoted by the pair (x, y).

It is often useful to denote a portfolio by the symbol φ so we write $\varphi = (x, y)$. Assumption 2.6 deserves a few words of explanation, because it reflects some simplifications of real markets. The fact that an investor may hold a fraction of a share or bond, is referred to as *perfect divisibility* of the assets. In reality, almost perfect divisibility is achieved whenever the volume of transactions is large in comparison to the unit prices.

The fact that no bounds are imposed on the size of x and y is related to another attribute of an ideal market, namely *liquidity*. This means that assets can be bought or sold on demand, in arbitrary quantities, at the market price. In practice there are obvious restrictions on the volume of trading and the possible size of a portfolio; for example, there is only a finite amount of currency in circulation, and a finite number of shares available for purchase. However, the essence of the following mathematical development is not affected if the sizes of holdings x and y are restricted.

Investors in this model are allowed to hold negative quantities of the bond; this is equivalent to borrowing money from the bank at the declared interest rate. This money may be spent provided it is returned with interest at time 1. As with a loan from a bank, a negative holding of bonds represents an obligation to repay or return these.

In the same way, investors in this model are also permitted to hold negative quantities of shares, which is equivalent to borrowing shares from someone else, and thus having an obligation towards them. Borrowed shares may be sold for cash provided that these (or others identical to them) are returned to the original owner at time 1. This is the device known as *short selling*. In practice the lender makes a small charge for this service, which for simplicity is omitted from our simple model—see Assumption 2.7 below.

If an investor holds a positive amount of a given asset, he is said to be in a *long position* (or just *long*) in that asset; if the holding is negative, then he is said to be in a *short position* (or just *short*). So a short position in either bond or stock means that the investor has borrowed some of this asset. The act of repaying a money loan, returning a bond, or returning a borrowed share is called *settling* or *closing* the short position. Similarly a long position in an asset is *closed* by selling it for its current value.

The next section explains how we model the buying and selling of assets. The following simplifying assumption is made in order to expose the main ideas of the subsequent theory.

Assumption 2.7 (No friction)

Trading is instantaneous, borrowing and lending rates are the same, and there are no transaction costs in trading (equivalently, buying and selling prices are the same).

This is a significant simplifying assumption, because in practice the interest rate for borrowing will reflect not only the change in value of the loan but also a charge for the service provided. That is why the return on a typical savings account is less than the interest that would be charged on a loan by the same bank. Similarly, since there are costs involved in trading, in practice there are differences between buying and selling prices of shares.

Section 8.3 gives a brief introduction to models with proportional transaction costs. Sophisticated models that include other costs can be developed, but these are beyond the scope of this book.

2.2 Viability

In this section we introduce the idea of *arbitrage*, which is the key concept used to identify realistic market models and also to define the notion of a fair price for an asset. The underlying idea is that in a realistic market no-one can guarantee a riskless profit; that is, there is no possibility of *arbitrage*.

To formulate the definition of arbitrage, we must first define the *value* of a portfolio $\varphi = (x, y)$ consisting of x bonds and y shares. This is denoted by V_t^{φ} for $t = 0, 1$ and is defined by

$$V_t^{\varphi} := xB_t + yS_t. \tag{2.2}$$

The *gain* of a portfolio φ is

$$G^{\varphi} := V_1^{\varphi} - V_0^{\varphi},$$

and if $V_0^{\varphi} \neq 0$, then the *return* is

$$R^{\varphi} := \frac{G^{\varphi}}{V_0^{\varphi}} = \frac{1}{V_0^{\varphi}} \left(V_1^{\varphi} - V_0^{\varphi} \right). \tag{2.3}$$

Observe that each of the quantities V_1^{φ}, G^{φ} and R^{φ} is a random variable, and therefore has two possible values depending on the scenario at time 1.

Example 2.2

There are two special cases of the return, namely when the portfolio consists of only bonds (so that $y = 0$) or only shares (so that $x = 0$). It is easy to verify that

$$R^{(x,0)} = r$$

whenever $x \neq 0$, and we refer to this by saying that the return on the bond is equal to $R^B := r$.

The return on the stock is

$$R^S := \frac{S_1}{S_0} - 1$$

since

$$R^{(0,y)} = \frac{S_1 - S_0}{S_0} = \frac{S_1}{S_0} - 1$$

whenever $y \neq 0$.

Example 2.3

Consider Model 2.2. The portfolio $\varphi = (-6, 1)$ consisting of $x = -6$ bonds and $y = 1$ share has initial value

$$V_0^\varphi = -6 \times 10 + 1 \times 75 = 15.$$

The final value, gain and return of this portfolio are given by

$$V_1^\varphi(\mathrm{u}) = -6 \times 12 + 1 \times 120 = 48, \qquad V_1^\varphi(\mathrm{d}) = -6 \times 12 + 1 \times 72 = 0,$$
$$G^\varphi(\mathrm{u}) = 48 - 15 = 33, \qquad G^\varphi(\mathrm{d}) = 0 - 15 = -15,$$
$$R^\varphi(\mathrm{u}) = \frac{33}{15} = 220\%, \qquad R^\varphi(\mathrm{d}) = \frac{-15}{15} = -100\%.$$

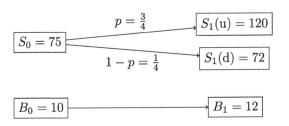

Model 2.2 One-period single-stock binary model in Example 2.3

Exercise 2.4

In Model 2.2, find the portfolio (x, y) with final value given by

$$V_1^{(x,y)}(\mathrm{u}) = 2148, \qquad V_1^{(x,y)}(\mathrm{d}) = 1236,$$

and calculate its return.

Exercise 2.5

(a) Suppose we know that the return on a portfolio (x, y) in Model 2.2 satisfies $V_1^{(x,y)}(\mathrm{u}) = 50\%$. Assuming that $y \neq 0$, find $\frac{x}{y}$, the ratio of shares to bonds in the portfolio.

(b) For a general binary model with one stock and a risk-free asset, show that if two portfolios (x, y) and (x', y') with $y \neq 0$, $y' \neq 0$, $V_0^{(x,y)} \neq 0$ and $V_0^{(x',y')} \neq 0$ satisfy $\frac{x}{y} = \frac{x'}{y'}$, then they have the same return.

Exercise 2.6

Suppose that the price of a bond in a single-period binary model is given by $B_0 = 15$ and $B_1 = 18$. A portfolio φ consisting of 30 bonds and 20 shares of a certain company has final value

$$V_1^\varphi = \begin{cases} 1140 & \text{with probability } p, \\ 1000 & \text{with probability } 1 - p. \end{cases}$$

Find the final stock price S_1 giving rise to these values.

The concept of *arbitrage* is fundamental to modern finance theory. Informally, this is a market condition that allows an investor to make a risk-free profit without initial investment, often referred to as a *free lunch*. Here is how to make it precise in our model.

Definition 2.7 (Arbitrage and viability)

(a) An *arbitrage opportunity* is a portfolio φ with $V_0^\varphi = 0$ and $V_1^\varphi \geq 0$, together with $V_1^\varphi(\omega) > 0$ for at least one $\omega \in \Omega$.

(b) A model is said to be *viable* if there are no arbitrage opportunities.

Strictly speaking we should call this an arbitrage opportunity *in bonds and stock* for reasons that we shall see later.

Exercise 2.8†

Decide whether a portfolio φ would be an arbitrage opportunity in each of the following situations:

(a) $V_0^\varphi > 0$, $V_1^\varphi(\mathrm{u}) > 0$ and $V_1^\varphi(\mathrm{d}) = 0$;

(b) $V_0^\varphi = 0$, $V_1^\varphi(\mathrm{u}) > 0$ and $V_1^\varphi(\mathrm{d}) < 0$;

(c) $V_0^\varphi = 0$, $G^\varphi(\mathrm{u}) = 0$ and $G^\varphi(\mathrm{d}) > 0$;

(d) $R^\varphi(\mathrm{u}) \geq r$, $R^\varphi(\mathrm{d}) \geq r$ and one of these inequalities is strict.

Arbitrage opportunities rarely exist in practice. If they can be found at all, they are generally beyond the reach of small investors, because the possible gain from an arbitrage opportunity is typically extremely small in comparison to the size of the transaction required to benefit from it. Moreover, in an *efficient market*, traders will move quickly to take advantage of an arbitrage opportunity. Other traders who find themselves being thus taken advantage of will move to prevent it by adjusting prices. Consequently, as well as being difficult to spot, a situation when arbitrage exists is generally short-lived because prices quickly change to eliminate arbitrage opportunities. For this reason real-world markets are effectively arbitrage-free.

The **No-Arbitrage Principle** is often referred to in the literature, but not always clearly defined. We take it to be the guiding principle that in a realistic market model there should be no arbitrage, for the reasons given above. In other words, it is only *viable* market models that are realistic. It follows that it is important to be able easily to identify viable models, and this is one major theme of this and later chapters.

The viability of a binary model is neatly characterized as follows.

Theorem 2.9

A single-period binary model with a single stock is viable if and only if

$$S_1(\mathrm{d}) < (1+r)S_0 < S_1(\mathrm{u}). \tag{2.4}$$

Remark 2.10

Equation (2.4) can equivalently be expressed as

$$\frac{S_1(\mathrm{d})}{1+r} < S_0 < \frac{S_1(\mathrm{u})}{1+r} \tag{2.5}$$

or

$$\frac{S_1(\mathrm{d})}{S_0} < 1+r < \frac{S_1(\mathrm{u})}{S_0},$$

which is in turn equivalent to

$$R^S(\mathrm{d}) < r < R^S(\mathrm{u}),$$

where R^S is the return on the stock (as computed in Example 2.2).

Proof (of Theorem 2.9)

Suppose that (2.4) holds, and let (x, y) be any portfolio with zero initial value; that is

$$0 = V_0^{(x,y)} = xB_0 + yS_0. \tag{2.6}$$

To see that this portfolio cannot be an arbitrage opportunity, examine the three possibilities for y. First, if $y = 0$, it follows from (2.7) that $x = 0$ and so $V_1^{(x,y)} = 0$. Thus (x, y) is not an arbitrage opportunity.

Now consider the case that $y > 0$. Multiplying inequality (2.4) by y and adding xB_1 leads to

$$xB_1 + yS_1(\mathrm{d}) < xB_1 + y(1 + r)S_0 < xB_1 + yS_1(\mathrm{u}).$$

The middle term in this inequality is equal to

$$x(1 + r)B_0 + y(1 + r)S_0 = (1 + r)V_0^{(x,y)} = 0,$$

while the other terms are the portfolio values in the two scenarios at time 1. Thus

$$V_1^{(x,y)}(\mathrm{d}) < 0 < V_1^{(x,y)}(\mathrm{u}),$$

and consequently (x, y) is not an arbitrage opportunity.

Finally, if $y < 0$, it follows by a similar argument that

$$V_1^{(x,y)}(\mathrm{u}) < 0 < V_1^{(x,y)}(\mathrm{d})$$

so again (x, y) is not an arbitrage opportunity. We conclude that the model is viable.

Conversely, suppose that the model is viable. We will show by contradiction that (2.4) holds. Suppose to the contrary that

$$(1 + r)S_0 \leq S_1(\mathrm{d}) < S_1(\mathrm{u}). \tag{2.7}$$

This means that the return on the stock in both scenarios matches or exceeds the return on the riskless investment, so that we may form an arbitrage opportunity by borrowing money and investing it in the stock. In detail, consider the portfolio φ consisting of 1 share and $-S_0/B_0$ bonds. The initial value of this portfolio is

$$V_0^\varphi = -\frac{S_0}{B_0}B_0 + S_0 = 0.$$

Computing the final values of this portfolio and using (2.7) gives

$$V_1^\varphi(\mathrm{u}) = -\frac{S_0}{B_0}B_1 + S_1(\mathrm{u}) = -(1 + r)S_0 + S_1(\mathrm{u}) > 0,$$

$$V_1^\varphi(\mathrm{d}) = -\frac{S_0}{B_0}B_1 + S_1(\mathrm{d}) = -(1 + r)S_0 + S_1(\mathrm{d}) \geq 0.$$

Consequently, the portfolio $(-S_0/B_0, 1)$ is an arbitrage opportunity. This contradicts the viability of the model, and so inequality (2.7) cannot hold true.

A similar construction using the portfolio $(S_0/B_0, -1)$ gives arbitrage in the case when

$$S_1(\mathrm{d}) < S_1(\mathrm{u}) \leq (1+r)S_0. \tag{2.8}$$

Since neither (2.7) nor (2.8) is true, we conclude that (2.4) holds. $\square$

Example 2.11

Inequality (2.4) holds for Model 2.2 because $S_0(1+r) = 75 \times 1.2 = 90$ and

$$S_1(\mathrm{d}) = 72 < 90 < 120 = S_1(\mathrm{u}).$$

Exercise 2.12

Explain why Model 2.3 is not viable, and construct an arbitrage opportunity. What would you change to make this model viable?

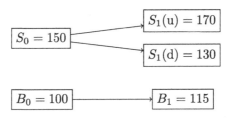

Model 2.3   One-period single-stock binary model in Exercise 2.12

Exercise 2.13

Construct an example of a viable model with $S_0 < S_1(\mathrm{d}) < S_1(\mathrm{u})$. What can you say about the interest rate in any such model?

2.3 Derivative Securities and the Pricing Problem

As explained in the introduction, this book is about pricing derivative securities. Before developing the theory it is useful to examine the pricing problem for a European call option, and show that an intuitive solution turns out to be wrong. This will motivate the need for the theory that ensues.

Example 2.14

Recall that the owner of a *European call option* (discussed briefly in Chapter 1) has the right (without obligation) to buy a share of the underlying stock S at time 1 at a *strike price*, say K, which is determined and written into the option contract when it is purchased at time 0. At time 1 when the actual future value of S is known, there are three possibilities. If the stock price S_1 is greater than the strike price K, we say that the option is *in the money*, and if $S_1 = K$, the model is *at the money*; otherwise, the option is *out of the money*.

At time 1, the decision of whether to exercise the option or not will depend on the stock price S_1. The owner of the option will exercise it (meaning he will exercise his right to buy a share at the fixed price K) at time 1 only if it is in the money; that is, if $S_1 > K$. In that case, the value of the option to the owner is $S_1 - K$ since it represents the right to buy a share worth S_1 for the lower price K. If, on the other hand, $S_1 \leq K$, then he is able to buy a share for K or less on the open market, and there would be no sense in exercising the option. The option is therefore worthless.

We summarize this by saying that the *payoff* of the option—that is, the value to the owner—at time 1 is

$$C_1 := \begin{cases} S_1 - K & \text{if } S_1 > K, \\ 0 & \text{if } S_1 \leq K \end{cases}$$

$$= [S_1 - K]_+$$

recalling that for any $a \in \mathbb{R}$, the *positive and negative parts* of a, denoted by a_+ and a_- are defined by $a_+ := \max\{a, 0\}$ and $a_- := \max\{-a, 0\}$. (Some important properties of a_+ and a_- are given in Exercise 2.17 below.)

Notice that the payoff is never negative and could be positive, since the owner of this option will only exercise it if it is profitable to do so. Such an option is therefore a valuable asset, and a market agent should expect to pay something, let us call it π, at time 0 to purchase it. Obtaining a call option for free would amount to a free lunch or riskless profit.

The pricing problem is to determine whether there is a fair price to pay for a European call option, and if so what it is. Of course we will need to define what is meant by a fair price.

First let us consider in detail this problem for a European call option in Model 2.2. Suppose that a trader wishes to write (that is, sell) a European call option with strike $K = 96$. What price should he charge for this at time 0? As a first guess, one might take the expected payoff of the option. We have

$$C_1 = \begin{cases} [120 - 96]_+ = 24 & \text{with probability } \frac{3}{4}, \\ [72 - 96]_+ = 0 & \text{with probability } \frac{1}{4}. \end{cases}$$

The expected value of the option is

$$\tfrac{3}{4} \times 24 + \tfrac{1}{4} \times 0 = 18,$$

which would seem to be a reasonable guess for the price of this option. However, we will see that this price is *too high*—since it allows the seller to make a risk free profit as follows.

The trader should simultaneously sell a call option for the price $\pi = 18$ and create the portfolio $\varphi = (-1.95, 0.5)$ in bonds and stock. The value of this *extended portfolio* (consisting of bonds, shares *and* the option) at time 0 is

$$V_0^\varphi - 18 = (-1.95 \times 10) + (0.5 \times 75) - 1 \times 18 = 0$$

so there is no net outlay. An alternative way to see this is to consider the cash flow: 0.5 shares (worth 37.5) are purchased by borrowing 1.95 bonds (worth 19.5) and the proceeds 18 from selling the option.

At time 1, there are two possibilities. If the stock price moves to 120, the call option will be exercised; consequently the trader must buy a share at 120 and then sell it to the owner of the option for 96, making a loss of 24. In this case, on closing the position in bonds and shares at the same time, the trader will receive

$$V_1^\varphi(\mathrm{u}) - 24 = (-1.95 \times 12) + (0.5 \times 120) - 24 = 12.6.$$

If the stock price moves to 72, the option is worthless, and is not exercised. The trader has nothing to pay out. Closing his position in shares, bonds and options gives

$$V_1^\varphi(\mathrm{d}) - 0 = (-1.95 \times 12) + (0.5 \times 72) - 0 = 12.6.$$

This means that if option is priced at 18, a trader selling at this price is able to make a risk-free profit of 12.6 in cash—*no matter what happens at time $t = 1$*. No market agent who is aware of this would be willing to buy a call option at a price of 18, as he would certainly not wish to assist the trader in making money out of nothing. We say that 18 is an *unfair* price for the option.

Remark 2.15

The argument presented here is quite typical in the construction of arbitrage opportunities. The idea is to sell those assets that are considered to be over-priced, and/or to buy those assets that are considered (or proven) to be too cheap.

The next exercise is designed to show that if the option price is too low, a trader may create an arbitrage opportunity, giving him a risk free profit, by following a similar strategy to the one above, but buying an option, instead of selling. Of course, no other market agent would be willing to sell a call option to our trader at such a price.

Exercise 2.16

Show that if the call option with strike $K = 96$ in Model 2.2 is priced at 6, then a trader can make a risk-free profit by creating the portfolio $(3.15, -0.5)$ and buying one call option. Deduce that the price 6 is too low.

Exercise 2.17

Prove the following for any real number a:

(a) $a_- = (-a)_+$.

(b) $a = a_+ - a_-$.

(c) $|a| = a_+ + a_-$.

2.4 Fair Pricing

The theory we are about to develop centres around extending the idea of arbitrage to portfolios that include assets other than the primary ones, as in the above example. In this way we will be able to define precisely what is meant by a price being "too high" or "too low", or "fair".

Let us suppose then that we have, in addition to the primary assets, a *derivative security* D (such as the European call option in the above example) that may be traded in the markets. The derivative will have a *payoff* or value at time 1, which depends on the way the stock price S_1 behaves. The stock price S_1 itself depends on the scenario followed by the model, and so we formally define a *derivative security*, usually referred to simply as a *derivative*, as follows.

Definition 2.18 (Derivative in a single period model)

A *derivative* (or *derivative security* or *contingent claim*) is an asset whose payoff at time 1 is given by a random variable $D : \Omega \to \mathbb{R}$.

Example 2.19

A European call option as above is a derivative show payoff at time 1 is the random variable C_1 where

$$C_1(\omega) = \left[S_1(\omega) - K\right]_+$$

for $\omega \in \Omega$.

Let π denote a price at which a derivative D is traded at time 0. The pricing problem is to find those values of π that are fair. To make this precise we will need the following definitions.

Definition 2.20 (Extended portfolio, extended arbitrage opportunity)

Let D be a derivative priced at π at time 0. An *extended portfolio* involving D is a triple $\psi = (x, y, z)$ that denotes a holding of x bonds, y shares and z units of the derivative D. The *value* V_t^ψ of the extended portfolio $\psi = (x, y, z)$ at $t = 0, 1$ is given by

$$V_0^\psi := xB_0 + yS_0 + z\pi, \qquad V_1^\psi := xB_1 + yS_1 + zD. \tag{2.9}$$

An *extended arbitrage opportunity* in bonds, shares and the derivative D is an extended portfolio ψ with $V_0^\psi = 0$ and $V_1^\psi \geq 0$, and $V_1^\psi(\omega) > 0$ for at least one $\omega \in \Omega$.

A rational seller will never agree to sell D at the price π if it allows an extended arbitrage opportunity (x, y, z) with $z > 0$, because that would mean that the buyer can make a risk-free profit by buying D and creating the portfolio $(x/z, y/z)$. Likewise, an extended arbitrage opportunity with $z < 0$ would allow a seller to make a risk free profit by selling D and simultaneously creating the portfolio $(-x/z, -y/z)$.

This leads to the extension of the No-Arbitrage Principle to also include derivatives: in a rational market model the prices of derivatives will be such as to prevent extended arbitrage opportunities. In other words, in a rational market all derivatives will be traded at *fair prices*, defined precisely as follows.

Definition 2.21 (Fair price)

A *fair* or *rational* or *arbitrage-free* price for a derivative D is any price π for it that prevents extended arbitrage opportunities involving D.

Finding fair prices is clearly very important. A natural question, which the theory will answer, is whether fair prices actually exist (they do in our model, if it is viable) and whether they are unique (yes in our simple model, but not always in more general models).

2.5 Replication and Completeness

The key to finding a fair price for a derivative D is the idea of *replication*, which enables us find a unique fair price at time 0 for any derivative in our simple model.

Definition 2.22 (Replication, attainability)

A portfolio φ *replicates* a derivative D if its value at time 1 is equal to D in all possible scenarios; that is.

$$V_1^\varphi(\omega) = D(\omega) \quad \text{for all } \omega \in \Omega,$$

or $V_1^\varphi = D$ for short. If a derivative D admits a replicating portfolio, then D is called *attainable*.

The main tool for finding a fair price by means of replication is the *Law of One Price*. In its general form, it states that if two extended portfolios have the same value at time 1 for all scenarios, then in a rational market the prices at time 0 must be such as to give the same value at time 0—otherwise, there would be arbitrage. The particular case that we need is in the next theorem.

Theorem 2.23 (Law of One Price in a simple model)

If a portfolio $\varphi = (x, y)$ replicates an attainable derivative D in a viable single-period single-stock model, then there is a unique fair price D_0 for D at time 0, which is equal to the initial value of the portfolio. That is

$$D_0 = V_0^\varphi. \tag{2.10}$$

Proof

Let $\varphi = (x, y)$ be a replicating portfolio for D. We begin by showing that any price π other than V_0^φ is unfair.

Suppose first that $\pi > V_0^\varphi$. Then at time 0 we may sell D at the price π, and with the proceeds buy the portfolio φ which costs V_0^φ and still have spare cash amounting to $\pi - V_0^\varphi$, which we can invest in bonds. This amounts to constructing the extended portfolio $(x', y, -1)$, where

$$x' = x + \frac{1}{B_0}\left(\pi - V_0^\varphi\right) = \frac{1}{B_0}(\pi - yS_0).$$

The value of this portfolio at time 0 is

$$V_0^{(x',y,-1)} = (\pi - yS_0) + yS_0 - \pi = 0.$$

At time 1, the value of this portfolio in any scenario $\omega \in \Omega$ is

$$\begin{aligned} V_1^{(x',y,-1)}(\omega) &= x'B_1 + yS_1(\omega) - D(\omega) \\ &= xB_1 + (1+r)\left(\pi - V_0^\varphi\right) + yS_1(\omega) - D(\omega) \\ &= V_1^\varphi(\omega) + (1+r)\left(\pi - V_0^\varphi\right) - D(\omega). \end{aligned}$$

Since φ replicates D, we have $V_1^\varphi(\omega) = D(\omega)$, and therefore

$$V_1^{(x',y,-1)}(\omega) = (1+r)\left(\pi - V_0^\varphi\right) > 0.$$

Thus we have constructed an extended arbitrage opportunity.

If $\pi < V_0^\varphi$, then we may construct an arbitrage opportunity by buying the derivative, selling the portfolio, and investing the difference $V_0^\varphi - \pi$ in bonds. This amounts to constructing the extended portfolio $(x'', -y, 1)$ where

$$x'' = -x + \frac{1}{B_0}\left(V_0^\varphi - \pi\right) = \frac{1}{B_0}(yS_0 - \pi).$$

Arguing as in the first case, this guarantees a profit of

$$(1+r)\left(V_0^\varphi - \pi\right) > 0.$$

So any price π for D other than V_0^φ is not fair. To complete the proof, we must show that $\pi = V_0^\varphi$ is a fair price. To see this, assume that D is priced at $\pi = V_0^\varphi$, and suppose that an extended arbitrage is achieved by the extended portfolio (u, v, w) consisting of u bonds, v shares and w units of D. The value at time 0 of this portfolio is

$$\begin{aligned} V_0^{(u,v,w)} &= uB_0 + vS_0 + w\pi = uB_0 + vS_0 + wV_0^\varphi \\ &= (u + wx)B_0 + (v + wy)S_0 = V_0^{(u+wx,v+wy)}. \end{aligned}$$

At time 1, the replication of D by φ gives

$$\begin{aligned} V_1^{(u,v,w)} &= uB_1 + vS_1 + wD = uB_1 + vS_1 + wV_1^\varphi \\ &= (u + wx)B_1 + (v + wy)S_1 = V_1^{(u+wx,v+wy)}. \end{aligned}$$

Thus $(u + wx, v + wy)$ is an arbitrage opportunity in bonds and shares, which contradicts the assumption that the model is viable. Thus there can be no extended arbitrage opportunity if $\pi = V_0^\varphi$, so this is the unique fair price for D at time 0, and we may denote it D_0. $\square$

This result is only applicable if the derivative D is attainable (that is, can be replicated), so we need to ask whether there is always a replicating portfolio for D, and if so how to find it. In a single-period binary model, the answer is very simple indeed: all derivative securities are attainable, and we say that the model is *complete*.

Theorem 2.24

In a single-period model with one stock, every derivative D is attainable and has a unique replicating portfolio (x, y) consisting of

$$x = \frac{1}{B_1} \frac{D(\mathrm{d})S_1(\mathrm{u}) - D(\mathrm{u})S_1(\mathrm{d})}{S_1(\mathrm{u}) - S_1(\mathrm{d})} \tag{2.11}$$

bonds and

$$y = \frac{D(\mathrm{u}) - D(\mathrm{d})}{S_1(\mathrm{u}) - S_1(\mathrm{d})} \tag{2.12}$$

shares.

Proof

At time 1, there are only two possibilities for the equation $V_1^{(x,y)} = D$, so we need to find a portfolio (x, y) satisfying the two equations

$$xB_1 + yS_1(\mathrm{u}) = D(\mathrm{u}), \qquad xB_1 + yS_1(\mathrm{d}) = D(\mathrm{d}).$$

For any D this system has a unique solution given by (2.11) and (2.12). $\square$

Example 2.25

Consider the European call option discussed in Example 2.14. The replicating portfolio for this option is

$$x = \frac{1}{12} \frac{0 \times 120 - 24 \times 72}{120 - 72} = -3, \qquad y = \frac{24 - 0}{120 - 72} = \frac{1}{2}.$$

Consequently, the fair price of this option is

$$V_0^{(x,y)} = -3 \times 10 + \tfrac{1}{2} \times 75 = 7.5.$$

This should be contrasted with the guess of 18 in Example 2.14, which was too high and the price 6 which was shown in Exercise 2.16 to be too low.

The theory we have just developed also shows just how we arrived at the extended portfolio $(-1.95, 0.5, -1)$ to give arbitrage when the price was too high. We know that the option is replicated by the portfolio $\varphi = (-3, 0.5)$ to give the fair price of 7.5; that is, this is the price for the option that does not allow arbitrage. If D is priced at 18 then by selling at that price instead of 7.5 the difference $18 - 7.5 = 10.5$ can be invested risk free in 1.05 bonds. So, if the bond holding is now increased to $-3 + 1.05 = -1.95$ this figure contains both the bond holding needed to replicate D *and* the extra element of investment for profit. This extra bond holding gives the guaranteed profit $1.05 \times 12 = 12.6$ no matter what happens at time $t = 1$.

The extended portfolio in Exercise 2.16 that gives a risk free profit if D is priced at 6 was found in the same way.

Exercise 2.26

Let D be the derivative given by $D(u) = 49$, $D(d) = 21$ in Model 2.4.

(a) Show that Model 2.4 is viable.

(b) Find the replicating portfolio for D and hence the fair price D_0.

(c) Construct extended arbitrage opportunities to show that 25 is too low a price, and 40 is too high to pay for this derivative.

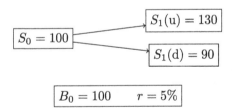

Model 2.4   One-period binary model in Exercises 2.26, 2.27 and 2.37

Exercise 2.27

Suppose that the fair price at time 0 of a derivative D in Model 2.4 is $D_0 = 100$, and you are told that $D(u) = 55$. What is the value of $D(d)$?

2.6 Risk-Neutral Probabilities

Finding the fair price of a derivative by first calculating the bond and stock holding in the replicating portfolio and then its initial value is quite laborious—

and becomes even more so when we consider models with more than one step. In this section we introduce the important concept of *risk-neutral probabilities*, which gives a much more efficient technique for finding fair prices, since it involves only the calculation of a new probability on the set Ω of possible scenarios. The fair price of any derivative is then simply its expected value under this new probability, after *discounting*.

The notion of the *discounted value* of any asset is an important one in mathematical finance. The idea is that of adjusting the future value of a random quantity to today's prices or *discounting the effect of inflation or interest rates*. This is made precise in the following definition.

Definition 2.28 (Discounted value)

Let $X : \Omega \to \mathbb{R}$ be a random variable representing a quantity (for example the payoff of a derivative) that is known at time 1. The *value of X discounted to time* 0 is the random variable

$$\bar{X} := \frac{X}{1+r}.$$

This is often referred to as simply the *discounted value* of X.

Recalling (2.5), we can say that a single-period binary model is viable if and only if the stock price at time 0 lies strictly between the two *discounted stock prices* at time 1; that is

$$\bar{S}_1(\mathrm{d}) < S_0 < \bar{S}_1(\mathrm{u}). \tag{2.13}$$

Thus, in a viable binary model, the point S_0 can be written as a weighted average (called a *convex combination*) of the two quantities $\bar{S}_1(\mathrm{u})$ and $\bar{S}_1(\mathrm{d})$. In other words, there is a number $q \in (0,1)$ such that

$$S_0 = q\bar{S}_1(\mathrm{u}) + (1-q)\bar{S}_1(\mathrm{d}), \tag{2.14}$$

where q is given explicitly by

$$q = \frac{S_0 - \bar{S}_1(\mathrm{d})}{\bar{S}_1(\mathrm{u}) - \bar{S}_1(\mathrm{d})}. \tag{2.15}$$

The key idea now is that the pair $(q, 1-q)$ can be interpreted as a new probability assignment for the scenarios u, d, and then equation (2.14) shows that S_0 is the expected value of the discounted price $\bar{S}_1$ under this new probability. That is, writing $\mathbb{Q} = (q, 1-q)$ we have

$$S_0 = \mathbb{E}_{\mathbb{Q}}(\bar{S}_1) := q\bar{S}_1(\mathrm{u}) + (1-q)\bar{S}_1(\mathrm{d}). \tag{2.16}$$

The new probability assignment $\mathbb{Q}$, called a *risk-neutral probability*, is in general *quite different from the original real-world probability* $(p, 1 - p)$: it is a purely artificial mathematical probability. The key defining property of $\mathbb{Q}$ is the equality (2.16); thus we have the following formal definition.

Definition 2.29 (Risk-neutral probability)

A *risk-neutral probability* is a probability assignment $\mathbb{Q} = (q, 1 - q)$ to the scenarios u, d such that (2.16) holds and $0 < q < 1$.

Notice that the original real-world probability $(p, 1 - p)$ has not featured at all in the above discussion, which is summarized as follows.

Theorem 2.30

A viable one-step binary model containing one stock admits a unique risk-neutral probability $\mathbb{Q}$.

Proof

We have already shown the existence of a probability assignment $\mathbb{Q}$ on Ω satisfying equation (2.16) if the model is viable. This probability assignment is unique because, as already observed, equation (2.16) gives

$$q = \frac{S_0 - \bar{S}_1(\mathrm{d})}{\bar{S}_1(\mathrm{u}) - \bar{S}_1(\mathrm{d})}, \qquad 1 - q = \frac{\bar{S}_1(\mathrm{u}) - S_0}{\bar{S}_1(\mathrm{u}) - \bar{S}_1(\mathrm{d})}. \tag{2.17}$$

$\square$

Example 2.31

For Model 2.2 we have

$$q = \frac{75 - 72/1.2}{120/1.2 - 72/1.2} = \frac{3}{8}$$

and we may easily verify that

$$q\bar{S}_1(\mathrm{u}) + (1 - q)\bar{S}_1(\mathrm{d}) = \frac{3}{8}\frac{120}{1.2} + \frac{5}{8}\frac{72}{1.2} = 75 = S_0.$$

The defining property (2.16) may be expressed in a number of equivalent ways. First consider the *increment of the discounted price*

$$\Delta\bar{S}_1 := \bar{S}_1 - S_0. \tag{2.18}$$

We may write (2.16) as

$$\mathbb{E}_{\mathbb{Q}}(\Delta \bar{S}_1) = 0,$$

since S_0 is constant. That is, the average movement of the price after discounting is zero.

Remarks 2.32

(1) The increment $\Delta \bar{S}_1$ of the discounted price is *not* the same as the *discounted price increment*, which would be

$$\overline{\Delta S_1} := \overline{S_1 - S_0} = \frac{S_1 - S_0}{1 + r}.$$

(2) In later development when there are several time steps we will again encounter this phenomenon, namely that there is on average no change in the discounted price provided the probability is chosen carefully. This is the defining property of an important concept—that of a *martingale*. In our context the discounted stock price is a one-step martingale.

A second alternative way to express equation (2.16) is

$$\mathbb{E}_{\mathbb{Q}}(S_1) = (1 + r)S_0,$$

which may be rearranged to yield

$$\mathbb{E}_{\mathbb{Q}}(R^S) = r,$$

where R^S is the return of the stock (see Example 2.2). This expression explains why the probability assignment $\mathbb{Q}$ is called *risk-neutral*: under the artificial probability $\mathbb{Q}$, the expected return from the risky investment is expected to be the same as the return from an investment in the risk free asset.

The existence of such a risk-neutral probability assignment actually characterizes a viable model; we have already proved one half of the following result in Theorem 2.30.

Theorem 2.33

A one-step binary model containing one stock is viable if and only if it allows a risk-neutral probability assignment. If a risk-neutral probability exists, then it is unique.

Exercise 2.34[‡]

Complete the proof of Theorem 2.33 by proving that if a one-step binary model with one stock has a risk-neutral probability, then it is viable.

Hint. Use Theorem 2.9.

Risk-neutral probabilities give us a quick and convenient way of calculating the fair price of a derivative. In view of results like Theorem 2.35 below, the fair price of a derivative is sometimes called its *risk-neutral price*.

Theorem 2.35

In a single-period binary model with risk-neutral probability $\mathbb{Q}$, the fair price of any derivative D is

$$D_0 = \frac{1}{1+r}\mathbb{E}_{\mathbb{Q}}(D) = \mathbb{E}_{\mathbb{Q}}(\bar{D}), \tag{2.19}$$

where $\bar{D}$ is the discounted value of D.

Proof

It follows from Theorem 2.24 that there exists a replicating portfolio (x, y) for D and we know from Theorem 2.23 that there is a unique fair price for D, namely $D_0 = V_0^{(x,y)}$. Since $D = V_1^{(x,y)} = xB_1 + yS_1$, then $\bar{D} = xB_0 + y\bar{S}_1$ and we conclude from (2.16) that

$$\mathbb{E}_{\mathbb{Q}}(\bar{D}) = \mathbb{E}_{\mathbb{Q}}(xB_0 + y\bar{S}_1) = xB_0 + y\mathbb{E}_{\mathbb{Q}}(\bar{S}_1) = xB_0 + yS_0 = V_0^{(x,y)} = D_0.$$

$\square$

Example 2.36

In Example 2.31, we found that the risk-neutral probability assignment for Model 2.2 is $\mathbb{Q} = (\frac{3}{8}, \frac{5}{8})$. The fair price of the European call option with strike 96 of Example 2.14 is then

$$\tfrac{3}{8} \times \tfrac{24}{1.2} + \tfrac{5}{8} \times \tfrac{0}{1.2} = 7.5.$$

As Theorem 2.35 tells us, this is the same as the initial value of the replicating portfolio for this option, given in Example 2.25.

Exercise 2.37[†]

Compute the fair prices for a European call option C_1 with strike 110 and a European call option C_1' with strike 120 in Model 2.4 using:

(a) Replication.

(b) Risk-neutral probabilities.

Check that these prices agree.

3
Single-Period Models

In the previous chapter, where we studied single-period binary models with one stock and one bond, we derived the following important properties:

- The model is viable if and only if it admits a risk-neutral probability. This risk-neutral probability is unique whenever it exists.

- Every attainable derivative has a unique fair price which is given by replication, and can also be obtained using the risk-neutral probability technique.

- The model is complete; that is, every derivative in this model is attainable. Thus every derivative has a unique fair price.

The aim of this chapter is to see the extent to which these properties continue to hold in more general one-step models. There are several natural ways in which the single-period binary model can be generalized to be more realistic:

- Increase the number of possible scenarios at time 1.

- Increase the number of risky assets.

- Increase the number of risk-free assets.

- Increase the number of time steps giving a so-called *multi-period* model.

Here we will consider only the first two possibilities, since these involve straightforward generalizations of the ideas of Chapter 2. The third possibility is not really a generalization, as can be seen from Exercise 3.1 below. The extension to the multi-period case will be developed in later chapters.

N.J. Cutland, A. Roux, *Derivative Pricing in Discrete Time*,
Springer Undergraduate Mathematics Series,
DOI 10.1007/978-1-4471-4408-3_3, © Springer-Verlag London 2012

Exercise 3.1[†]

Suppose that a one-step market model has two bonds, called B and B', with associated interest rates $r = B_1/B_0 - 1$ and $r' = B_1'/B_0' - 1$, and that investors may invest or borrow as many of these bonds as they wish. Show that if $r \neq r'$, then the model allows arbitrage. Explain why this makes it redundant to increase the number of risk-free assets.

Each of the key ideas in the previous chapter (arbitrage, viability, risk-neutral probability, attainable derivative) extends naturally to *any* single-period model. We will see that any such model is viable if and only if it admits a risk-neutral probability and this may be used to find the (unique) fair price of an attainable derivative. It turns out however that in general these models are incomplete, so there are non-attainable derivatives and these do not have a unique fair price.

We begin by examining what happens when the number of scenarios is increased, still with just one risky asset. The extension to models with several risky assets is considered from Section 3.2 onwards.

Several of the assumptions in Chapter 2, with minor amendments, remain in force throughout this chapter. We summarize the unchanged assumptions (2.1–2.2, 2.7 and part of 2.5) here for convenience.

Assumptions 3.1

(1) Trading takes place at time 0 and time 1.

(2) The model contains a *bond* with initial price $B_0 > 0$. The value of the bond at time 1 is

$$B_1 = (1 + r)B_0,$$

where the interest rate $r > -1$ is known in advance.

(3) A portfolio may consist of any number (positive or negative) of bonds and shares.

(4) Trading is instantaneous, borrowing and lending rates are the same, and there are no transaction costs in trading (equivalently, buying and selling prices are the same).

The assumption regarding the number of scenarios is now relaxed to give the following, which is in force throughout this chapter.

Assumption 3.2 (Discreteness of scenarios)

The set Ω of possible future scenarios has a finite number of elements (at least two). The probability of any scenario $\omega \in \Omega$ occurring is denoted by $\mathbb{P}(\omega)$; we assume that $\mathbb{P}(\omega) > 0$ for each ω, which means that we disregard any scenarios that are impossible (defined as having zero probability). The probability $\mathbb{P}$ is called the *real-world probability* or the *market probability*, and we have

$$\sum_{\omega \in \Omega} \mathbb{P}(\omega) = 1.$$

It is often convenient to assume that the different scenarios are numbered so that $\Omega = \{\omega_1, \ldots, \omega_n\}$ for some $n \geq 2$.

3.1 Single-Stock Models

First we investigate what happens when the number of scenarios can be more than two but there is still only one stock or risky asset. There is still just one bond B.

The single stock, which is the other primary asset and is again denoted by S, has share price S_0 at time 0 and S_1 at time 1. The latter is a random variable so the possible future values are $S_1(\omega)$ for $\omega \in \Omega$ (or $S_1(\omega_j)$ for $j = 1, \ldots, n$) and we assume that each of these values is different. We may suppose that the labelling of the scenarios is such that

$$S_1(\omega_1) > \cdots > S_1(\omega_n). \tag{3.1}$$

This may be illustrated by a tree (on its side) with n branches; see for example Model 3.1 below. We assume that $S_0 > 0$ and $S_1(\omega) > 0$ for all ω.

This section will show that the main results of Chapter 2 for the binary model extend, with some restrictions, to a model with two assets but possibly more than two scenarios. For such models the definitions of *arbitrage opportunity*, *viability*, *derivative*, *fair price*, *replicating portfolio* and *attainable derivative* are exactly the same as in Chapter 2 (Definitions 2.7, 2.18, 2.21 and 2.22) but bearing in mind that the set of scenarios may now be bigger.

To illustrate the complications that can arise, we begin with an example.

3.1.1 Example: A European Put Option in a Ternary Model

In the context of a single-period model with one bond B and stock S, a *European put option* on the stock S gives its owner the right (without any obligation) to *sell* one share of S at time 1 (the *exercise date*) at a predetermined strike price K. By contrast with a European call option, this has positive value to the owner at time $t = 1$ only if K is bigger than the market price of the stock, so the payoff of this option at time 1 is

$$P_1 := \max\{K - S_1, 0\} = [K - S_1]_+ = [S_1 - K]_-. \tag{3.2}$$

If, at time 1, the stock price S_1 is *less* than the strike price K, we say that the option is *in the money*; if $S_1 = K$, the option is *at the money*. If $S_1 > K$, the option is *out of the money* (it is worthless).

Consider the following model with $\Omega = \{\omega_1, \omega_2\}$, final stock price given by $S_1(\omega_1) = 230$ and $S_1(\omega_2) = 210$, and initial stock price $S_0 = 220$. For simplicity we also suppose that $B_0 = B_1 = 1$ and $r = 0$. Applying the theory of Chapter 2, it is easy to see that $q = \frac{1}{2}$ is the unique risk-neutral probability; that is, the solution to the equation

$$220 = S_0 = q S_1(\omega_1) + (1 - q) S_2(\omega_2) = 230q + 210(1 - q). \tag{3.3}$$

So this model is viable.

Consider now a put option with strike $K = 220$; then we have $P_1(\omega_1) = 0$ and $P_1(\omega_2) = 10$. The risk-neutral pricing theory of Chapter 2 shows that the fair price of this option is

$$P_0 = \tfrac{1}{2} \times 0 + \tfrac{1}{2} \times 10 = 5. \tag{3.4}$$

The option admits the unique replicating portfolio $(x, y) = (115, -0.5)$ which is found by solving the equations

$$x + 230y = 0, \qquad x + 210y = 10.$$

This gives the fair price

$$P_0 = V_0^{(115, -0.5)} = 115 \times 1 - 0.5 \times 220 = 5,$$

which is the same as (3.4), in confirmation of the theory of Chapter 2.

Suppose now that this model is modified by adding an additional scenario ω_3, with $S_1(\omega_3) = 190$, giving Model 3.1. The value of the put option at the new scenario is $P_1(\omega_3) = 30$ and the quest for a replicating portfolio must take the new scenario into account. Any such portfolio (x, y) must satisfy the three equations

$$x + 230y = 0, \qquad x + 210y = 10, \qquad x + 190y = 30.$$

The portfolio $(115, -0.5)$ calculated in the binary model above is the only one that satisfies the first two equations in this system. However, this portfolio does not satisfy the third equation, since $115 - 190 \times 0.5 = 20 \neq 30$. We conclude that there is *no* replicating portfolio for this derivative. In other words, this put option is not attainable, so the ternary Model 3.1 is *incomplete*.

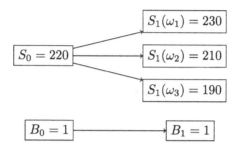

Model 3.1 One-period single-stock ternary model

Complications can also arise when searching for a risk-neutral probability in a ternary model. In such a model a risk-neutral probability will take the form $\mathbb{Q} = (q_1, q_2, q_3)$ with each $q_i > 0$ and $\mathbb{E}_{\mathbb{Q}}(\bar{S}_1) = S_0$. So in Model 3.1, since $r = 0$, we need to solve the equations

$$230q_1 + 210q_2 + 190q_3 = 220, \qquad q_1 + q_2 + q_3 = 1, \tag{3.5}$$

for the probabilities $q_1, q_2, q_3 \in (0, 1)$. This is a system of *two* equations in *three* unknowns; such a system, if it can be solved, has infinitely many solutions.

Exercise 3.2[†]

Show that Model 3.1 has infinitely many risk-neutral probabilities, each given by

$$\mathbb{Q}_\lambda = \left(\lambda, \tfrac{3}{2} - 2\lambda, \lambda - \tfrac{1}{2} \right) \tag{3.6}$$

for some $\lambda \in \left(\tfrac{1}{2}, \tfrac{3}{4} \right)$.

Calculating the expected payoff of a put option with strike 220 using any of the risk-neutral probabilities obtained in Exercise 3.2, we find that

$$\mathbb{E}_{\mathbb{Q}_\lambda}(P_1) = 0\lambda + 10\left(\tfrac{3}{2} - 2\lambda\right) + 30\left(\lambda - \tfrac{1}{2}\right) = 10\lambda \in (5, 7.5).$$

We may guess (correctly, as later theory will show) that this gives different 'fair prices' for different risk-neutral probabilities. Two traders might find it difficult to agree on which risk-neutral probability to use for fixing a price at which to trade this option: factors other than mathematical fairness, such as utility and attitude to risk, will come into play.

The above discussion raises an important question: which derivatives *are* attainable in Model 3.1? A replicating portfolio (x, y) for any derivative D in this model must satisfy

$$x + 230y = D(\omega_1), \qquad x + 210y = D(\omega_2), \qquad x + 190y = D(\omega_3). \qquad (3.7)$$

We can now express y in two ways. From the first two equations we must have

$$y = \tfrac{1}{20}\big(D(\omega_1) - D(\omega_2)\big), \qquad (3.8)$$

and from the second two

$$y = \tfrac{1}{20}\big(D(\omega_2) - D(\omega_3)\big). \qquad (3.9)$$

Combining (3.8) and (3.9), it follows that we need

$$D(\omega_1) - 2D(\omega_2) + D(\omega_3) = 0 \qquad (3.10)$$

for a replicating portfolio to exist. Conversely, it is easy to check that if (3.10) holds then the system (3.7) has a solution. We deduce that a derivative D is attainable in this ternary model if and only if (3.10) holds.

We will see that in an incomplete model the fair price of an *attainable* derivative is unique and is given by any one of the infinitely many risk-neutral probabilities. To see this in Model 3.1, for any $\lambda \in (\tfrac{1}{2}, \tfrac{3}{4})$, the expectation of a derivative D under the probability $\mathbb{Q}_\lambda$ is

$$\mathbb{E}_{\mathbb{Q}_\lambda}(D) = \lambda D(\omega_1) + \big(\tfrac{3}{2} - 2\lambda\big) D(\omega_2) + \big(\lambda - \tfrac{1}{2}\big) D(\omega_3)$$
$$= \lambda\big(D(\omega_1) - 2D(\omega_2) + D(\omega_3)\big) + \tfrac{3}{2}D(\omega_2) - \tfrac{1}{2}D(\omega_3). \qquad (3.11)$$

This expression is independent of λ precisely when the attainability criterion (3.10) is satisfied. For any *attainable* derivative D, equation (3.11) reduces to

$$\mathbb{E}_{\mathbb{Q}_\lambda}(D) = \tfrac{3}{2}D(\omega_2) - \tfrac{1}{2}D(\omega_3). \qquad (3.12)$$

The following exercise illustrates the pricing theory for attainable derivatives that will be established in Section 3.1.3 (Theorems 3.10 and 3.11).

Exercise 3.3

In the example above, suppose that D is a derivative satisfying (3.10). Show that D is attainable and that if φ is a any replicating portfolio then

$$V_0^\varphi = \tfrac{3}{2} D(\omega_2) - \tfrac{1}{2} D(\omega_3),$$

the price given by any risk-neutral probability as in (3.12).

This example illustrates two important points, which we will address in the coming sections. In general a model may be incomplete; that is, it may give rise to non-attainable derivatives. Further, in general there will be more than one risk-neutral probability in a viable model, and we will see that risk-neutral pricing gives a unique fair price only for attainable derivatives. It will emerge that these facts are all closely connected.

3.1.2 Viability, Risk Neutral Probabilities and Completeness

As suggested by the example, the definition of a *risk-neutral probability* now takes the following form, taking into account the fact that there are $n \geq 2$ scenarios. Recall that $\bar{X} := X/(1+r)$ is the discounted value of any random variable X and $r = B_1/B_0 - 1$ is the interest rate on the bond.

Definition 3.4 (Risk-neutral probability)

A *risk-neutral probability* $\mathbb{Q}$ for a single-period single-stock model with n scenarios is a family $(q_1, \ldots, q_n)$ with $q_j > 0$ for each $j = 1, \ldots, n$ and

$$S_0 = \mathbb{E}_{\mathbb{Q}}(\bar{S}_1) = \sum_{j=1}^{n} q_j \bar{S}_1(\omega_j), \qquad (3.13)$$

together with

$$\sum_{j=1}^{n} q_j = 1. \qquad (3.14)$$

For future reference, note that the defining property (3.13) of a risk-neutral probability extends to the value of any portfolio—a fact that is important in the development of derivative pricing theory.

Theorem 3.5

If $\mathbb{Q}$ is a risk-neutral probability and φ is any portfolio, then

$$\mathbb{E}_{\mathbb{Q}}(\bar{V}_1^\varphi) = V_0^\varphi.$$

Proof

Writing $\varphi = (x, y)$, we have

$$\mathbb{E}_{\mathbb{Q}}(\bar{V}_1^\varphi) = \mathbb{E}_{\mathbb{Q}}\left(\frac{xB_1 + yS_1}{1+r}\right) = \mathbb{E}_{\mathbb{Q}}(xB_0 + y\bar{S}_1) = xB_0 + yS_0 = V_0^\varphi.$$

$\square$

We are now in a position to extend Theorems 2.9 and 2.33 to this model, giving what is known as the *First Fundamental Theorem of Asset Pricing* for a one-period model with a single stock.

Theorem 3.6 (*First Fundamental Theorem of Asset Pricing*)

In a single-period model with one stock and $n \geq 2$ scenarios, the following statements are equivalent:

(1) The model is viable.

(2) The stock and bond prices satisfy

$$S_1(\omega_n) < S_0(1+r) < S_1(\omega_1) \tag{3.15}$$

or, equivalently,

$$\bar{S}_1(\omega_n) < S_0 < \bar{S}_1(\omega_1). \tag{3.16}$$

(3) The model admits a risk-neutral probability.

Proof

Exercise 3.7 below establishes the equivalence of (1) and (2), so it is sufficient to show that (2) and (3) are equivalent. Suppose first that (3) holds, that is, there is a risk-neutral probability $\mathbb{Q} = (q_1, \ldots, q_n)$. Since each $q_j > 0$, it follows from (3.1) and (3.14) that

$$\bar{S}_1(\omega_n) = \sum_{j=1}^n q_j \bar{S}_1(\omega_n) < \sum_{j=1}^n q_j \bar{S}_1(\omega_j) = S_0 < \sum_{j=1}^n q_j \bar{S}_1(\omega_1) = \bar{S}_1(\omega_1).$$

This gives (2).

Conversely, suppose that the model is viable, that is (2) holds. From assumption (3.1), there exists k with $1 \le k < n$ such that

$$\bar{S}_1(\omega_n) < \cdots < \bar{S}_1(\omega_{k+1}) \le S_0 \le \bar{S}_1(\omega_k) < \cdots < \bar{S}_1(\omega_1).$$

Taking the averages

$$a := \frac{1}{k}\big(\bar{S}_1(\omega_1) + \cdots + \bar{S}_1(\omega_k)\big), \qquad b := \frac{1}{n-k}\big(\bar{S}_1(\omega_{k+1}) + \cdots + \bar{S}_1(\omega_n)\big),$$

we obtain $b < S_0 < a$. Hence there is $q \in (0,1)$ such that $S_0 = qa + (1-q)b$, that is,

$$S_0 = \frac{q}{k}\big(\bar{S}_1(\omega_1) + \cdots + \bar{S}_1(\omega_k)\big) + \frac{1-q}{n-k}\big(\bar{S}_1(\omega_{k+1}) + \cdots + \bar{S}_1(\omega_n)\big).$$

If we let

$$q_j := \begin{cases} \dfrac{q}{k} & \text{if } 1 \le j \le k, \\[2mm] \dfrac{1-q}{n-k} & \text{if } k+1 \le j \le n, \end{cases}$$

then $q_j > 0$ for all j and

$$S_0 = \sum_{j=1}^{n} q_j \bar{S}_1(\omega_j).$$

Moreover,

$$\sum_{j=1}^{n} q_j = k \times \frac{q}{k} + (n-k) \times \frac{1-q}{n-k} = 1.$$

Therefore $\mathbb{Q} = (q_1, \ldots, q_n)$ is a risk-neutral probability for this model, and (3) holds true. $\qquad\square$

Exercise 3.7[‡]

Show that a single-period model with $n \ge 2$ scenarios containing a bond and a single stock contains no arbitrage if and only if the stock and bond prices satisfy inequality (3.15).

Hint. Generalize the proof of Theorem 2.9.

Having established necessary and sufficient conditions for a single-period single-stock model to be viable, we now turn to the issue of completeness. We noted that the simple model in Chapter 2 was complete. Here is the general definition, bearing in mind that a derivative can be *any* random variable D, representing a payoff at time 1.

Definition 3.8 (Completeness)

A market model is called *complete* if every derivative it gives rise to is attainable.

The next result is often referred to as the *Second Fundamental Theorem of Asset Pricing* for a single-period single-stock model.

Theorem 3.9 (Second Fundamental Theorem of Asset Pricing)

In a viable single-period model with one stock and $n \geq 2$ scenarios, the following statements are equivalent:

(1) The model is complete.

(2) The risk-neutral probability is unique.

(3) $n = 2$.

Proof

Theorem 2.24 shows that every derivative in the binary model is attainable, so binary models are complete. It also follows from Theorem 2.30 that there can only be one risk-neutral probability in any viable binary model. Thus (3) implies (1) and (2).

To complete the proof, we show that if (3) is false, then both (1) and (2) are false. Suppose then that $n > 2$. Consider the derivative with payoff

$$D(\omega_1) = S_1(\omega_2), \qquad D(\omega) = S_1(\omega) \quad \text{for all } \omega \neq \omega_1.$$

To find a replicating portfolio (x, y) for D, we need to find a simultaneous solution (x, y) to the n equations

$$xB_1 + yS_1(\omega) = D(\omega) \quad \text{for } \omega \in \Omega.$$

The first three equations are

$$xB_1 + yS_1(\omega_1) = D(\omega_1) = S_1(\omega_2),$$
$$xB_1 + yS_1(\omega_2) = D(\omega_2) = S_1(\omega_2),$$
$$xB_1 + yS_1(\omega_3) = D(\omega_3) = S_1(\omega_3).$$

The first two give

$$y = \frac{D(\omega_2) - D(\omega_1)}{S_1(\omega_2) - S_1(\omega_1)} = \frac{S_1(\omega_2) - S_1(\omega_2)}{S_1(\omega_2) - S_1(\omega_1)} = 0, \tag{3.17}$$

while the second and third equations yield

$$y = \frac{D(\omega_3) - D(\omega_2)}{S_1(\omega_3) - S_1(\omega_2)} = \frac{S_1(\omega_3) - S_1(\omega_2)}{S_1(\omega_3) - S_1(\omega_2)} = 1. \tag{3.18}$$

This shows that there is no solution and hence no replicating portfolio (x, y) for this particular D. That is, the derivative D is not attainable, so that the model is incomplete. Thus if (3) is false, then so is (1).

Finally we show that if $n > 2$, then (2) cannot be true either. A risk-neutral probability is a solution $\mathbb{Q} = (q_1, \ldots, q_n)$ to the two equations

$$q_1 \bar{S}_1(\omega_1) + \cdots + q_n \bar{S}_1(\omega_n) = S_0, \qquad q_1 + \cdots + q_n = 1,$$

together with $q_j \in (0, 1)$ for $j = 1, \ldots, n$. We are assuming that the model is viable, so Theorem 3.6 tells that there is a solution. Thus we have two equations in $n > 2$ unknowns, with at least one solution (that is, the system is consistent) so there must be infinitely many solutions and the risk-neutral probability is not unique. □

3.1.3 Pricing Attainable Derivatives

The pricing theory developed in Chapter 2 for the binary model extends easily to the pricing of *attainable* derivatives in more general models. The basic result is the Law of One Price, which reads exactly as in Chapter 2 (see Theorem 2.23) for a binary model.

Theorem 3.10 (Law of One Price in single period one stock model)

If a portfolio φ replicates an attainable derivative D in a viable single-stock, single-period model, then there is a unique fair price D_0 for D given by $D_0 = V_0^\varphi$, the initial φ.

The proof of Theorem 3.10 is exactly the same as the proof of Theorem 2.23 (which did not assume that Ω has only two elements), so it is not repeated here.

Pricing of attainable derivatives using risk-neutral probabilities $\mathbb{Q}$ must now take into account that there may be more than one such $\mathbb{Q}$. Fortunately this is not restrictive, as the next result shows.

Theorem 3.11

The unique fair price at time 0 for an attainable derivative D in a viable single-period single-stock model is

$$D_0 = \mathbb{E}_{\mathbb{Q}}(\bar{D}),$$

where $\mathbb{Q}$ is *any* risk-neutral probability.

Proof

Suppose that the portfolio φ replicates D. The Law of One Price (Theorem 3.10) together with Theorem 3.5 gives

$$D_0 = V_0^{\varphi} = \mathbb{E}_{\mathbb{Q}}(\bar{V}_1^{\varphi}) = \mathbb{E}_{\mathbb{Q}}(\bar{D}).$$

$\square$

Theorem 3.11 shows that in an incomplete model (where there are many risk-neutral probabilities), it doesn't matter which risk-neutral probability we use to price an attainable derivative. The following corollary makes this explicit.

Corollary 3.12

Suppose that D is an attainable derivative in a viable single-period single-stock model. For any two risk-neutral probabilities $\mathbb{Q}$ and $\mathbb{Q}'$,

$$\mathbb{E}_{\mathbb{Q}}(\bar{D}) = \mathbb{E}_{\mathbb{Q}'}(\bar{D}).$$

Proof

Both $\mathbb{E}_{\mathbb{Q}}(\bar{D})$ and $\mathbb{E}_{\mathbb{Q}'}(\bar{D})$ are equal to the fair price of D. $\square$

Theorem 3.11 only applies to the pricing of *attainable* derivatives. In order to price a derivative with no replicating portfolio, one needs to consider super-replicating portfolios instead; the final value of a super-replicating portfolio for a derivative is greater than or equal to the payoff of the derivative. We discuss this in the next section.

Exercise 3.13[†]

(a) Find a general form for the risk-neutral probability in Model 3.2.

(b) Show that any attainable derivative D in Model 3.2 must satisfy

$$3D(\omega_1) - 7D(\omega_2) + 4D(\omega_3) = 0, \qquad (3.19)$$

and compute its fair price in terms of $D(\omega_1)$ and $D(\omega_2)$.

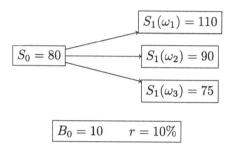

Model 3.2 One-period ternary model with a single stock in Exercises 3.13, 3.22 and 3.26

3.1.4 Pricing Non-attainable Derivatives

The pricing theory above is applicable only to attainable derivatives, and tells us nothing as yet about pricing of a *non*-attainable derivative D in a viable but incomplete single-stock model. Is there a fair price for such a D—and if so is it unique? Could there be a range of fair prices, and if so how can we identify them? The preceding theory suggests two approaches to these questions.

– Consider portfolios that give good approximations to D. This leads to the ideas of *super-replication* and *sub-replication* that we explore below.

– Consider the prices $\mathbb{E}_\mathbb{Q}(\bar{D})$ for the different possible risk-neutral probabilities $\mathbb{Q}$.

We will see that these two approaches coincide and give a well-defined range of fair prices for D. Recall that the definition of a *fair price* for a derivative D is any price that does not give extended arbitrage.

Let us denote by F_D the set of all possible fair prices for a derivative D. We know from the theory above that if D is attainable then F_D has a single element D_0 where

$$D_0 = \mathbb{E}_\mathbb{Q}(\bar{D}) \quad \text{for any risk neutral probability } \mathbb{Q}$$
$$= V_0^\varphi \quad \text{for any replicating portfolio } \varphi.$$

Beginning with the first approach suggested above we make the following definitions (applicable to any derivative).

Definition 3.14 (Super- and sub-replication)

(a) A portfolio φ *super-replicates* a derivative D if $V_1^\varphi \geq D$ (meaning that $V_1^\varphi(\omega) \geq D(\omega)$ for all ω).

(b) A portfolio φ *sub-replicates* a derivative D if $V_1^\varphi \leq D$.

Remark 3.15

If a portfolio φ replicates a derivative D, then $V_1^\varphi = D$, so that φ is both a sub- and super-replicating portfolio for D.

A portfolio φ that super-replicates a derivative D can be used by a trader *selling* D to hedge his liability. If the trader is able to sell D at the price $\pi = V_0^\varphi$ at time 0 and invests the proceeds in the portfolio φ, then he is guaranteed not to make a loss, because at time 1 he will own a portfolio worth V_1^φ which will at least cover his liability D, which can thus be delivered without risk. If he sells at any price greater than V_0^φ he can make a risk free profit.

Thus

$$\pi_D^a := \inf\{V_0^\varphi : \varphi \text{ super-replicates } D\} \tag{3.20}$$

intuitively represents an upper bound on prices that a *buyer* would find acceptable, because any greater price gives a risk-free profit for the seller. Putting this another way, the value π_D^a is an upper bound on selling prices that the market would find acceptable, and so it is sometimes called the *ask price* or *seller's price* for D. It is an upper limit on the prices that a seller could reasonably ask for D.

Similar considerations apply to

$$\pi_D^b := \sup\{V_0^\varphi : \varphi \text{ sub-replicates } D\}. \tag{3.21}$$

This is a lower bound on the price that a seller would consider selling D for, because any lower price would give a free lunch to the buyer. For this reason π^b is sometimes called the *bid price* or *buyer's price* for D; it is a lower limit on the prices that a buyer should reasonably offer to pay for D. Any lower price would make it too cheap.

Exercise 3.16[‡]

(a) Show that in a viable model with derivative D, if φ super-replicates D and φ' sub-replicates D, then $V_0^{\varphi'} \leq V_0^{\varphi}$.

Hint. Use Theorem 3.5.

(b) Deduce that $\pi_D^b \leq \pi_D^a$, and that if D is attainable then

$$\pi_D^a = \pi_D^b = D_0, \tag{3.22}$$

which is the unique fair price for D.

Exercise 3.17[†]

Let D be a non-attainable derivative in a viable model with one stock. Show that:

(a) If φ super-replicates D and $\pi \geq V_0^{\varphi}$, then π is not a fair price for D.

(b) If φ sub-replicates D and $\pi \leq V_0^{\varphi}$, then π is not a fair price for D.

The key to characterizing F_D for a non-attainable derivative by means of super- and sub-replication is the following result.

Theorem 3.18

Let D be a non-attainable derivative in a viable incomplete single-stock model. There exist super- and sub-replicating portfolios φ^a and φ^b for D such that $V_0^{\varphi^a} = \pi_D^a$ and $V_0^{\varphi^b} = \pi_D^b$. Hence if $\pi \geq \pi_D^a$ or $\pi \leq \pi_D^b$ then π is not fair, and $\pi_D^b < \pi_D^a$.

Remark 3.19

Theorem 3.18 shows that in the case of a non-attainable derivative D, the names *ask price* and *bid price* are confusing misnomers: π_D^a is *not* a fair price to ask, since it allows extended arbitrage for the seller, and π_D^b is *not* a fair price to offer, as it leads to extended arbitrage for the buyer.

Before we prove Theorem 3.18 let us see how we can quickly use it to pin down the set of fair prices for a non-attainable derivative.

Theorem 3.20

The set F_D of fair prices for a non-attainable derivative D in a viable single-stock model is a non-empty interval given by

$$F_D = \left(\pi_D^b, \pi_D^a\right).$$

Proof

Theorem 3.18 shows that any $\pi \geq \pi_D^a$ is not fair, and similarly for any price $\pi \leq \pi_D^b$. Hence $F_D \subseteq (\pi_D^b, \pi_D^a)$.

Now suppose that π is an unfair price for D. To complete the proof we must show that $\pi \geq \pi_D^a$ or $\pi \leq \pi_D^b$. There is an extended portfolio (x, y, z) with

$$xB_0 + yS_0 + z\pi = 0, \tag{3.23}$$

$$xB_1 + yS_1 + zD \geq 0 \tag{3.24}$$

and $xB_1 + yS_1(\omega) + zD(\omega) > 0$ for at least one $\omega \in \Omega$. By scaling we may assume that $z = \pm 1$ since the viability of the model implies that $z \neq 0$. If $z = 1$ then rearrangement in (3.24) yields $-xB_1 - yS_1 \leq D$, so the portfolio $(-x, -y)$ sub-replicates D. It follows from (3.23) that

$$\pi = -xB_0 - yS_0 = V_0^{(-x,-y)} \leq \pi_D^b. \tag{3.25}$$

If $z = -1$, a similar argument shows that $xB_1 + yS_1 \geq D$, and so the portfolio (x, y) super-replicates D. Thus

$$\pi \geq \pi_D^a. \tag{3.26}$$

It follows that $F_D = (\pi_D^b, \pi_D^a)$.   □

Proof (of Theorem 3.18)

We give an intuitive geometric proof of the existence of φ^a for π_D^a. We may assume that $B_0 = 1$ without any loss of generality.

Consider Figure 3.3 that plots above each discounted stock price $\bar{S}_1(\omega)$ the corresponding value of the discounted derivative $\bar{D}(\omega)$. Suppose we are now given a portfolio $\varphi = (\xi, \eta)$. The discounted value $\bar{V}_1^\varphi$ for any given value of the discounted stock $\bar{S}_1 = s$ can be illustrated as in Figure 3.3 by plotting the line $v = \xi + \eta s$, where we have s running along the horizontal ("discounted stock") axis and v along the vertical ("discounted value") axis. Note that here the values ξ and η are fixed, and it is s and v that are the variables.

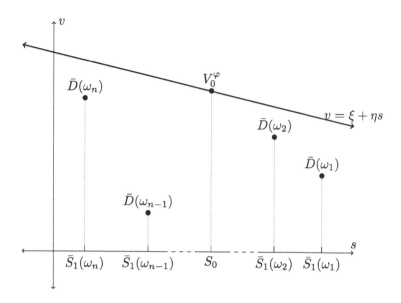

Figure 3.3 Super-replicating portfolio $\varphi = (\xi, \eta)$ in a single-stock model

The portfolio $\varphi = (\xi, \eta)$ super-replicates D if and only if

$$\xi + \eta \bar{S}_1(\omega) \geq \bar{D}(\omega) \quad \text{for each } \omega \in \Omega;$$

that is, if the line $v = \xi + \eta s$ lies on or above each of the points $(\bar{S}_1(\omega), \bar{D}(\omega))$ that we have plotted. Such a line is illustrated in the diagram. The point where the line $v = \xi + \eta s$ intercepts the vertical line $s = S_0$ is the value $\xi + \eta S_0 = V_0^\varphi$.

Thus, to find a super-replicating portfolio φ with *least* value $V_0^\varphi = \pi_D^a$ means finding a line $v = \xi + \eta s$ on or above each of the points $(\bar{S}_1(\omega), \bar{D}(\omega))$, with the lowest possible intercept with the vertical line $s = S_0$. It is easy to see that such a line exists, and in fact (as we see in Example 3.21 below) it will join two of the plotted points, one on either side of the line $s = S_0$. Then we let $\varphi^a := (\xi, \eta)$ for this line.

The portfolio φ^b can be found similarly, using lines $v = \xi + \eta s$ that lie on or below the points that plot the values of $\bar{D}$, and taking the line with maximum intercept on the vertical line $s = S_0$.

Since φ^b sub-replicates D and φ^a super-replicates D, Exercise 3.17 shows that any price π with $\pi \leq \pi_D^b$ or $\pi \geq \pi_D^a$ is not fair.

It remains to check that $\pi_D^b < \pi_D^a$. We have

$$V_1^{\varphi^b} \leq D \leq V_1^{\varphi^a}$$

and, since D is not attainable, this means that

$$V_1^{\varphi^b}(\omega) < D(\omega) \quad \text{and} \quad D(\omega') < V_1^{\varphi^a}(\omega')$$

for some $\omega, \omega' \in \Omega$. Taking any risk-neutral probability $\mathbb{Q}$, it follows that

$$\mathbb{E}_{\mathbb{Q}}(\bar{V}_1^{\varphi^b}) < \mathbb{E}_{\mathbb{Q}}(\bar{D}) < \mathbb{E}_{\mathbb{Q}}(\bar{V}_1^{\varphi^a}). \qquad (3.27)$$

But Theorem 3.5 gives

$$\mathbb{E}_{\mathbb{Q}}(\bar{V}_1^{\varphi^b}) = V_0^{\varphi^b} = \pi_D^b, \qquad \mathbb{E}_{\mathbb{Q}}(\bar{V}_1^{\varphi^a}) = V_0^{\varphi^a} = \pi_D^a,$$

and therefore $\pi_D^b < \pi_D^a$. $\qquad\qquad\qquad\qquad\qquad\qquad\qquad\qquad\quad \square$

The following example illustrates Theorems 3.18 and 3.20.

Example 3.21

Consider again the pricing problem of the non-attainable put option with strike 220 in Model 3.1. As in Section 3.1.1 its payoff is the random variable

$$P_1 = [S_1 - 220]_+.$$

Let us compute the bid and ask prices of this option using the above geometrical idea.

Recall that $r = 0$, so that $B_0 = B_1 = 1$, $\bar{S}_1 = S_1$ and $\bar{P}_1 = P_1$. Any super-replicating portfolio (ξ, η) for the put option satisfies

$$\xi B_0 + \eta \bar{S}_1(\omega) = \xi B_1 + \eta S_1(\omega) \geq P_1(\omega) = \bar{P}_1(\omega) \quad \text{for } \omega \in \Omega.$$

In other words, if we consider the three points

$$\left.\begin{aligned}(\bar{S}_1(\omega_1), P_1(\omega_1)) &= (230, 0), \\ (S_1(\omega_2), P_1(\omega_2)) &= (210, 10), \\ (S_1(\omega_3), P_1(\omega_3)) &= (190, 30)\end{aligned}\right\} \qquad (3.28)$$

in the (s, v) plane in Figure 3.4, then any super-replicating portfolio (ξ, η) corresponds to a straight line with equation $v = \xi + \eta s$ that is *above* these points, that is

$$v = \xi + \eta s \geq P_1(\omega_k) \quad \text{whenever} \quad s = S_1(\omega_k)$$

for $k = 1, 2, 3$. To compute the ask price, we need to find a line $v = \xi + \eta s$ above these points and such that $\xi + 220\eta$ is a minimum. This can be done

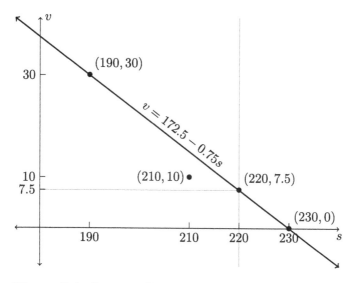

Figure 3.4 Super-replicating portfolios in Example 3.21

graphically: it is easy to see that the line connecting $(190, 30)$ and $(230, 0)$ satisfies this property. The equation of this line is

$$v = 172.5 - 0.75s,$$

which corresponds to the portfolio $(x^a, y^a) := (172.5, -0.75)$. It is clear that

$$x^a B_1 + y^a S_1(\omega_1) = x^a + 230y^a = 0 = P_1(\omega_1),$$
$$x^a B_1 + y^a S_1(\omega_2) = x^a + 210y^a = 15 > 10 = P_1(\omega_2),$$
$$x^a B_1 + y^a S_1(\omega_3) = x^a + 190y^a = 30 = P_1(\omega_3).$$

That is, (x^a, y^a) replicates the put option exactly in the scenarios ω_1 and ω_3, while its value in scenario ω_2 is strictly greater than the payoff of the put, exactly as in Figure 3.4. So the ask price of the put option is

$$\pi_{P_1}^a = V_0^{(x^a, y^a)} = 172.5 - 0.75 \times 220 = 7.5.$$

A similar argument can be used to compute the bid price of the put option. A portfolio (ξ, η) sub-replicates the put option if

$$\xi B_0 + \eta \bar{S}_1(\omega) = \xi B_1 + \eta S_1(\omega) \leq P_1(\omega) = \bar{P}_1(\omega) \quad \text{for } \omega \in \Omega.$$

That is, if it corresponds to a line $v = \xi + \eta s$ lying *below* the points in (3.28) in the (s, v) plane in Figure 3.5. The bid price can be found by finding the line in

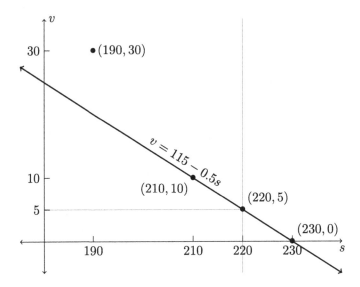

Figure 3.5 Sub-replicating portfolios in Example 3.21

Figure 3.5 lying below the given points with the maximum value when $s = 220$. This line is

$$v = 115 - \tfrac{1}{2}s,$$

which corresponds to the portfolio $(x^b, y^b) = (115, -0.5)$ that replicates the put option exactly in the scenarios ω_2 and ω_3. It follows that the bid price of the put option is

$$\pi_{P_1}^b = V_0^{(x^b, y^b)} = x^b + 220 y^b = 5.$$

Hence $F_D = (5, 7.5)$.

Exercise 3.22[†]

Consider a call option with strike 80 in Model 3.2.

(a) Show that it is not attainable.

   Hint. Use Exercise 3.13(b).

(b) Compute its ask price, and find a super-replicating portfolio that realizes this price.

(c) Compute its bid price, and find a sub-replicating portfolio that realizes this price.

Exercise 3.23[‡]

For enthusiasts only. Prove the first part of Theorem 3.18 for π_D^a analytically as follows. Recall from (2.18) that $\Delta \bar{S}_1 = \bar{S}_1 - S_0$.

(a) Show that a portfolio $\varphi = (x, y)$ super-replicates D if and only if

$$y\Delta \bar{S}_1(\omega) \geq \bar{D}(\omega) - V_0^\varphi \quad \text{for all } \omega \in \Omega. \tag{3.29}$$

(b) By the definition of infimum, there exists a sequence $(\varphi_k)_{k \in \mathbb{N}}$ of super-replicating portfolios such that $V_0^{\varphi_k} \to \pi_D^a$. Show that there exists a number $M \geq 0$ such that

$$y_k \Delta \bar{S}_1(\omega) \geq -M \quad \text{for all } \omega \in \Omega \tag{3.30}$$

for $k \in \mathbb{N}$, where $\varphi_k = (x_k, y_k)$.

Hint. A convergent sequence is bounded.

(c) Show that

$$\Delta \bar{S}_1(\omega_n) < 0 < \Delta \bar{S}_1(\omega_1) \tag{3.31}$$

if the model is viable. Use this to show that $(y_k)_{k \in \mathbb{N}}$ is bounded, and hence has a convergent subsequence.

(d) Taking a subsequence if necessary, let $y := \lim_{k \to \infty} y_k$ and show that $x_k \to x$, where $x := \frac{1}{B_0}(\pi_D^a - yS_0)$. Show that $V_0^{(x,y)} = \pi_D^a$.

(e) Verify that (x, y) super-replicates D.

The other natural approach to pricing a non-attainable derivative D is to use risk-neutral probabilities as at the end of Example 3.21. We might guess that if $\mathbb{Q}$ is a risk-neutral probability then $\mathbb{E}_\mathbb{Q}(\bar{D})$ is a fair price—although this may not be unique since there is more than one (in fact infinitely many) such $\mathbb{Q}$ in an incomplete model. This guess is correct, as we now show.

Theorem 3.24

Let D be a non-attainable derivative in a viable incomplete model with one stock. The following statements are equivalent:

(1) π is a fair price for D.

(2) $\pi = \mathbb{E}_\mathbb{Q}(\bar{D})$ for some risk-neutral probability $\mathbb{Q}$.

Hence the set F_D of fair prices for D is

$$F_D = (\pi_D^b, \pi_D^a) = \{\mathbb{E}_\mathbb{Q}(\bar{D}) : \mathbb{Q} \text{ is a risk neutral probability}\}.$$

Proof

Suppose that $\pi = \mathbb{E}_{\mathbb{Q}}(\bar{D})$ for some risk-neutral probability $\mathbb{Q}$. It follows from (3.27) that $\pi_D^b < \pi < \pi_D^a$, and from Theorem 3.18 that π is a fair price for D. Thus (2) implies (1).

The proof of the converse is deferred to Theorem 3.68 in the next section, since it uses the extension of Theorem 3.6 to the multi-stock model. □

Example 3.25

In Example 3.21, the bid-ask interval for a put option with strike 220 in Model 3.1 was shown to be

$$\left(\pi_{P_1}^b, \pi_{P_1}^a\right) = (5, 7.5).$$

Exercise 3.2 and the discussion following it showed that every risk-neutral probability in Model 3.1 takes the form

$$\mathbb{Q} = \left(\lambda, \tfrac{3}{2} - 2\lambda, \lambda - \tfrac{1}{2}\right)$$

for some $\lambda \in (\tfrac{1}{2}, \tfrac{3}{4})$, and that

$$\mathbb{E}_{\mathbb{Q}}(\bar{P}_1) = \mathbb{E}_{\mathbb{Q}}(P_1) = 10\lambda \in (5, 7.5).$$

Thus the bid-ask interval $(\pi_{P_1}^b, \pi_{P_1}^a)$ coincides exactly with the set of expected values of the put option under equivalent risk-neutral probabilities; that is,

$$F_{P_1} = \left(\pi_{P_1}^b, \pi_{P_1}^a\right) = (5, 7.5) = \left\{\mathbb{E}_{\mathbb{Q}}(\bar{P}_1) : \mathbb{Q} \text{ a risk-neutral probability}\right\}.$$

Exercise 3.26

Consider Model 3.2.

(a) Show that the set F_{C_1} of fair prices found by means of Theorem 3.24 for a call option C_1 with strike 80 in this model is the same as $(\pi_{C_1}^b, \pi_{C_1}^a)$ found in Exercise 3.22.

   Hint. Use Exercise 3.13(a).

(b) Find the set of fair prices for a put option with strike 100 in this model.

The pricing theory for a derivative in a single period one stock model can now be summarized as follows.

Theorem 3.27

If D is a derivative in a viable single period model with a single stock, then

$$\pi_D^a := \inf\{V_0^\varphi : \varphi \text{ super-replicates } D\}$$

$$= \sup\{\mathbb{E}_\mathbb{Q}(\bar{D}) : \mathbb{Q} \text{ is a risk neutral probability}\},$$

$$\pi_D^b := \sup\{V_0^\varphi : \varphi \text{ sub-replicates } D\}$$

$$= \inf\{\mathbb{E}_\mathbb{Q}(\bar{D}) : \mathbb{Q} \text{ is a risk neutral probability}\}.$$

(1) The following statements are equivalent:

    (i) D is attainable.

    (ii) $\pi_D^b = \pi_D^a$.

    (iii) π_D^a is a fair price for D.

    (iv) π_D^b is a fair price for D.

    (v) There is a unique fair price for D.

(2) The following statements are equivalent:

    (i) D is not attainable.

    (ii) $\pi_D^b < \pi_D^a$.

    (iii) The set of fair prices for D is $F_D = (\pi_D^b, \pi_D^a)$.

3.2 Models with Several Stocks

We now turn our attention to models that may contain several stocks, denoted by $S^1, \ldots, S^m$, where $m \geq 1$. We assume that the initial prices $S_0^1, \ldots, S_0^m$ are known at time 0, and as before that the stocks do not pay dividends. The price of each stock S^i at time 1 is a random variable; that is, a function $S_1^i : \Omega \to \mathbb{R}$. Again Ω is the set of all possible scenarios, having $n \geq 2$ elements. We continue to assume that the bond price is always strictly positive; that is, $B_t > 0$ for $t = 0, 1$. We *do not* need to assume that stock prices are always positive, because the theory developed below applies even if $S_t^i \leq 0$ for some i and t. This will be useful later when we need to apply the theory to risky assets other than stock that may take negative values. Of course all the results in this section apply to the single-stock case already treated (taking $m = 1$), including the binary model discussed in Chapter 2.

The stock prices at time t can be combined into a single *price vector*

$$S_t := (S_t^1, \ldots, S_t^m).$$

Here and below we will switch freely between the row and column representation of vectors, so this is equivalent to

$$S_t := \begin{bmatrix} S_t^1 \\ \vdots \\ S_t^m \end{bmatrix}.$$

We assume that the future price vectors $S_1(\omega)$ are different (in at least one component) for different $\omega \in \Omega$. Any model of this kind may be represented as a tree with n branches (one for each scenario ω). Model 3.6 below gives an example of this.

Example 3.28

Model 3.6 contains two risky assets. Despite the fact that

$$S_1^1(\omega_2) = 154 = S_1^1(\omega_3),$$

we have

$$S_1(\omega_2) = \begin{bmatrix} 154 \\ 220 \end{bmatrix} \neq \begin{bmatrix} 154 \\ 275 \end{bmatrix} = S_1(\omega_3),$$

equivalently

$$S_1(\omega_2) = (154, 220) \neq (154, 275) = S_1(\omega_3),$$

and therefore Model 3.6 has four distinct scenarios.

We continue to allow unrestricted trading in all the primary assets. An investor or trader may form any *portfolio* φ consisting of $x \in \mathbb{R}$ bonds and $y = (y^1, \ldots, y^m) \in \mathbb{R}^m$ shares. The value of the portfolio $(x, y) \in \mathbb{R}^{m+1}$ is

$$V_t^\varphi = V_t^{(x,y)} := xB_t + y \cdot S_t = xB_t + \sum_{i=1}^m y^i S_t^i.$$

Here $\cdot$ is the Euclidean inner product.

In addition to the primary assets, a model may contain a derivative with a random payoff D at time 1, and trading at a price π at time 0. In this case, an investor may form any *extended portfolio* of $x \in \mathbb{R}$ bonds, $y = (y^1, \ldots, y^m) \in \mathbb{R}^m$

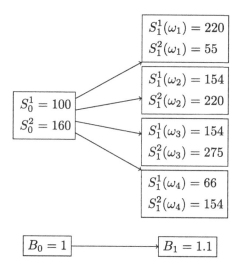

Model 3.6 One-period model with two stocks and four outcomes in Examples 3.28 and 3.40

shares and z derivatives. The values of the extended portfolio $\psi = (x, y, z) \in \mathbb{R}^{m+2}$ are then as expected:

$$V_0^\psi := xB_0 + y \cdot S_0 + z\pi = xB_0 + \sum_{i=1}^m y^i S_0^i + z\pi,$$

$$V_1^\psi := xB_1 + y \cdot S_1 + zD = xB_1 + \sum_{i=1}^m y^i S_1^i + zD.$$

The notions of arbitrage, extended arbitrage and fair pricing in models with multiple risky assets and derivatives are the natural extension of the single stock case, and are defined as follows.

Definition 3.29 (Arbitrage, extended arbitrage, fair price)

(a) An *arbitrage opportunity* is a portfolio $\varphi = (x, y^1, \ldots, y^m)$ with $V_0^\varphi = 0$ and $V_1^\varphi \geq 0$, together with $V_1^\varphi(\omega) > 0$ for at least one $\omega \in \Omega$. A model is *viable* if there are no arbitrage opportunities.

(b) An *extended arbitrage opportunity* is an extended portfolio

$$\psi = (x, y^1, \ldots, y^m, z)$$

with $V_0^\psi = 0$ and $V_1^\psi \geq 0$, together with $V_1^\psi(\omega) > 0$ for at least one $\omega \in \Omega$.

(c) For a derivative D, a price π is a *fair price* if it does not allow extended arbitrage involving D.

Note that an arbitrage opportunity $\varphi = (x, y)$ is specified completely by the stock holding y, because if this is known then the property $V_0^\varphi = 0$ means that $x = -\frac{1}{B_0} y \cdot S_0$. Thus we make the following definition.

Definition 3.30 (Implicit arbitrage opportunity)

An *implicit arbitrage opportunity* is any stock holding y such that the portfolio

$$\varphi := \left(-\frac{1}{B_0} y \cdot S_0, y \right)$$

is an arbitrage opportunity.

The lemma below will be useful later; first we need some notation. The idea of discounting is the same as before. Define the discounted stock price at time 1 as $\bar{S}_1 := \frac{1}{1+r} S_1$, and the increment of the discounted stock price vector as

$$\Delta \bar{S}_1 := \bar{S}_1 - S_0. \tag{3.32}$$

Similarly, the discounted value of a portfolio φ at time 1 is $\bar{V}_1^\varphi := \frac{1}{1+r} V_1^\varphi$ and the increment of the discounted value of φ is

$$\Delta \bar{V}_1^\varphi := \bar{V}_1^\varphi - V_0^\varphi. \tag{3.33}$$

Notice for any portfolio $\varphi = (x, y)$ that

$$\Delta \bar{V}_1^\varphi = x\bar{B}_1 + y \cdot \bar{S}_1 - (xB_0 + y \cdot S_0) = y \cdot (\bar{S}_1 - S_0) = y \cdot \Delta \bar{S}_1. \tag{3.34}$$

Lemma 3.31 (Implicit Arbitrage Lemma)

(1) The following statements are equivalent for a stock holding y:

   (i) y is an implicit arbitrage opportunity.

   (ii) $y \cdot \Delta \bar{S}_1(\omega) \geq 0$ for all ω, and $y \cdot \Delta \bar{S}_1(\omega) > 0$ for at least one ω.

(2) A model is viable if and only if there is no implicit arbitrage opportunity.

Proof

To establish (1), take any $y \in \mathbb{R}^m$ and let $\varphi := (-\frac{1}{B_0} y \cdot S_0, y)$. Then

$$V_0^\varphi = xB_0 + y \cdot S_0 = 0$$

and so (3.33) and (3.34) give

$$\bar{V}_1^\varphi = \Delta \bar{V}_1^\varphi = y \cdot \Delta \bar{S}_1.$$

The result then follows because $V_1^\varphi \geq 0$ if and only if $\bar{V}_1^\varphi \geq 0$, and similarly for $V_1^\varphi > 0$.

Statement (2) holds true because every arbitrage opportunity gives rise to an implicit arbitrage opportunity, and vice versa.   □

The observation (1) in the above lemma accords with the intuition that an arbitrage opportunity is given by any stock holding (financed by a loan from the bank) that gives a risk free profit after discounting the effect of inflation. The quantity $y \cdot \Delta \bar{S}_1(\omega)$ is precisely the gain, after discounting, from a holding of y shares.

3.3 Viability with Several Stocks

In a model with a single stock, we saw that lack of arbitrage was equivalent to having a risk-neutral probability. In mathematical terms this can be explained by saying that the initial stock price can be expressed as a *convex combination* or *weighted average* of the possible discounted future stock prices. The aim now is to extend this result to the case of more than one stock.

The proof of the First Fundamental Theorem in a single-stock model used the fact that the future prices of the stock can be set out in increasing order, as in (3.1), leading to the expression of S_0 as a convex combination of the future discounted prices $\bar{S}_1(\omega)$ in a viable model. This can be done because there is just one stock, and the future values can be set out in order on a line, which is 1-dimensional.

If the number of stocks is $m > 1$ then the vectors of future prices are "m-dimensional numbers" and are naturally represented in m-dimensional space. For example, if there are 2 stocks then the future prices would be the set of vectors $(S_1^1(\omega), S_1^2(\omega))$ that lie in the plane. Clearly it is not possible in general to order vectors in dimensions higher than 1; however, the idea of viewing risk-neutral probabilities as the weights of a convex combination does extend to models with more than one risky asset, and this is one way to understand the First Fundamental Theorem for many stocks below.

First, we need the definition of a risk-neutral probability, which is essentially the same as before—but bearing in mind that we are dealing with price *vectors* S_t rather than single prices. Writing $\Omega = \{\omega_1, \ldots, \omega_n\}$ for convenience, we have the following.

Definition 3.32 (Risk-neutral probability)

A *risk-neutral probability* $\mathbb{Q}$ for a single-period model with m stocks and n scenarios is a family $(q_1, \ldots, q_n)$ satisfying

$$S_0 = \mathbb{E}_{\mathbb{Q}}(\bar{S}_1) = \sum_{i=1}^{n} q_i \bar{S}_1(\omega_i), \tag{3.35}$$

together with

$$\sum_{i=1}^{n} q_i = 1 \tag{3.36}$$

and $q_i \in (0,1)$ for $i = 1, \ldots, n$.

Remarks 3.33

(1) Equation (3.35) is actually a *vector equation* and is equivalent to

$$S_0^i = \mathbb{E}_{\mathbb{Q}}(\bar{S}_1^i) = \sum_{j=1}^{n} q_j \bar{S}_1^i(\omega_j) \quad \text{for } i = 1, \ldots, m.$$

(2) A vector v is called a *convex combination* of the vectors $w_1, \ldots, w_p$ with associated weights $\lambda_1, \ldots, \lambda_p$ if

$$y = \sum_{k=1}^{p} \lambda_k w_k,$$

where $\lambda_k \geq 0$ for all k and $\lambda_1 + \cdots + \lambda_p = 1$. Thus $\mathbb{Q} = (q_1, \ldots, q_n)$ is a risk-neutral probability if and only if $q_i \in (0,1)$ for all i and S_0 is a convex combination of $\bar{S}_1(\omega_1), \ldots, \bar{S}_1(\omega_n)$ with weights $q_1, \ldots, q_n$.

(3) The condition $q_j \in (0,1)$ for all j is equivalent to $q_j > 0$ for all i because of (3.36).

(4) Strictly speaking (and in more advanced financial mathematics), a risk-neutral probability $\mathbb{Q}$ is required to be *equivalent* to the real-world probability $\mathbb{P}$; that is, for every set $A \subseteq \Omega$, we have $\mathbb{Q}(A) > 0$ if and only if $\mathbb{P}(A) > 0$. Since we have assumed that $\mathbb{P}(\omega) > 0$ for all $\omega \in \Omega$, it is therefore sufficient to require that $q_j > 0$ for $j = 1, \ldots, n$.

Exercise 3.34[‡]

(a) Show that the set of risk-neutral probabilities is convex.

• (b) Show that if D is any random variable on Ω, the set

$$\{\mathbb{E}_{\mathbb{Q}}(D) : \mathbb{Q} \text{ a risk-neutral probability}\}$$

is convex.

Hint. If $x = \lambda\mathbb{E}_{\mathbb{Q}}(D) + (1-\lambda)\mathbb{E}_{\mathbb{Q}'}(D)$, consider the probability

$$\mathbb{Q}^\lambda = \lambda\mathbb{Q} + (1-\lambda)\mathbb{Q}'. \tag{3.37}$$

There several ways to recognize a risk neutral probability, summarized in the following theorem, part of which we saw in the single-stock case in Theorem 3.5.

Theorem 3.35

If $\mathbb{Q} = (q_1, \ldots, q_n)$ is any probability with $q_j > 0$ for all j, then the following are equivalent:

(1) $\mathbb{Q}$ is a risk neutral probability.

(2) $\mathbb{E}_{\mathbb{Q}}(\Delta\bar{S}_1) = 0$.

(3) $\mathbb{E}_{\mathbb{Q}}(\bar{V}_1^\varphi) = V_0^\varphi$ for every portfolio φ.

(4) $\mathbb{E}_{\mathbb{Q}}(\Delta\bar{V}_1^\varphi) = 0$ for every portfolio φ.

(5) $\mathbb{E}_{\mathbb{Q}}(y \cdot \Delta\bar{S}_1) = 0$ for any $y \in \mathbb{R}^m$.

Proof

The equivalence of (1) and (2) comes from the definition $\Delta\bar{S}_1 := \bar{S}_1 - S_0$; the equivalence of (3), (4) and (5) comes from the definition $\Delta\bar{V}_1^\varphi := \bar{V}_1^\varphi - V_0^\varphi$ and the equation (3.34).

We now show that (2) is equivalent to (5). First, if $\mathbb{E}_{\mathbb{Q}}(\Delta\bar{S}_1) = 0$, then

$$\mathbb{E}_{\mathbb{Q}}(y \cdot \Delta\bar{S}_1) = y \cdot \mathbb{E}_{\mathbb{Q}}(\Delta\bar{S}_1) = 0$$

for all $y \in \mathbb{R}^m$. Thus (2) implies (5). For the converse, assuming (5) we may take $y = \mathbb{E}_{\mathbb{Q}}(\Delta\bar{S}_1)$, which gives

$$0 = \mathbb{E}_{\mathbb{Q}}\big(\mathbb{E}_{\mathbb{Q}}(\Delta\bar{S}_1) \cdot \Delta\bar{S}_1\big) = \mathbb{E}_{\mathbb{Q}}(\Delta\bar{S}_1) \cdot \mathbb{E}_{\mathbb{Q}}(\Delta\bar{S}_1).$$

It follows that $\mathbb{E}_{\mathbb{Q}}(\Delta\bar{S}_1) = 0$, which is (2). $\qquad\square$

We can now state the *First Fundamental Theorem of Asset Pricing* for a general discrete single-period model.

Theorem 3.36 (First Fundamental Theorem of Asset Pricing)

A single-period model with a finite number of scenarios and an arbitrary number of stocks is viable if and only if it admits a risk-neutral probability.

One half of this is easy to prove, so we do so immediately.

Proposition 3.37

If a single-period model with a finite number of scenarios and an arbitrary number of stocks admits a risk-neutral probability, then it is viable.

Proof

Suppose that the model admits a risk-neutral probability $\mathbb{Q} = (q_1, \ldots, q_n)$. If φ is any portfolio satisfying $V_1^\varphi \geq 0$ and $V_1^\varphi(\omega) > 0$ for some $\omega \in \Omega$, then

$$\mathbb{E}_\mathbb{Q}\big(\bar{V}_1^\varphi\big) = \sum_{i=1}^n q_i \bar{V}_1^\varphi(\omega_i) > 0.$$

It follows from Theorem 3.35(3) that $V_0^\varphi > 0$, and consequently there is no possibility of arbitrage. $\square$

Exercise 3.38[†]

Extend Model 3.1 by adding another stock S' with values $S_0' = 85$, $S_1'(\omega_1) = 80$, $S_1'(\omega_2) = 70$ and $S_1'(\omega_3) = 90$. Show that there is no risk-neutral probability assignment in this model, and then construct an arbitrage opportunity.

To complete the proof of Theorem 3.36 we have the following, which is the converse of Proposition 3.37.

Proposition 3.39

A viable single-period model with a finite number of scenarios and an arbitrary number of stocks admits a risk-neutral probability.

We will give two different proofs of Proposition 3.39. The first is easier to visualize, but the second is important from a theoretical point of view because it generalizes more readily to more complex models (especially where time is continuous).

The first proof of Proposition 3.39 hinges on the fact that the price vectors S_0 and S_1 for m stocks lie in $\mathbb{R}^m$, and it involves the idea of a *convex hull*. The *convex hull* of a finite collection of points in $\mathbb{R}^m$ is defined as the set of convex combinations of those points, or equivalently as the smallest convex set that contains those points.

Example 3.40

Consider Model 3.6. The discounted stock prices and discounted stock price increments are

$$\bar{S}_1(\omega) = \begin{cases} (200, 50) & \text{if } \omega = \omega_1, \\ (140, 200) & \text{if } \omega = \omega_2, \\ (140, 250) & \text{if } \omega = \omega_3, \\ (60, 140) & \text{if } \omega = \omega_4, \end{cases} \qquad \Delta\bar{S}_1(\omega) = \begin{cases} (100, -110) & \text{if } \omega = \omega_1, \\ (40, 40) & \text{if } \omega = \omega_2, \\ (40, 90) & \text{if } \omega = \omega_3, \\ (-40, -20) & \text{if } \omega = \omega_4. \end{cases}$$

It is quite natural to represent these increments as points on a two-dimensional plane, as in see Figure 3.7. The shaded region in this figure is

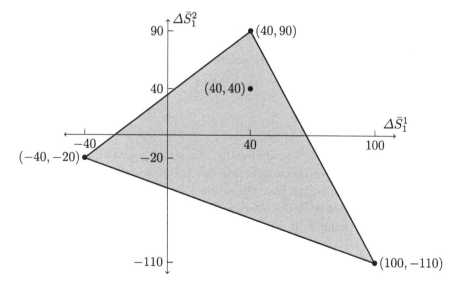

Figure 3.7 Convex hull of discounted price increments in Example 3.40

the convex hull of the discounted increments; that is, the smallest convex set that contains all the increments. Observe that the origin $(0,0)$ is in the interior of this convex hull, and therefore it can be written as a convex combination of the discounted increments; that is, there exist positive numbers q_1, q_2, q_3, q_4 such that

$$q_1 \Delta \bar{S}_1(\omega_1) + q_2 \Delta \bar{S}_1(\omega_2) + q_3 \Delta \bar{S}_1(\omega_3) + q_4 \Delta \bar{S}_1(\omega_4) = 0,$$

$$q_1 + q_2 + q_3 + q_4 = 1.$$

This is exactly what is meant by a risk-neutral probability, and therefore Model 3.6 must be viable.

Exercise 3.41[‡]

Sketch the convex hull of $\{\Delta \bar{S}_1(\omega) : \omega \in \Omega\}$ for the non-viable model in Exercise 3.38. Is 0 is an element of this convex hull?

Convexity and the existence of risk-neutral probabilities are connected in the following way.

Lemma 3.42

In a single-period model with m stocks and scenarios $\omega_1, \ldots, \omega_n$ the following are equivalent:

(1) There is a risk-neutral probability.

(2) The origin $0 \in \mathbb{R}^m$ belongs to the *relative interior* of the convex hull of the discounted increments $\Delta \bar{S}_1(\omega_1), \ldots, \Delta \bar{S}_1(\omega_n)$.

Remark 3.43

The *interior* of a set in $\mathbb{R}^m$ is, informally, the set of points that are not on its boundary. The idea of the *relative interior* involves a slight subtlety. It is possible for the convex hull to have dimension k smaller than m, in which case the relative interior refers to the interior of the convex hull when viewed as a subset of $\mathbb{R}^k$. For example, if $m = 2$, the convex hull of two points is simply the line segment joining them (so $k = 1$), and this has no interior as a set in the plane. However, since it is a one-dimensional topological space its relative interior is the line segment without its endpoints. Of course, if the dimension of the convex hull is m, then its relative interior coincides with its interior.

Remark 3.44

Both proofs of Proposition 3.39 involve the notion of a *hyperplane* in $\mathbb{R}^m$ which is a set of the form

$$\{v \in \mathbb{R}^m : a \cdot v = b\}$$

for some $a \in \mathbb{R}^m \setminus \{0\}$ and $b \in \mathbb{R}$. It is clear that a hyperplane in $\mathbb{R}$ is a point, in $\mathbb{R}^2$ it is a line and a hyperplane in $\mathbb{R}^3$ is a plane. In general it is a subspace of dimension one less than the whole space, but not necessarily through the origin.

It is helpful to view a hyperplane in $\mathbb{R}^m$ geometrically as a linear (affine) set that divides $\mathbb{R}^m$ into two parts—it is this interpretation that makes the first proof of Proposition 3.39 easy to visualize. The second proof of Proposition 3.39 uses the notion of a hyperplane in $\mathbb{R}^n$ together with the Separating Hyperplane Theorem (Lemma 3.46).

Here is the first proof of Proposition 3.39.

Proof (first proof of Proposition 3.39)

Suppose that the model does not admit a risk-neutral probability. We know from Lemma 3.42 that this is equivalent to 0 not being in the relative interior of the convex hull of the increments $\Delta \bar{S}_1(\omega_n), \ldots, \Delta \bar{S}_1(\omega_n)$. Then we can find an arbitrage opportunity (x, y) as follows, so that the model is not viable.

We need to consider two possibilities. Suppose first that 0 is not on the boundary of the convex hull. It is possible to find a hyperplane ℓ that separates 0 from the convex hull of the stock price increments; if $m = 2$ then ℓ is a line as in Figure 3.8, and if $m = 3$ then ℓ is a plane. Let $y = (y^1, \ldots, y^m)$ be any non-zero vector from the origin, perpendicular to ℓ and pointing towards it. Simple geometry gives

$$y \cdot \Delta \bar{S}_1(\omega) > 0 \quad \text{for all } \omega \in \Omega.$$

Then by the Implicit Arbitrage Lemma (Lemma 3.31) y is an implicit arbitrage opportunity, and the model is not viable.

The other possibility is that 0 lies on the boundary of the convex hull but not in its relative interior; a typical situation for $m = 2$ is given in Figure 3.9. In this case it is possible to find a hyperplane ℓ that intersects the convex hull on its boundary, and with the relative interior of the convex hull all to one side of ℓ. Take any non-zero $y = (y^1, \ldots, y^m)$ perpendicular to ℓ and pointing in the direction of the convex hull. There will be at least one ω such that $\Delta \bar{S}_1(\omega)$ does *not* lie on ℓ and, and for any such ω we have $y \cdot \Delta \bar{S}_1(\omega) > 0$. For those

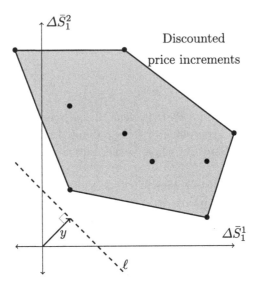

Figure 3.8 Finding arbitrage when the convex hull of discounted price increments does not contain the origin

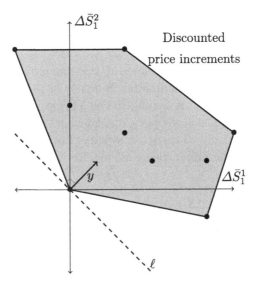

Figure 3.9 Finding arbitrage when the origin is on the boundary of the convex hull of discounted price increments

ω such that $\Delta \bar{S}_1(\omega)$ lies on ℓ we have $y \cdot \Delta \bar{S}_1(\omega) = 0$. Again by the Implicit Arbitrage Lemma, y is an implicit arbitrage opportunity, and the model is not viable. $\qquad \square$

Exercise 3.45[‡]

From your sketch of the convex hull in Exercise 3.41 find an arbitrage opportunity using the idea of the first proof of Proposition 3.39 and compare it with the one you found in Exercise 3.38.

The second proof of Proposition 3.39 makes use of the fact that any random variable X on Ω can be represented as the vector $(X(\omega_1), \ldots, X(\omega_n)) \in \mathbb{R}^n$, as can any probability measure $\mathbb{Q} = (q_1, \ldots, q_n)$ on Ω.

We will need the following general result from linear algebra (which we used informally in the first proof).

Lemma 3.46 (Separating Hyperplane Theorem)

If $\mathcal{A}$ is a subspace of $\mathbb{R}^n$ and $\mathcal{R}$ is a compact convex subset of $\mathbb{R}^n$ disjoint from $\mathcal{A}$, then there is a vector $\theta \in \mathbb{R}^n$ such that

$$\theta \cdot z = 0 \quad \text{for all } z \in \mathcal{A}, \qquad \theta \cdot z > 0 \quad \text{for all } z \in \mathcal{R}.$$

Proof (second proof of Proposition 3.39)

Since there is no arbitrage, the Implicit Arbitrage Lemma (Lemma 3.31) shows that there is no stock holding y such that the gain $y \cdot \Delta \bar{S}_1(\omega)$ after discounting is always non-negative, and positive for at least one scenario. Let $\mathcal{A}$ be the collection of all such gains, represented as a subset of $\mathbb{R}^n$:

$$\mathcal{A} := \left\{ (y \cdot \Delta \bar{S}_1(\omega_1), y \cdot \Delta \bar{S}_1(\omega_2), \ldots, y \cdot \Delta \bar{S}_1(\omega_n)) : y \in \mathbb{R}^m \right\}. \qquad (3.38)$$

Then the opening remark above means that in a viable model the set $\mathcal{A}$ does not have any points in common with

$$\mathbb{R}_+^n := \left\{ z \in \mathbb{R}^n : z_j \geq 0 \text{ for } j = 1, \ldots, n \text{ and } z_j > 0 \text{ for at least one } j \right\}$$

which is just $[0, \infty)^n$ with the origin 0 removed; that is, $\mathcal{A} \cap \mathbb{R}_+^n = \emptyset$.

It is easy to check that $\mathcal{A}$ is a *subspace* of $\mathbb{R}^n$; that is, if $v, w \in \mathcal{A}$ and $\lambda \in \mathbb{R}$, then $v + w \in \mathcal{A}$ and $\lambda v \in \mathcal{A}$.

Now the aim is to find a probability $\mathbb{Q} = q = (q_1, \ldots, q_n) \in \mathbb{R}^n$ such that $\mathbb{E}_\mathbb{Q}(\Delta \bar{S}_1) = 0$, which we have seen (Theorem 3.35) is equivalent to

$$\mathbb{E}_\mathbb{Q}(y \cdot \Delta \bar{S}_1) = 0 \quad \text{for all } y \in \mathbb{R}^m.$$

This can be written as

$$q \cdot z = 0 \quad \text{for all } z \in \mathcal{A}.$$

These comments motivate the application of the Separating Hyperplane Theorem to the subspace $\mathcal{A}$ as follows.

The set $\mathcal{R}$ defined by

$$\mathcal{R} := \left\{ z \in \mathbb{R}_+^n : \sum_{j=1}^n z_j = 1 \right\}$$

is a subset of $\mathbb{R}_+^n$ and so $\mathcal{A} \cap \mathcal{R} = \varnothing$ also; moreover, $\mathcal{R}$ is compact and convex, so the Separating Hyperplane Theorem applies. Thus there is a vector $\theta = (\theta_1, \ldots, \theta_n) \in \mathbb{R}^n$ such that

$$\theta \cdot z = 0 \quad \text{for all } z \in \mathcal{A},$$

$$\theta \cdot z > 0 \quad \text{for all } z \in \mathcal{R}.$$

For each j the vector $e_j = (0, \ldots, 1, \ldots 0)$ with 1 in the jth position is in $\mathcal{R}$ and so $\theta_j = \theta \cdot e_j > 0$.

Now scale θ to give the vector $q = (q_1, \ldots, q_n) \in \mathbb{R}^n$ by

$$q := \frac{\theta}{\sum_{l=1}^n \theta_l} \tag{3.39}$$

so that $\sum_{j=1}^n q_j = 1$ and each $q_j > 0$. Thus $\mathbb{Q} = q$ is a candidate for a risk neutral probability. What we know about q (since it is a scaled version of θ) is that

$$q \cdot z = 0 \quad \text{for all } z \in \mathcal{A}.$$

The discussion above shows that this is another way of writing

$$\mathbb{E}_\mathbb{Q}(y \cdot \Delta \bar{S}_1) = 0 \quad \text{for all } y \in \mathbb{R}^m$$

and so $\mathbb{Q}$ is indeed a risk-neutral probability. $\square$

It is useful to illustrate the above proof with an example.

Example 3.47

Consider Model 3.10. Then

$$\Delta \bar{S}_1(\omega_1) = (-30, 20, -10),$$

$$\Delta \bar{S}_1(\omega_2) = (15, -10, 5) = -\tfrac{1}{2} \Delta \bar{S}_1(\omega_1).$$

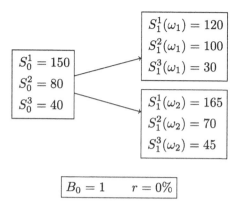

$$S_0^1 = 150$$
$$S_0^2 = 80$$
$$S_0^3 = 40$$

$$S_1^1(\omega_1) = 120$$
$$S_1^2(\omega_1) = 100$$
$$S_1^3(\omega_1) = 30$$

$$S_1^1(\omega_2) = 165$$
$$S_1^2(\omega_2) = 70$$
$$S_1^3(\omega_2) = 45$$

$$B_0 = 1 \qquad r = 0\%$$

Model 3.10   One-period binary model with three stocks in Example 3.47

The proof above tells us in advance that the set $\mathcal{A}$ of all possible discounted gains is a (non-trivial) subspace of $\mathbb{R}^2$, so it will either be all of $\mathbb{R}^2$ (in which case the model is not viable) or a line through the origin. In the second alternative, the model is viable provided the line $\mathcal{A}$ only intersects the positive quadrant $[0, \infty)^2$ at the origin. Examining $\mathcal{A}$ in detail for this example we have

$$\mathcal{A} = \big\{ \big(y \cdot (-30, 20, -10), y \cdot (15, -10, 5)\big) : y \in \mathbb{R}^3 \big\}$$
$$= \big\{ (z_1, z_2) : z_1 \in \mathbb{R}, z_2 = -\tfrac{1}{2} z_1 \big\}$$

which is a line in $\mathbb{R}^2$ through the origin with slope $-\tfrac{1}{2}$ so the model is viable.

Now take a vector θ perpendicular to the line $\mathcal{A}$ (so that $\theta \cdot z = 0$ for all $z \in \mathcal{A}$) and pointing in the direction of $\mathcal{R}$, for example $\theta = (\tfrac{1}{2}, 1)$, as in Figure 3.11. The above proof guaranteed the fact that $\theta_1, \theta_2 > 0$. Defining $\mathbb{Q} = q = (q_1, q_2)$ as in (3.39), we get

$$q = \tfrac{2}{3}\theta = \big(\tfrac{1}{3}, \tfrac{2}{3}\big)$$

since $\theta_1 + \theta_2 = \tfrac{3}{2}$, so that $q_1 + q_2 = 1$ and $q \cdot z = 0$ for all $z \in \mathcal{A}$. To verify that $q = (\tfrac{1}{3}, \tfrac{2}{3})$ is indeed a risk neutral probability for this model (as the proof guarantees) we have

$$q_1 \bar{S}_1(\omega_1) + q_2 \bar{S}_1(\omega_2) = \tfrac{1}{3} \begin{bmatrix} 120 \\ 100 \\ 30 \end{bmatrix} + \tfrac{2}{3} \begin{bmatrix} 165 \\ 70 \\ 45 \end{bmatrix} = \begin{bmatrix} 150 \\ 80 \\ 40 \end{bmatrix} = S_0.$$

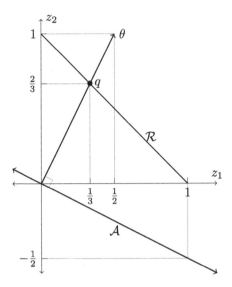

Figure 3.11   Separating disjoint sets in Example 3.47

3.4 Risk-Neutral Pricing with Several Stocks

Now that we have the First Fundamental Theorem, the theory of derivative pricing developed earlier extends in a straightforward fashion, giving an easy way to price attainable derivatives in general viable models using risk-neutral probabilities.

The first task is to generalize the Law of One Price from the one-stock model to the case of many stocks. In fact the proof is identical to that of Theorem 2.23, provided we bear in mind that in general the share holding y is now a vector $y = (y^1, \ldots, y^m)$.

We give below a slight variation of that proof. This involves a simple new idea, that of *strong arbitrage*, which makes the proof itself a little more straightforward.

Definition 3.48 (Strong arbitrage)

A portfolio φ is a *strong arbitrage opportunity* in the bond and stock if $V_0^\varphi < 0$ and $V_1^\varphi \geq 0$. An extended portfolio ψ involving a derivative D is an *extended strong arbitrage opportunity* in the bond, stock and D if $V_0^\psi < 0$ and $V_1^\psi \geq 0$.

A strong arbitrage opportunity is, informally, an investment that allows us to *receive* money at time 0 without anything to pay back at time 1, which is a desirable but rather unlikely state of affairs. The usefulness of the idea of strong arbitrage stems from the following, utilized in the proof of Theorem 3.51 below and elsewhere.

Lemma 3.49

(1) If (x, y) is a strong arbitrage opportunity, then the portfolio (x', y) with

$$x' := x - \frac{1}{B_0} V_0^{(x,y)} \tag{3.40}$$

is an arbitrage opportunity.

(2) If (x, y, z) is an extended strong arbitrage opportunity, then the extended portfolio (x', y, z) with

$$x' := x - \frac{1}{B_0} V_0^{(x,y,z)}$$

is an extended arbitrage opportunity.

Proof

The proof of (1) is left as Exercise 3.50 below. For (2), clearly $V_0^{(x',y,z)} = 0$ and

$$V_1^{(x',y,z)} = V_1^{(x,y,z)} - (1+r)V_0^{(x,y,z)} > 0$$

since $-V_0^{(x,y,z)} > 0$ and $V_1^{(x,y,z)} \geq 0$.   □

Exercise 3.50‡

(a) Prove Lemma 3.49(1).

(b) Deduce that if a model contains a strong arbitrage opportunity, then it is not viable.

(c) Give an example to show that the converse to (b) does not hold; that is, an example of a non-viable model that does not contain a strong arbitrage opportunity.

Here is the Law of One Price for a general single period model, proved using the idea of strong extended arbitrage.

Theorem 3.51 (Law of One Price in a general single-period model)

If a portfolio φ replicates an attainable derivative D in a viable single-period model, then there is a unique fair price D_0 for D at time 0, given by the initial value of φ; that is

$$D_0 = V_0^{\varphi}.$$

Proof

Let D be a derivative with replicating portfolio $\varphi = (x, y)$, traded at time 0 for the price π. If $\pi > V_0^\varphi$, an extended strong arbitrage opportunity is created by short-selling D at the price π and investing in the portfolio φ. This is gives the extended portfolio $\psi = (\varphi, -1) = (x, y, -1)$ with initial value

$$V_0^\psi = V_0^\varphi - \pi < 0$$

and final value

$$V_1^\psi = V_1^\varphi - D = 0.$$

So ψ is an extended strong arbitrage opportunity, and the model is not viable.

If $\pi < V_0^\varphi$, then an extended strong arbitrage opportunity is constructed by buying the derivative, and investing in the portfolio $-\varphi$, giving the extended portfolio $(-\varphi, 1)$.

If $\pi = V_0^\varphi$, then there can be no arbitrage in the bond, stock and the derivative D. The proof here is exactly as in the simple model of Chapter 2 (Theorem 2.23) but we repeat it here for good measure. Suppose $\psi = (u, v, w)$ is an extended arbitrage opportunity. At time 0, we have

$$V_0^\psi = uB_0 + v \cdot S_0 + w\pi = uB_0 + v \cdot S_0 + wV_0^\varphi$$
$$= (u + wx)B_0 + (v + wy) \cdot S_0 = V_0^{(u+wx, v+wy)}.$$

At time 1, the replication of D by φ gives

$$V_1^\psi = uB_1 + v \cdot S_1 + wD = uB_1 + v \cdot S_1 + wV_1^\varphi$$
$$= (u + wx)B_1 + (v + wy) \cdot S_1 = V_1^{(u+wx, v+wy)}.$$

Thus $(u + wx, v + xy)$ is an arbitrage opportunity, which contradicts the viability of the model. Thus there is no extended arbitrage opportunity, and $\pi = V_0^\varphi$ is the unique fair price of the derivative at time 0. $\square$

Exercise 3.52[†]

Show that in Model 3.10 the derivative D with $D(\omega_1) = 20$ and $D(\omega_2) = 0$ admits at least three different replicating portfolios. Compute the fair price of this derivative.

Theorem 3.51 has the following corollary.

Corollary 3.53

If two portfolios φ and φ' replicate a derivative D in a viable one-step model, then their initial values are the same; that is, $V_0^\varphi = V_0^{\varphi'}$.

Proof

Both V_0^φ and $V_0^{\varphi'}$ are equal to the unique fair price of D. $\quad\square$

The following exercise says that, in a single-stock model, the assumption of viability may be removed in Corollary 3.53.

Exercise 3.54

In a one-period model with a single stock, suppose that two portfolios φ and φ' have the same value at time 1; that is, $V_1^\varphi = V_1^{\varphi'}$. Show that $\varphi = \varphi'$ and hence $V_0^\varphi = V_0^{\varphi'}$.

It is now an easy task to extend Theorems 2.35 and Theorem 3.11 to the present situation. The proofs look the same, but again remember that in a portfolio $\varphi = (x, y)$, the stock holding y is a *vector*.

Theorem 3.55

In a viable single-period model, the unique fair price D_0 of any attainable derivative D is given by

$$D_0 = \mathbb{E}_\mathbb{Q}(\bar{D}),$$

where $\mathbb{Q}$ is any risk-neutral probability.

Proof

Suppose that the portfolio φ replicates D. For any risk-neutral probability $\mathbb{Q}$, Theorems 3.35 and 3.51 give

$$\mathbb{E}_\mathbb{Q}(\bar{D}) = \mathbb{E}_\mathbb{Q}(\bar{V}_1^\varphi) = V_0^\varphi = D_0.$$

$\quad\square$

Exactly as in the case of a single stock, we have the following corollary.

Corollary 3.56

Suppose that D is an attainable derivative in a general viable single-period model. For any two risk-neutral probabilities $\mathbb{Q}$ and $\mathbb{Q}'$

$$\mathbb{E}_{\mathbb{Q}}(\bar{D}) = \mathbb{E}_{\mathbb{Q}'}(\bar{D}).$$

Proof

Both $\mathbb{E}_{\mathbb{Q}}(\bar{D})$ and $\mathbb{E}_{\mathbb{Q}'}(\bar{D})$ are equal to the fair price of D. $\square$

3.5 Completeness with Several Stocks

We now consider the issue of completeness for a *viable* single-period model with m stocks and n scenarios. We have seen that any such model admits a risk-neutral probability $\mathbb{Q}$ (which may not be unique).

Recall that a model is complete if and only if every derivative is attainable. Continuing to write $\Omega = \{\omega_1, \ldots, \omega_n\}$, this means that the model is complete if and only if the system

$$xB_1 + y \cdot S_1(\omega_1) = D(\omega_1),$$

$$\vdots$$

$$xB_1 + y \cdot S_1(\omega_n) = D(\omega_n)$$

admits a solution (x, y) for each derivative D. Linear algebra immediately tells us that the $n \times (m+1)$ matrix

$$A := \begin{bmatrix} B_1 & S_1^1(\omega_1) & \cdots & S_1^m(\omega_1) \\ \vdots & \vdots & \ddots & \vdots \\ B_1 & S_1^1(\omega_n) & \cdots & S_1^m(\omega_n) \end{bmatrix}$$

has rank n, or, equivalently, has n linearly independent columns, or n linearly independent rows. Thus a single-period model with n scenarios is complete if and only if it contains at least n *independent assets* (both risky and risk-free). By the *linear independence of a collection of assets* we mean that the future price of any one of them (including the bond) cannot be written as a linear combination of the prices of the others. For the model to have n independent assets, there must be at least n assets (that is, the matrix A must have at least n columns). Consequently, if the model is complete, then we must have

$$m + 1 \geq n.$$

Example 3.57

Model 3.1 in Section 3.1.1 is not complete, because it has $m + 1 = 2$ assets and $n = 3$ scenarios, so that $m + 1 \not\geq n$. Recall from Exercise 3.2 that the risk-neutral probability in this model is not unique; we will see that this is another characteristic of incomplete models.

Putting this discussion more succinctly, a derivative D is attainable if the equation

$$\begin{bmatrix} B_1 \; S_1^1(\omega_1) & \cdots & S_1^m(\omega_1) \\ \vdots & \ddots & \vdots \\ B_1 \; S_1^1(\omega_n) & \cdots & S_1^m(\omega_n) \end{bmatrix} \begin{bmatrix} x \\ y^1 \\ \vdots \\ y^m \end{bmatrix} = \begin{bmatrix} D(\omega_1) \\ \vdots \\ D(\omega_n) \end{bmatrix}$$

has a solution $(x, y^1, \ldots, y^m) \in \mathbb{R}^{1+m}$. Completeness requires that every derivative D should be attainable. If we denote the vector

$$\begin{bmatrix} D(\omega_1) \\ \vdots \\ D(\omega_n) \end{bmatrix}$$

by $b \in \mathbb{R}^n$, then the model is complete if and only if the equation

$$A \begin{bmatrix} x \\ y^1 \\ \vdots \\ y^m \end{bmatrix} = b$$

has a solution for each $b \in \mathbb{R}^n$, which is the case if and only if A has rank n.

The matrix A also features in the characterization of a risk-neutral probability. If we write $q = (q_1, \ldots, q_n)$, then q represents a risk-neutral probability assignment if and only if

$$q_1 + \cdots + q_n = 1,$$

$$q_1 S_1^1(\omega_1) + \cdots + q_n S_1^1(\omega_n) = (1 + r)S_0^1,$$

$$\vdots$$

$$q_1 S_1^m(\omega_1) + \cdots + q_n S_1^m(\omega_n) = (1 + r)S_0^m,$$

together with $q_j > 0$ for $j = 1, \ldots, n$. Writing this in matrix form, we see that q is a risk-neutral probability if and only if

$$
A^T q = \begin{bmatrix} B_1 \\ (1+r)S_0^1 \\ \vdots \\ (1+r)S_0^m \end{bmatrix} = (1+r) \begin{bmatrix} B_0 \\ S_0 \end{bmatrix}
\tag{3.41}
$$

and $q_j > 0$ for $j = 1, \ldots, n$.

The above discussion leads to the following characterization of viable single-period models that are complete.

Theorem 3.58

For a viable single-period model with $n \geq 2$ scenarios, a bond and m stocks, the following statements are equivalent:

(1) The model is complete.

(2) The rank of the matrix A above is n.

(3) $m + 1 \geq n$, and the model contains n independent assets.

(4) The model admits a unique risk-neutral probability.

Remarks 3.59

(1) The equivalence between (1) and (4) of this theorem is often called the *Second Fundamental Theorem of Asset Pricing* for single-period models.

(2) The condition of completeness is quite restrictive. For example, as we have seen in Section 3.1, a model with a bond and a single stock is incomplete whenever it has more than binary branching.

(3) Since the values of each asset in the n different scenarios form a vector of length n (of the form $(B_1, \ldots, B_1)$ or $(S_1^i(\omega_1), \ldots, S_1^i(\omega_n)))$, there can be *at most* n independent assets.

Proof (of Theorem 3.58)

We have already shown above that (1)–(3) are all equivalent.

For the equivalence of (1) and (4) first note that viability of the model implies that (3.41) has a solution for q; that is, it is consistent. If the model is complete, it follows from (2) that the matrix A above has rank n. Basic linear algebra then tells us that the solution q to (3.41) is unique. In other words, there is a unique risk-neutral probability. So (1) implies (4).

Conversely, if the model is incomplete then the rank of A is less than n. Basic linear algebra tells us that since equation (3.41) has at least one solution q (guaranteed by viability), there must be infinitely many solutions; thus there are infinitely many risk-neutral probability assignments.   □

Example 3.60

Model 3.10 in Example 3.47 has $m + 1 = 4$ assets (including the bond), and only $n = 2$ scenarios. It is complete since the rows of the matrix

$$\begin{bmatrix} B_1 & S_1^1(\omega_1) & S_1^2(\omega_1) & S_1^3(\omega_1) \\ B_1 & S_1^1(\omega_2) & S_1^2(\omega_2) & S_1^3(\omega_2) \end{bmatrix} = \begin{bmatrix} 1 & 120 & 100 & 30 \\ 1 & 165 & 70 & 45 \end{bmatrix}$$

are independent. Moreover, we showed in Example 3.47 that the unique risk-neutral probability is $(\frac{1}{3}, \frac{2}{3})$.

Example 3.61

Model 3.12 has $m + 1 = 3$ assets and $n = 3$ scenarios, but it is not complete because the 3 assets it contains are not independent. Indeed, it is straightforward to verify that

$$S_1^2 = 2S_1^1 + 50B_1,$$

and therefore the rank of the matrix

$$\begin{bmatrix} 1 & S_1^1(\omega_1) & S_1^2(\omega_1) \\ 1 & S_1^1(\omega_2) & S_1^2(\omega_2) \\ 1 & S_1^1(\omega_3) & S_1^2(\omega_3) \end{bmatrix} = \begin{bmatrix} 1 & 100 & 250 \\ 1 & 70 & 190 \\ 1 & 30 & 110 \end{bmatrix}$$

is equal to $2 < n$. The discounted increments

$$\Delta \bar{S}_1(\omega) = \begin{cases} (40, 80) & \text{if } \omega = \omega_1, \\ (10, 20) & \text{if } \omega = \omega_2, \\ (-30, -60) & \text{if } \omega = \omega_3 \end{cases}$$

are collinear in $\mathbb{R}^2$ and therefore the model has more than one risk-neutral probability.

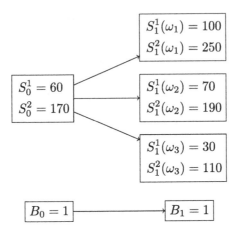

Model 3.12 One-period ternary model with two stocks in Example 3.61

Exercise 3.62[†]

Explain why Model 3.13 is viable and complete. Find a replicating portfolio for the derivative D with payoff $D(\omega_1) = 20$, $D(\omega_2) = 10$ and $D(\omega_3) = 0$, and calculate its fair price.

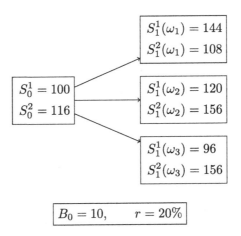

Model 3.13 One-period ternary model with two stocks in Exercise 3.62

Exercise 3.63[†]

Consider Model 3.14.

(a) Find the general form for a risk-neutral probability for this model. Explain why it is viable but not complete.

(b) Derive a necessary and sufficient condition for a derivative D to be attainable in this model, and give a formula for its fair price in terms of $D(\omega_1)$, $D(\omega_2)$ and $D(\omega_3)$.

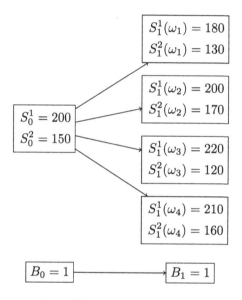

Model 3.14 One-period model with four scenarios and two stocks in Exercises 3.63 and 3.71

3.6 Derivative Pricing in an Incomplete Model

The ideas and results concerning pricing an non-attainable derivative in an incomplete model with several stocks are essentially the same as for single-stock models, detailed in Section 3.1.4.

The definitions of super- and sub-replication of a derivative D, and the ask and bid prices π_D^a, π_D^b are the same as in the single-stock case, which we repeat here for convenience.

Definition 3.64 (Super- and sub-replication)

Let D be a derivative in a single-period model.

(a) A portfolio φ *super-replicates* D if $V_1^\varphi \geq D$.

(b) A portfolio φ *sub-replicates* D if $V_1^\varphi \leq D$.

Definition 3.65 (Ask and bid price)

Let D be a derivative in a single-period model. Its *ask price* π_D^a and its *bid price* π_D^b is defined as

$$\pi_D^a := \inf\{V_0^\varphi : \varphi \text{ super-replicates } D\},$$
$$\pi_D^b := \sup\{V_0^\varphi : \varphi \text{ sub-replicates } D\}.$$

For attainable D we know that $\pi_D^a = \pi_D^b = D_0$. For non-attainable D, as in the single stock case we have an interval F_D of fair prices given by $F_D = (\pi_D^b, \pi_D^a)$. The steps to proving this are the same as before, but with a different proof for the following generalization of Theorem 3.18.

Theorem 3.66

Let D be a non-attainable derivative in a viable incomplete model with m stocks. There exist super- and sub-replicating portfolios φ^a and φ^b for D such that $V_0^{\varphi^a} = \pi_D^a$ and $V_0^{\varphi^b} = \pi_D^b$. Hence if $\pi \geq \pi_D^a$ or $\pi \leq \pi_D^b$ then π is not fair, and $\pi_D^b < \pi_D^a$.

Proof

We give an analytic proof that generalizes the proof for the single stock case sketched in Exercise 3.23. First notice that any portfolio $\varphi = (x, y)$ super-replicates D if and only if

$$x B_0 + y \cdot \bar{S}_1 \geq \bar{D} \quad \text{for all } \omega,$$

which is equivalent to

$$y \cdot \Delta \bar{S}_1(\omega) \geq \bar{D}(\omega) - V_0^\varphi \quad \text{for all } \omega.$$

Now take a sequence of super-replicating portfolios $(\varphi_k)_{k \in \mathbb{N}} = ((x_k, y_k))_{k \in \mathbb{N}}$ with $V_0^{\varphi_k} \to \pi^a$. Then the sequence $V_0^{\varphi_k}$ is bounded and

$$y_k \cdot \Delta \bar{S}_1(\omega) \geq \bar{D}(\omega) - V_0^{\varphi_k} \quad \text{for all } \omega.$$

Since Ω is finite this means that $y_k \cdot \Delta \bar{S}_1(\omega)$ is bounded below; that is, there exists a number $M \geq 0$ such that

$$y_k \cdot \Delta \bar{S}_1(\omega) \geq -M \tag{3.42}$$

for all $k \in \mathbb{N}$ and $\omega \in \Omega$.

We would like to obtain an upper bound on these numbers as well. For this, and writing $\Omega = \{\omega_1, \ldots, \omega_n\}$, let $\mathbb{Q} = (q_1, \ldots, q_n)$ be any risk-neutral probability. For each $k \in \mathbb{N}$, the property

$$\mathbb{E}_{\mathbb{Q}}(y_k \cdot \Delta \bar{S}_1) = 0$$

implies that

$$q_i y_k \cdot \Delta \bar{S}_1(\omega_i) = -\sum_{j \neq i} q_j y_k \cdot \Delta \bar{S}_1(\omega_j)$$

for all i. Inequality (3.42) then implies that for each i

$$q_i y_k \cdot \Delta \bar{S}_1(\omega_i) \leq \sum_{j \neq i} q_j M \leq M,$$

from which we see that

$$y_k \cdot \Delta \bar{S}_1(\omega_i) \leq \frac{M \cdot}{q_i}.$$

Hence

$$y_k \cdot \Delta \bar{S}_1(\omega) \leq M' \quad \text{for all } \omega, \tag{3.43}$$

where $M' := M / \min\{q_1, \cdots, q_n\}$. Combining (3.42) and (3.43) it follows that the set

$$\left\{ y_k \cdot \Delta \bar{S}_1(\omega) : k \in \mathbb{N}, \omega \in \Omega \right\}$$

is bounded.

We would like to conclude that the sequence y_k is bounded and thus has a convergent subsequence, but this may not be the case because the space W of $\mathbb{R}^m$ spanned by the vectors $\Delta \bar{S}_1(\omega_1), \ldots, \Delta \bar{S}_1(\omega_n)$ may be smaller than $\mathbb{R}^m$ itself. For this reason we modify each φ_k as follows. Let y_k' be the projection of y_k onto W, and define

$$x_k' := \frac{1}{B_0} \left(V_0^{\varphi_k} - y_k' \cdot S_0 \right).$$

Writing $\varphi_k' = (x_k', y_k')$, it follows that

$$V_0^{\varphi_k'} = V_0^{\varphi_k} \longrightarrow \pi_D^a. \tag{3.44}$$

Moreover,

$$y'_k \cdot \Delta\bar{S}_1(\omega) = y_k \cdot \Delta\bar{S}_1(\omega) \qquad (3.45)$$

for all k and ω so the sequence $(y'_k \cdot \Delta\bar{S}_1(\omega))_{k\in\mathbb{N}}$ is bounded for all ω. Since the set $\{\Delta\bar{S}_1(\omega) : \omega \in \Omega\}$ is finite and spans W, and each $y'_k \in W$, it follows from linear algebra that the sequence $(y'_k)_{k\in\mathbb{N}}$ itself is bounded, so that it has a convergent subsequence, again denoted $(y'_k)_{k\in\mathbb{N}}$.

Let $y \in \mathbb{R}^m$ be the limit of $(y'_k)_{k\in\mathbb{N}}$; then $(x'_k)_{k\in\mathbb{N}}$ converges to

$$x := \frac{1}{B_0}\left(\pi_D^a - y \cdot S_0\right)$$

and, writing $\varphi = (x, y)$, the sequence $(V_0^{\varphi'_k})_{k\in\mathbb{N}}$ converges to

$$V_0^\varphi = xB_0 + y \cdot S_0 = \pi_D^a.$$

It remains to verify that φ super-replicates D. Equations (3.44) and (3.45) together with the super-replicating property of $(y_k)_{k\in\mathbb{N}}$ show that

$$y'_k \cdot \Delta\bar{S}_1(\omega) \geq \bar{D}(\omega) - V_0^{\varphi'_k} \qquad \text{for all } \omega$$

for $k \in \mathbb{N}$ and so, taking limits,

$$y \cdot \Delta\bar{S}_1(\omega) \geq \bar{D}(\omega) - V_0^\varphi \qquad \text{for all } \omega.$$

Thus φ super-replicates D and we may set $\varphi^a := \varphi$.

The portfolio φ^b can be found similarly beginning with a sequence of sub-replicating portfolios. The proof that any price π with $\pi \leq \pi_D^b$ or $\pi \geq \pi_D^a$ is not fair is exactly as in the single-stock case. $\qquad\qquad\square$

The previous theorem is the key to the proof of the characterization of F_D when D is non-attainable, which is proved exactly as in the single-stock case (Theorem 3.20) except that all stock holdings and stock prices are now vectors.

Theorem 3.67

If D is a non-attainable derivative in a viable model then

$$F_D = \left(\pi_D^b, \pi_D^a\right).$$

Now let us turn to the other natural approach to pricing a non-attainable derivative D, using risk-neutral probabilities. This gives an easy direct characterization of the set of fair prices for any derivative, already noted for the single-stock case (Theorem 3.24).

Theorem 3.68

The set F_D of fair prices of a derivative D in a viable model with m stocks is given by

$$F_D = \{\mathbb{E}_\mathbb{Q}(\bar{D}) : \mathbb{Q} \text{ is a risk-neutral probability}\}.$$

Proof

Suppose that $\pi = \mathbb{E}_\mathbb{Q}(\bar{D})$ for some risk-neutral probability $\mathbb{Q}$ and $\psi = (x, y, z)$ is an extended portfolio with D priced at π. If $V_1^\psi \geq 0$ and $V_1^\psi(\omega) > 0$ for at least one ω then $\mathbb{E}_\mathbb{Q}(V_1^\psi) > 0$ and hence $\mathbb{E}_\mathbb{Q}(\bar{V}_1^\psi) > 0$. This means that

$$V_0^\psi = xB_0 + yS_0 + z\pi = \mathbb{E}_\mathbb{Q}(x\bar{B}_1 + y\bar{S}_1 + z\bar{D}) = \mathbb{E}_\mathbb{Q}(\bar{V}_1^\psi) > 0,$$

and so ψ is not an extended arbitrage. Thus π is a fair price for D.

For the converse, suppose that π is a fair price for D. This means that the model consisting of the bond, the m stocks and the derivative (with price π at time 0) is viable. (We may assume without any loss of generality that $\pi > 0$ and $D(\omega) > 0$ for all ω; if not add an appropriate constant and note Exercise 3.72.) This is mathematically just the same as a model with $m + 1$ risky assets, so the First Fundamental Theorem gives a risk-neutral probability $\mathbb{Q}$ such that

$$\mathbb{E}_\mathbb{Q}\big((\bar{S}_1, \bar{D})\big) = (S_0, \pi).$$

In particular,

$$\mathbb{E}_\mathbb{Q}(\bar{S}_1) = S_0,$$

so $\mathbb{Q}$ is a risk-neutral probability for the basic model consisting of bond and stock only, and in addition

$$\mathbb{E}_\mathbb{Q}(\bar{D}) = \pi.$$

□

Remark 3.69

Theorem 3.68 provides the missing proof of part of Theorem 3.24, the corresponding result for the single-stock case.

It is natural to define the numbers

$$\pi_D^+ := \sup\{\mathbb{E}_\mathbb{Q}(\bar{D}) : \mathbb{Q} \text{ is a risk neutral probability}\},$$

$$\pi_D^- := \inf\{\mathbb{E}_\mathbb{Q}(\bar{D}) : \mathbb{Q} \text{ is a risk neutral probability}\}.$$

Then pricing theory for any derivative in a single period model with finitely many stocks can be summarized as follows, as for the one stock case.

Theorem 3.70

If D is a derivative in a viable model with a single stock, then

$$\pi_D^a = \pi_D^+, \qquad \pi_D^b = \pi_D^-.$$

(1) The following statements are equivalent:

   (i) D is attainable.

   (ii) $\pi_D^b = \pi_D^a$.

   (iii) π_D^a is a fair price for D.

   (iv) π_D^b is a fair price for D.

   (v) There is a unique fair price for D.

(2) The following statements are equivalent:

   (i) D is not attainable.

   (ii) $\pi_D^b < \pi_D^a$.

   (iii) The set of fair prices for D is $F_D = (\pi_D^b, \pi_D^a)$.

Exercise 3.71[†]

Consider the derivative $D = |S_1^2 - 150|$ in Model 3.14.

(a) Show that D is not attainable and find the general form of $\mathbb{E}_\mathbb{Q}(D)$ for a risk-neutral probability $\mathbb{Q}$.

   Hint. Use Exercise 3.63.

(b) Find the set F_D of fair prices for D at time 0.

Exercise 3.72[‡]

Show that for any derivative D and constant c, the price π is fair for D if and only if $\pi + c$ is a fair price for the derivative with payoff $D + c(1 + r)$.

Exercise 3.73[‡]

Find an example to show that if π, π' are fair prices for derivatives D, D' respectively, then $\pi + \pi'$ need not be a fair price for $D + D'$.

Hint. Consider a non-attainable European put P_1 and a European call C_1 with the same strike price K on a stock S in a viable model, and let $D = P_1$ and $D' = -C_1$.

Exercise 3.74[‡]

Assuming only Theorem 3.68, show directly that if a derivative D in a viable model has a unique fair price then it is attainable, as follows. Consider the matrix

$$A = \begin{bmatrix} B_1 & S_1^1(\omega_1) & \cdots & S_1^m(\omega_1) \\ \vdots & \vdots & \ddots & \vdots \\ B_1 & S_1^1(\omega_n) & \cdots & S_1^m(\omega_n) \end{bmatrix}$$

in Section 3.5, and let D_0 be the unique fair price of D.

(a) Show that a vector $q = (q_1, \ldots, q_n)$ with $\sum_i q_i = 1$ and $q_i > 0$ for all i satisfies the system

$$A^T q = (1 + r) \begin{bmatrix} B_0 \\ S_0 \end{bmatrix} \tag{3.46}$$

if and only if

$$\begin{bmatrix} A^T \\ D \end{bmatrix} q = (1 + r) \begin{bmatrix} B_0 \\ S_0 \\ D_0 \end{bmatrix}. \tag{3.47}$$

(b) Deduce from the theory of linear equations that

$$\mathrm{rank} \begin{bmatrix} A^T \\ D \end{bmatrix} = \mathrm{rank}\, A^T. \tag{3.48}$$

From this conclude that D is attainable.

4
Multi-Period Models: No-arbitrage Pricing

Any realistic financial market will have more than one time step. In this chapter we extend the no-arbitrage pricing theory of the previous chapter to models that have any finite number of possible trading dates. The models are still discrete, but the theory developed in this and the next chapter will pave the way for generalization to continuous time models.

There will be a fixed finite set of trading dates $\mathbb{T} = \{0, 1, \dots, T\}$. A multi-period model can be regarded as a succession of single-period models, so that the theory is in many respects a simple extension of that of the previous chapter. There is, however, one important new feature. When there is more than one time step a trader will be allowed to change his portfolio at any of the time points $t = 1, 2, 3, \dots, T - 1$, so that the portfolio of bonds and shares may be different for different periods $(t, t+1]$. This means that the concept of a viable market is a little more complicated, and likewise the fair price of a derivative. However, the underlying idea is the same as before: a market consisting of bond and stock is viable if there is no arbitrage ("free lunch") involving trading in these assets, and a fair price for a derivative is any price that prevents extended arbitrage (that is, a "free lunch" obtained by trading in bonds, stock and the derivative).

We begin by describing the multi-period models that we will consider, and the idea of a *trading strategy*, which can be thought of as a portfolio that changes over time. This allows the definition of *arbitrage* and *viability*, and leads to the appropriate extension of the definition of a *fair price* for a derivative in this setting. Pricing of derivatives by *replication* is then routine; the chapter

N.J. Cutland, A. Roux, *Derivative Pricing in Discrete Time*,
Springer Undergraduate Mathematics Series,
DOI 10.1007/978-1-4471-4408-3_4, © Springer-Verlag London 2012

concludes with a discussion of completeness. Extending the idea of risk-neutral probabilities and their use in pricing is deferred to the next chapter, where we also discuss pricing in an incomplete multi-period market.

Here are the basic assumptions for the models that will be discussed in this and the next chapter.

Assumptions 4.1

(1) There is a finite set of times $\mathbb{T} = \{0, 1, \ldots, T\}$ at which trading can take place.

(2) The model contains a *bond* with initial price $B_0 > 0$ at time 0. The value of the bond at any time t is

$$B_t = (1+r)^t B_0$$

where the interest rate $r > -1$ is fixed and known at time 0.

(3) There is a finite number of stocks $\{S^1, S^2, \ldots, S^m\}$ and the vector of stock prices is denoted $S = (S^1, S^2, \ldots, S^m)$. Stock prices change over time; the vector of stock prices at time t is denoted by $S_t = (S_t^1, S_t^2, \ldots, S_t^m)$.

(4) An investor or trader can hold on each time interval $(t-1, t]$ a portfolio $\varphi_t = (x_t, y_t)$, where $x_t \in \mathbb{R}$ denotes the holding in bonds and the vector $y_t = (y_t^1, \ldots, y_t^m) \in \mathbb{R}^m$ denotes the stock holding. There is no restriction on the values of x_t, y_t^i.

(5) Trading is instantaneous, borrowing and lending rates are the same, and there are no transaction costs in trading; equivalently, the buying and selling prices of shares are the same.

(6) There is a finite number of scenarios Ω, each of which describes a possible history of the stock price evolution as described below. The probability of any scenario $\omega \in \Omega$ occurring is denoted by $\mathbb{P}(\omega)$. It is assumed that $\mathbb{P}(\omega) > 0$ for each ω, which means that impossible scenarios (that is, zero-probability scenarios) are disregarded. The probability $\mathbb{P}$ is called the *real-world probability* or the *market probability*, and

$$\sum_{\omega \in \Omega} \mathbb{P}(\omega) = 1.$$

Remarks 4.1

(1) The time between trading dates represents a fixed unit τ of real time, for example one second, one minute, one month or one year. Since r denotes the interest accrued over time τ, if this is not one year then r is *not* the annual rate that would be quoted by a lender.

(2) The assumption that interest rates are constant over time is only for notational convenience. The theory that will be developed can easily be adapted to work in the slightly more general model that allows interest rates to vary over time, that is, where the interest rate on the interval $[t-1, t]$ is denoted by r_t, and $B_t = (1 + r_t)B_{t-1}$ for each $t > 0$. See Section 8.2 for further details.

(3) As with single-period models (see the opening paragraph of Section 3.2), the theory below works even if $S_t^i \leq 0$ for some i and t, which means that it applies to a very general class of risky assets (including derivatives). We will nevertheless continue to think of S as a vector of stock prices for the sake of convenience.

(4) We will see that as in single-period models the real world probability $\mathbb{P}$ is irrelevant to the pricing theory that is developed.

4.1 Stock Price Evolution

As noted above, the set of scenarios Ω represents the collection of possible histories, or states of the world that could evolve by time T. In reality, the market will follow just one such scenario, but we cannot say in advance which one it will be. As time progresses, more information about the history of stock prices becomes known, so the number of possible final outcomes decreases.

We make the fundamental assumption that each scenario is simply one of the possible price histories of the stock prices. Some notation is needed to explain this in detail. As above S denotes the stock price vector $(S^1, S^2, \ldots, S^m)$. The initial stock prices $S_0 = (S_0^1, S_0^2, \ldots, S_0^m)$ are known at time 0. The price S_t^k of stock k at time $t > 0$ is a random variable, whose value only becomes known at time t and depends on the particular scenario that the market has followed. We write $S_t(\omega) = (S_t^1(\omega), S_t^2(\omega), \ldots, S_t^m(\omega))$ to denote the *stock price vector* at time t for each scenario ω.

Each scenario $\omega \in \Omega$ thus generates a *stock price history*

$$\big(S_0, S_1(\omega), S_2(\omega), \ldots, S_T(\omega)\big).$$

The assumption above means that different scenarios produce different price histories. That is, we may identify each scenario ω with its price history, so that

$$\omega = \big(S_0, S_1(\omega), S_2(\omega), \ldots, S_T(\omega)\big).$$

Since Ω has a finite number of elements, it follows that the price of any stock at any time can take only a finite number of values.

Just as a single-period model may be represented as a collection of branches, a multi-period model may be represented as a *tree*. The branches of the tree are the scenarios (or possible price histories). A *node* at time t is any *partial history* up to time t. If

$$\omega = \big(S_0, S_1(\omega), S_2(\omega), \ldots, S_T(\omega)\big) \in \Omega$$

is a scenario or *complete price history*, the *node at time t* on ω is the partial history (or partial scenario)

$$\omega \uparrow t := \big(S_0, S_1(\omega), S_2(\omega), \ldots, S_t(\omega)\big).$$

This should be read as "*the scenario* (or *price history*) ω *up to time t*".

There is a single node at time 0, namely $(S_0) = \omega \uparrow 0$ for any ω. This is called the *root node* and denoted by $\emptyset$. Any time-T node is called a *terminal node*; a terminal node is none other than a scenario because clearly

$$\omega \uparrow T = \omega$$

for any scenario $\omega \in \Omega$.

The stock price moves randomly, so it is reasonable to expect that any node at time $t < T$ belongs to more than one scenario (otherwise the future stock price movements from that point on would be known). In other words, if $\omega \uparrow t$ is a time-t node with $t < T$ there will be at least one other scenario (or history) $\omega' \neq \omega$ that is the same up to time t; that is $\omega \uparrow t = \omega' \uparrow t$. So there is no one-to-one correspondence between scenarios and *non-terminal nodes*.

Note that if ω and ω' are different scenarios having the same time-t node λ say and $s \leq t$ then we have $S_s(\omega) = S_s(\omega')$ for all $s \leq t$. Thus, if λ is a time-t node and $s \leq t$ we may define

$$S_s(\lambda) := S_s(\omega)$$

for any scenario ω with $\lambda = \omega \uparrow t$. In particular we say that $S_t(\lambda)$ is the *stock price vector at the node* λ.

If λ is a time-t node with $t < T$, a *successor node* to λ is any node $\omega \uparrow (t+1)$ at time $t+1$ such that $\omega \uparrow t = \lambda$. For any such non-terminal node λ, the collection of its successor nodes is denoted by

$$\mathrm{succ}\,\lambda := \big\{\omega \uparrow (t+1) : \omega \in \Omega, \omega \uparrow t = \lambda\big\}.$$

The above discussion of randomness is encapsulated in the following assumption, which is the natural generalization to multi-period models of the Assumption 3.2 for single-period models.

Assumption 4.2 (Randomness)

Every non-terminal node on the pricing tree has at least two distinct successor nodes.

Example 4.2

Consider the price evolution of a single stock $S \equiv S^1$ in a two-period model with four possible scenarios w_1, w_2, w_3 and w_4 given in the following table.

w	S_0	$S_1(w)$	$S_2(w)$
w_1	100	120	140
w_2	100	120	110
w_3	100	90	120
w_4	100	90	80

The root node is clearly $\emptyset = (100)$, and there are two nodes at time 1, namely

$$w_1 \uparrow 1 = w_2 \uparrow 1 = (100, 120),$$

$$w_3 \uparrow 1 = w_4 \uparrow 1 = (100, 90).$$

There are as many terminal nodes as scenarios, and they are given by

$$w_1 = w_1 \uparrow 2 = (100, 120, 140),$$

$$w_2 = w_2 \uparrow 2 = (100, 120, 110),$$

$$w_3 = w_3 \uparrow 2 = (100, 90, 120),$$

$$w_4 = w_4 \uparrow 2 = (100, 90, 80).$$

Both $w_1 \uparrow 1 = w_2 \uparrow 1$ and $w_3 \uparrow 1 = w_4 \uparrow 1$ are successor nodes to the root node $\emptyset$. In turn, the successor nodes of $w_1 \uparrow 1 = w_2 \uparrow 1$ are w_1 and w_2, and the successor nodes to $w_3 \uparrow 1 = w_4 \uparrow 1$ are w_3 and w_4.

This is a (single stock) *binary model*, because each node has exactly two successors. In such a model it is customary to label nodes according to the up and down movements of the (discounted) stock price at successive times. Thus the time-1 nodes are

$$u = (100, 120),$$

$$d = (100, 90),$$

and the time-2 nodes are

$$uu = (100, 120, 140) = \omega_1,$$

$$ud = (100, 120, 110) = \omega_2,$$

$$du = (100, 90, 120) = \omega_3,$$

$$dd = (100, 90, 80) = \omega_4.$$

Using this notation, the stock price movement may be represented by the tree in Model 4.1; arrows indicate succession.

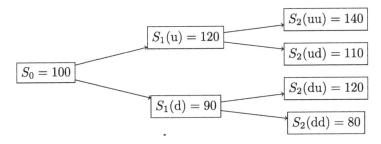

Model 4.1   Two-step single-stock binary model in Examples 4.2, 4.3 and 4.43

As already noted, a multi-period model may be viewed as a succession of single-period models, so the theory of single-period models developed in Chapter 3 will be used to inform the study of multi-period models in this and later chapters. For any non-terminal node λ at time $t < T$, we refer to the one-step model that can be constructed from the node λ and its successors at time $t+1$ as the *single-step submodel at* λ.

Example 4.3

Model 4.1 contains three single-step submodels, namely at the root node (together with its successors u and d), at u (with uu and ud) and at d (with du and dd).

Example 4.4

In a model with more than one stock, the randomness assumption does not mean that *every* stock has at least two distinct prices at the next time step. In

Model 4.2, we have $S_1^2(\mathrm{u}) = S_1^2(\mathrm{d}) = 110$, but since

$$S_1(\mathrm{u}) = \begin{bmatrix} 132 \\ 110 \end{bmatrix} \neq \begin{bmatrix} 77 \\ 110 \end{bmatrix} = S_1(\mathrm{d}),$$

the root node has two different successor nodes and therefore Model 4.2 satisfies Assumption 4.2. It is convenient to denote the five terminal nodes in this model by uu, um, ud, du, dd. Since terminal nodes are scenarios this means, for example, that ud is the price history

$$\mathrm{ud} = \left(\begin{bmatrix} 100 \\ 100 \end{bmatrix}, \begin{bmatrix} 132 \\ 110 \end{bmatrix}, \begin{bmatrix} 157.3 \\ 96.8 \end{bmatrix} \right)$$

and $\Omega = \{\mathrm{uu}, \mathrm{um}, \mathrm{ud}, \mathrm{du}, \mathrm{dd}\}$. In another context these scenarios might be labelled as $\omega_1, \ldots, \omega_5$.

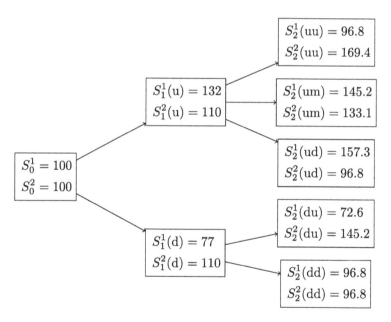

Model 4.2 Two-step two-stock binary model in Example 4.4 and Exercise 4.31

Exercise 4.5

Find a tree representation of the stock price evolution of a single stock given by the following table.

ω	S_0	$S_1(\omega)$	$S_2(\omega)$
ω_1	55	66	75
ω_2	55	66	65
ω_3	55	40	65
ω_4	55	40	50
ω_5	55	40	30

4.2 Trading

Traders in a multi-period model are allowed to adjust their portfolios as time progresses, and in doing so may take account of the way that each scenario evolves, as next described. Recall that we are assuming that the market places no restrictions on the amount that a trader can invest in any of the assets at any time.

In single-period models, trading consists of buying a portfolio at time $t = 0$, holding it during the time period $[0, 1]$ and then clearing it at the final time $t = 1$. By contrast, in models with several time steps, a trader's portfolio may be changed at intermediate time steps $t = 1, 2, \ldots$ by buying or selling shares and bonds. A trader would typically form an initial portfolio $\varphi_1 = (x_1, y_1)$ at time 0 which is held up to time 1, and then adjust it at time $t = 1$ giving a new portfolio $\varphi_2 = (x_2, y_2)$ that is held for the time period $(1, 2]$, and so on. In general, for each $t > 0$ a trader chooses a portfolio φ_t at time $t - 1$, and holds it during the time period $(t - 1, t]$. The portfolio that is denoted φ_{t+1} is the result of adjusting the portfolio φ_t at time t. For a graphical illustration, see Figure 4.3.

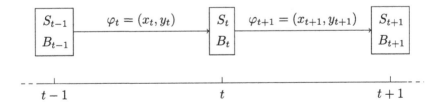

Figure 4.3   Adjustment of trading strategy

The resulting sequence of portfolios $\Phi \equiv (\varphi_t)_{t=1}^T = (x_t, y_t)_{t=1}^T$ is called a *trading strategy* provided it meets the requirement of *predictability* explained below. For convenience of notation, we assume that there is no change of port-

folio at time $t = 0$, which means that writing φ_0 for the holding at $t = 0$ we
have $\varphi_0 = \varphi_1$.

Naturally, a trader should be allowed to take into account the known history
of stock prices to date, so the choice of a new portfolio at any time could be
different for different scenarios. Mathematically this means that the portfolio
at time t is random, so we need to write $\varphi_t(w)$ or $(x_t(w), y_t(w))$ to indicate
dependence on $w \in \Omega$. However, the model should assume that traders in the
market have no knowledge of future events on which to base the choice of
portfolio, and in particular any information about the future evolution of stock
prices; in other words there should be no 'insider dealing'. This assumption is
made mathematically precise as follows, using the notion of partial histories or
nodes. A *stochastic* or *random process* $X = (X_t)_{t=0}^{T}$ or $X = (X_t)_{t=1}^{T}$ is a family
of random variables indexed by time, which can be thought of alternatively as
a random quantity that changes over time.

Definition 4.6 (Predictability)

A stochastic process $(X_t)_{t=1}^{T}$ is called *predictable* if for each $t > 0$ the value
$X_t(w)$ depends only on $w \uparrow (t-1)$; that is, if $w \uparrow (t-1) = w' \uparrow (t-1)$ then
$X_t(w) = X_t(w')$.

If $(X_t)_{t=1}^{T}$ is predictable, then for any time t, and any node λ at the earlier
time $t - 1$ we may write $X_t(\lambda)$ to mean $X_t(w)$ for any w with $w \uparrow (t-1) = \lambda$.

Remark 4.7

If X is a predictable process, then X_1 is non-random, and the value X_0 is not
necessarily defined. It is often convenient to define $X_0 = X_1$.

The way to build in the assumption of no insider trading is to allow only
predictable choices of portfolio on the time interval $(t-1, t]$. This leads to the
following definition.

Definition 4.8 (Trading strategy)

A *trading strategy* is a sequence $\Phi \equiv (\varphi_t)_{t=0}^{T}$ of portfolios such that:

(a) $\varphi_0 = \varphi_1$.

(b) The sequence of portfolios $(\varphi_t)_{t=1}^{T}$ is predictable.

Property (a) is included because mathematically it is convenient to have an initial portfolio φ_0 at time $t = 0$ that continues to be held over the time interval $(0, 1]$ where it is denoted φ_1.

A trading strategy Φ gives rise to a *value process* V^Φ that tracks the value of the portfolio as time changes. Here is the definition.

Definition 4.9 (Value process)

If $\Phi \equiv (x_t, y_t)_{t=0}^T$ is a trading strategy its *value process* $V^\Phi = (V_t^\Phi)_{t=0}^T$ is given for all t by

$$V_t^\Phi := x_t B_t + y_t \cdot S_t. \tag{4.1}$$

Note that since both the stock prices and the portfolios in Φ are in general random variables, then so is the value V_t^Φ for each $t > 0$. To emphasize this the above definition would be written in full as

$$V_t^\Phi(\omega) = x_t(\omega) B_t + y_t(\omega) \cdot S_t(\omega)$$

for $t > 0$ and $\omega \in \Omega$. The *initial value* of the strategy is

$$V_0^\Phi = x_0 B_0 + y_0 \cdot S_0, \tag{4.2}$$

which is non-random. For later times $t > 0$ note that although the portfolio $(x_t(\omega), y_t(\omega))$ depends only on $\omega \uparrow (t-1)$, the value $V_t^\Phi(\omega)$ depends on $\omega \uparrow t$ since it also involves the current value of the stock. This motivates the following general definition.

Definition 4.10 (Adaptedness)

A stochastic process $(X_t)_{t=0}^T$ is *adapted* or *non-anticipating* if, for each t, the value $X_t(\omega)$ depends only on $\omega \uparrow t$; that is, if $\omega \uparrow t = \omega' \uparrow t$ then $X_t(\omega) = X_t(\omega')$.

If $(X_t)_{t=0}^T$ is an adapted process, then for any node λ at time t we may write $X_t(\lambda)$ to mean $X_t(\omega)$ for any ω with $\omega \uparrow t = \lambda$.

Example 4.11

The stock price process $(S_t)_{t=0}^T$ and the value process $(V_t^\Phi)_{t=0}^T$ of any trading strategy Φ are adapted processes.

Trading strategies that allow investors to make only trades that they can afford form an important class called *self-financing trading strategies*. When following a self-financing trading strategy, all changes in an investor's wealth come entirely from the market movement and interest; after his initial investment the investor does not inject or withdraw any funds for any purpose.

To express this idea mathematically, consider the changes in the value process due to trading using a strategy $\Phi \equiv (\varphi_t)_{t=0}^T = (x_t, y_t)_{t=0}^T$. At the instant $t > 0$ the investor adjusts his portfolio from φ_t, which was held on the interval $(t-1, t]$, to φ_{t+1}, which will be held on the interval $(t, t+1]$. Before the change at time t the value is

$$V_t^\Phi = x_t B_t + y_t \cdot S_t,$$

and after the change of portfolio it is

$$V_{t+}^\Phi := x_{t+1} B_t + y_{t+1} \cdot S_t. \tag{4.3}$$

If the change in portfolios is to be self-financing (that is, with no injection or withdrawal of funds) then these values must be the same, although the way they are made up can differ. The precise definition of a self-financing trading strategy is as follows.

Definition 4.12 (Self-financing trading strategy)

Let $\Phi \equiv (x_t, y_t)_{t=0}^T$ be a trading strategy. Then Φ is *self-financing* if

$$V_{t+}^\Phi = V_t^\Phi \tag{4.4}$$

for $t < T$.

Remark 4.13

By the convention adopted at $t = 0$, namely $(x_0, y_0) = (x_1, y_1)$, equation (4.4) always holds for $t = 0$.

Remark 4.14

The simplest example of a self-financing trading strategy is when no changes are made; that is, $\varphi_t = \varphi_0$ for all t. Note also that if at any particular time t and time-t node λ no changes are made to the portfolio (that is, if $\varphi_{t+1}(\lambda) = \varphi_t(\lambda)$), then the self-financing condition automatically holds at that node.

We write

$$\Delta X_t := X_t - X_{t-1}$$

for $t > 0$ to denote the *increments* of any random process X. In particular, for a trading strategy Φ the increment ΔV_t^Φ is called the *incremental gain of Φ at time t*. Note that ΔV_t^Φ can be negative if Φ loses value over the time period $[t-1, t)$. The following shows that a strategy is self-financing precisely when the incremental gain is equal to the change in the values of the assets.

Theorem 4.15

For any trading strategy $\Phi \equiv (\varphi_t)_{t=0}^T = (x_t, y_t)_{t=0}^T$ the following statements are equivalent:

(1) Φ is self-financing.

(2) If $t > 0$ then

$$\Delta V_t^\Phi = x_t \Delta B_t + y_t \cdot \Delta S_t.$$

(3) If $t > 0$ then

$$V_t^\Phi = V_0^\Phi + \sum_{s=1}^t x_s \Delta B_s + \sum_{s=1}^t y_s \cdot \Delta S_s.$$

Proof

Suppose that Φ is self-financing. For any $t > 0$ we have $V_{t-1}^\Phi = V_{(t-1)+}^\Phi$ and therefore

$$\begin{aligned}
\Delta V_t^\Phi = V_t^\Phi - V_{t-1}^\Phi &= V_t^\Phi - V_{(t-1)+}^\Phi \\
&= V_t^\Phi - (x_t B_{t-1} + y_t \cdot S_{t-1}) \\
&= x_t(B_t - B_{t-1}) + y_t \cdot (S_t - S_{t-1}) \\
&= x_t \Delta B_t + y_t \cdot \Delta S_t.
\end{aligned}$$

So (1) implies (2).

Conversely, if $\Delta V_t^\Phi = x_t \Delta B_t + y_t \cdot \Delta S_t$ for each $t > 0$ then by definition of ΔV_t^Φ, ΔB_t and ΔS_t we can rewrite this as

$$\begin{aligned}
V_t^\Phi - V_{t-1}^\Phi &= x_t(B_t - B_{t-1}) + y_t \cdot (S_t - S_{t-1}) \\
&= x_t B_t + y_t \cdot S_t - x_t B_{t-1} - y_t \cdot S_{t-1} \\
&= V_t^\Phi - V_{(t-1)+}^\Phi.
\end{aligned}$$

It follows that

$$V_{t-1}^\Phi = V_{(t-1)+}^\Phi$$

for each $t > 0$ so Φ is self-financing. Thus (2) implies (1).

The equivalence of (2) and (3) is routine since by definition

$$V_t^\Phi = V_0 + \sum_{s=1}^{t} \Delta V_s^\Phi$$

for $t > 0$ so (2) implies (3). Conversely, from (3) we obtain that

$$V_t^\Phi - V_{t-1}^\Phi = x_t \Delta B_t + y_t \cdot \Delta S_t,$$

which is (2). $\qquad\qquad\qquad\qquad\qquad\qquad\qquad\qquad\qquad\qquad\qquad\qquad$ □

Example 4.16

For Model 4.1 (with $B_0 = 1$ and $r = 10\%$), consider the trading strategy $\Phi = (x_t, y_t)_{t=0}^2$ with

$$(x_1, y_1) = (100, 2), \qquad (x_2(u), y_2(u)) = (50, 3), \qquad (x_2(d), y_2(d)) = (200, 1).$$

The value and gains processes of Φ are straightforward to calculate. For example, for the scenario uu, the value process is

$$V_0^\Phi = 100 + 2 \times 100 = 300,$$

$$V_1^\Phi(u) = 100 \times 1.1 + 2 \times 120 = 350,$$

$$V_2^\Phi(uu) = 50 \times 1.21 + 3 \times 140 = 480.5.$$

To see whether Φ is self-financing we need to check the changes in value at *all* nodes at the intermediate time steps. In this case the only intermediate time is $t = 1$ and, the relevant nodes are u and d. For u we have

$$V_1^\Phi(u) = 350,$$

$$V_{1+}^\Phi(u) = 50 \times 1.1 + 3 \times 120 = 415,$$

so this strategy is not self-financing.

To complete the picture, check that at d we have

$$V_1^\Phi(d) = 100 \times 1.1 + 2 \times 90 = 290,$$

$$V_{1+}^\Phi(d) = 200 \times 1.1 + 1 \times 90 = 310,$$

so this strategy is not self-financing at either of the intermediate nodes.

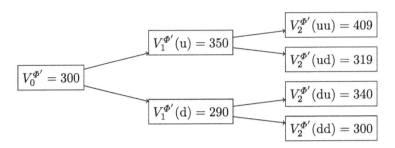

Figure 4.4   Value of adjusted strategy in Model 4.1

To illustrate Theorem 4.15 above we have, for example

$$x_2(u)\Delta B_2 + y_2(u)\Delta S_2(uu) = 50 \times (1.21 - 1.1) + 3 \times (140 - 120) = 65.5$$

but $\Delta V_2^{\Phi}(uu) = 130.5$, which gives an alternative verification that Φ is not self-financing.

Notice now that the bond holding in this strategy may be adjusted to create a strategy that *is* self-financing. Consider the trading strategy $\Phi' \equiv (x_t', y_t)$ with

$$x_1' := x_1 = 100,$$

$$x_2'(u) := \tfrac{1}{1+r}\big(V_1^{\Phi}(u) - y_2(u)S_1(u)\big) = \tfrac{1}{1.1}(350 - 3 \times 120) \approx -9.0909,$$

$$x_2'(d) := \tfrac{1}{1+r}\big(V_1^{\Phi}(d) - y_2(d)S_1(d)\big) = \tfrac{1}{1.1}(290 - 1 \times 90) \approx -181.8182.$$

It is straightforward to check that this is self-financing. The value process of Φ' appears in Figure 4.4.

Remark 4.17

As this example shows there is less work involved in using the basic definition (equation (4.4)) rather than Theorem 4.15 to check whether a strategy is self-financing. However, Theorem 4.15 (and its consequence Theorem 4.20) is important from a theoretical point of view, as we will see later.

Exercise 4.18

In Model 4.5 below, consider the trading strategy $\Phi \equiv (x_t, y_t)_{t=0}^2$, where

$$(x_1, y_1) = (-10, 11),$$

$$\big(x_2(u), y_2(u)\big) = (290, 5.5), \qquad \big(x_2(d), y_2(d)\big) = (1115, -16.5).$$

Construct the value process of this strategy, and show that it is self-financing.

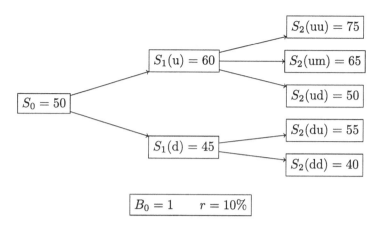

Model 4.5 Two-step model in Exercises 4.18 and 4.23

The next theorem gives further insight into what it means for a trading strategy to be self-financing. First we need to extend the idea of discounting to the multi-period setting.

Definition 4.19 (Discounted value)

Given a stochastic process $X = (X_t)_{t=0}^T$ the *discounted process* $\bar{X}$ corresponding to X is defined by

$$\bar{X}_t := \frac{X_t}{(1+r)^t}$$

for all t. The value $\bar{X}_t$ is referred to as the *discounted* (to time 0) *value* of the random variable X_t.

For $t > 0$ the *increment of the discounted process* is

$$\Delta \bar{X}_t = \bar{X}_t - \bar{X}_{t-1}.$$

Note that in the particular case $X = B$ we have $\bar{B}_t = B_0$ and $\Delta \bar{B}_t = 0$ for all t. The following should be compared with Theorem 4.15.

Theorem 4.20

If $\Phi = (x_t, y_t)_{t=0}^T$ is a trading strategy, then the following statements are equivalent.

(1) Φ is self-financing.

(2) If $t > 0$ then

$$\Delta \bar{V}_t^\Phi = y_t \cdot \Delta \bar{S}_t.$$

(3) If $t > 0$ then

$$\bar{V}_t^\Phi = \bar{V}_0^\Phi + \sum_{s=1}^t y_s \cdot \Delta \bar{S}_s. \tag{4.5}$$

Proof

Observe that the trading strategy Φ is self-financing if and only if $V_t^\Phi = V_{t+}^\Phi$, equivalently $\bar{V}_t^\Phi = \bar{V}_{t+}^\Phi$ for all t.

Consider now the discounted multi-period model with stock price $\bar{S}_t$ and bond price $\bar{B}_t$ for all t. In the discounted model, the value process of Φ in the discounted model is $\bar{V}^\Phi = (\bar{V}_t^\Phi)_{t=0}^T$, and its value after changing portfolio at time t is $\bar{V}_{t+}^\Phi$, so by our observation Φ is self-financing in the original model if and only if Φ is self-financing in the discounted model. The result follows by applying Theorem 4.15 to the discounted model, and noting that $\Delta \bar{B}_t = 0$ for all t.   □

Remarks 4.21

(1) Theorem 4.20 cannot be proved by simply discounting in parts (2) and (3) of Theorem 4.15. This is because the increment of a discounted process is not the same as the discounted increment of that process. For example,

$$\overline{\Delta S_t} = \frac{\Delta S_t}{(1+r)^t} = \frac{S_t}{(1+r)^t} - \frac{S_{t-1}}{(1+r)^t}$$

$$\neq \frac{S_t}{(1+r)^t} - \frac{S_{t-1}}{(1+r)^{t-1}} = \bar{S}_t - \bar{S}_{t-1} = \Delta \bar{S}_t.$$

(2) The above proof raises an important point about the self-financing property. A trading strategy Φ could be self-financing in a given model with stock price S, but if the asset prices are changed (even if the tree structure of the price process remains the same) then Φ might not be self-financing for the new model.

The next theorem shows that it is always possible to transform a trading strategy that is not self-financing into one that is self-financing, by adjusting only the bond holding as in Example 4.16. The idea (and the idea of the proof) is that a trader may at any time modify his portfolio to have any desired stock holding at any time, but to do so in a self-financing way the holding in bonds must be adjusted appropriately. If, for example, stock holdings are increased, then there must be an increase in borrowing from the bank to finance this.

The formal statement is as follows.

Theorem 4.22

Suppose that a predictable sequence $(y_t)_{t=1}^T$ of stock holdings is given, together with an initial value v. There is a unique predictable sequence of bond positions $(x_t)_{t=1}^T$ such that the trading strategy $\Phi = (x_t, y_t)_{t=0}^T$ with $x_0 = x_1$ and $y_0 = y_1$ is self-financing and has initial value $V_0^\Phi = v$.

Proof

In order to have $V_0^\Phi = v$, and in view of the definition (4.2), we must have

$$x_1 := \frac{1}{B_0}(v - y_1 \cdot S_0). \tag{4.6}$$

The bond holdings x_t for $t > 1$ are now constructed by recursion on $t \leq T$.

Suppose that $x_1, \ldots, x_t$ have already been chosen in a predictable way so that Φ is self-financing for all re-balancing times before time t. To maintain the self-financing property at time t, the random variable x_{t+1} must be chosen to ensure that the self-financing property (4.4) holds at time t. That is, we require that for any $\omega \in \Omega$

$$V_t^\Phi(\omega) = V_{t+}^\Phi(\omega) = x_{t+1}(\omega)B_t + y_{t+1}(\omega) \cdot S_t(\omega).$$

Solving for $x_{t+1}(\omega)$ (which is the only unknown in this equation) gives

$$x_{t+1}(\omega) := \frac{1}{B_t}\left(V_t^\Phi(\omega) - y_{t+1}(\omega) \cdot S_t(\omega)\right) \tag{4.7}$$

as the unique value that maintains the self-financing property at time t. Since the right hand side depends only on $\omega \uparrow t$ it follows that the trading strategy Φ thus constructed is predictable. ∎

Exercise 4.23

Given stock holdings $y_1 = 30$, $y_2(u) = 41$, $y_2(d) = 8$ and an initial value $v = 2000$, find a predictable sequence $(x_t)_{t=1}^2$ of bond holdings in Model 4.5 so that the trading strategy $\Phi := (x_t, y_t)_{t=0}^2$ is self-financing and has initial value v.

Exercise 4.24[†]

(a) Show that if $\Phi = (x_t, y_t)_{t=0}^T$ is self-financing and c is a constant (positive or negative), then the trading strategy $c\Phi \equiv (cx_t, cy_t)_{t=0}^T$ is self-financing and $V_t^{c\Phi} = cV_t^{\Phi}$ for all t.

(b) Show that if $\Phi = (x_t, y_t)_{t=0}^T$ and $\Phi' = (x_t', y_t')_{t=0}^T$ are self-financing, then the trading strategy $\Phi + \Phi' \equiv (x_t + x_t', y_t + y_t')_{t=0}^T$ is self-financing and $V_t^{\Phi+\Phi'} = V_t^{\Phi} + V_t^{\Phi'}$ for all t.

In the pricing and replication of derivatives, the only trading strategies of importance are the self-financing ones. From here on, unless explicitly stated otherwise, we will use the term "trading strategy" (or "strategy" for short) to mean "self-financing trading strategy".

4.3 Viability

We are now in a position to define what is meant by an *arbitrage opportunity* in a model with several time steps. The idea is the same as in a one-step model: an arbitrage opportunity is a way to obtain a riskless profit, or a "free lunch"; here this will mean a trading strategy rather than a single portfolio. As before, realistic market models will not have any arbitrage opportunities: they will be *viable*. Here are the definitions.

Definition 4.25 (Arbitrage, viability)

(a) An *arbitrage opportunity* is a self-financing trading strategy Φ with $V_0^{\Phi} = 0$, such that $V_T^{\Phi} \geq 0$ (meaning that $V_T^{\Phi}(\omega) \geq 0$ for all $\omega \in \Omega$) and $V_T^{\Phi}(\omega) > 0$ for at least one ω.

(b) A market model is *viable* if there are no arbitrage opportunities.

In other words, an arbitrage opportunity is a self-financing strategy for buying and selling shares and bonds at any times in the period $[0, T]$, with zero initial outlay, that provides a positive probability of profit, and no risk of loss.

Exercise 4.26[‡]

Suppose that, in a model with time horizon T, there exists a self-financing strategy $\Phi = (\varphi_t)_{t=0}^T$ with the following properties:

(1) There is a time $s < T$ such that $V_s^{\Phi} = 0$.

(2) There is a time $u > s$ such that $V_u^{\Phi} \geq 0$ and $V_u^{\Phi}(\omega) > 0$ for at least one $\omega \in \Omega$.

Show that the model is not viable.

Note that an arbitrage opportunity Φ is specified completely by the strategy of stock holdings $y = (y_t)_{t=1}^T$, because if this is known then Theorem 4.22 tells us that the property $V_0^{\Phi} = 0$ specifies a unique (self-financing) trading strategy with those values for y. Thus we make the following extension of the definition of implicit arbitrage.

Definition 4.27 (Implicit arbitrage)

An *implicit arbitrage opportunity* is any predictable sequence of stock holdings $(y_t)_{t=1}^T$ such that the unique self-financing trading strategy Φ with $V_0^{\Phi} = 0$ given by Theorem 4.22 is an arbitrage opportunity.

The following result extends the Implicit Arbitrage Lemma (Lemma 3.31) to multi-period models.

Lemma 4.28 (Implicit Arbitrage Lemma)

(1) The following statements are equivalent for any predictable sequence of stock holdings $(y_t)_{t=1}^T$:

(i) $(y_t)_{t=1}^T$ is an implicit arbitrage opportunity.

(ii) $\sum_{s=1}^T y_s \cdot \Delta \bar{S}_s \geq 0$, and $\sum_{s=1}^T y_s(\omega) \cdot \Delta \bar{S}_s(\omega) > 0$ for some $\omega \in \Omega$.

(2) A model is viable if and only if there is no implicit arbitrage opportunity.

Proof

The proof is almost identical to the proof of Lemma 3.31. Take any predictable sequence $(y_t)_{t=1}^T$ and let Φ be the unique trading strategy with $V_0^{\Phi} = 0$ guaranteed by Theorem 4.22. Then from Theorem 4.20 we have

$$\bar{V}_t^{\Phi} = \sum_{s=1}^{t} y_s \cdot \Delta \bar{S}_s$$

from which (1) follows.

For (2), every arbitrage opportunity gives rise to an implicit arbitrage opportunity, and vice versa. □

By viewing a multi-period model as a succession of single steps, we now show that a multi-period model is viable if and only if each single-step submodel is viable. (Recall that by a *single-step submodel* we mean any non-terminal node together with its successor nodes.) Consequently we will be able to use the theory of viability in single-period models to capture the notion of viability in terms of the *convexity* of discounted prices at successor nodes in a multi-period model: the model will be viable if and only if, at each non-terminal node on the stock price tree, the stock price can be expressed as a strictly convex combination of the discounted prices of its successors.

Here are the details.

Theorem 4.29

A finite multi-period market model is viable if and only if each single-step submodel is viable.

Although this result and the idea underlying it is simple, the detailed proof is somewhat lengthy. However, in view of its frequent use in the sequel, the reader is encouraged to persevere. Before embarking on the proof of Theorem 4.29 it is worth illustrating its usefulness by showing how it applies to Model 4.1.

Example 4.30

Theorem 4.29 (together with what we know about viability in single step binary models) tells us that a binary tree model is viable provided that at each non-terminal node, in discounted terms, each upward price movement has a corresponding downward counterpart. Recalling that $B_0 = 1$ and $r = 10\%$, Model 4.1 is viable, since

$$90 < 100 \times 1.1 < 120,$$

$$80 < 90 \times 1.1 < 120,$$

$$110 < 120 \times 1.1 < 140.$$

Proof (of Theorem 4.29)

First, suppose that each single-step submodel is viable and that $\Phi \equiv (\varphi_t)_{t=0}^{T}$ is a self-financing trading strategy.

We will verify the following for any node λ at any time $t < T$:

(1) If $V_{t+1}^{\Phi}(\mu) \geq 0$ for all $\mu \in \text{succ}\,\lambda$, then $V_t^{\Phi}(\lambda) \geq 0$.

(2) If $V_{t+1}^{\Phi}(\mu) \geq 0$ for all $\mu \in \text{succ}\,\lambda$ and $V_{t+1}^{\Phi}(\mu) > 0$ for at least one $\mu \in \text{succ}\,\lambda$, then $V_t^{\Phi}(\lambda) > 0$.

Fix any node λ at time $t < T$ and consider the single-step submodel with root node λ. Considering the portfolio $(x, y) := \varphi_{t+1}(\lambda)$ in this submodel, note that

$$V_t^{\Phi}(\lambda) = V_{t+}^{\Phi}(\lambda) = xB_t + y \cdot S_t(\lambda)$$

and

$$V_{t+1}^{\Phi}(\mu) = xB_{t+1} + y \cdot S_{t+1}(\mu)$$

for all $\mu \in \text{succ}\,\lambda$. The submodel with root node λ is viable, so if $V_{t+1}^{\Phi}(\mu) \geq 0$ for all $\mu \in \text{succ}\,\lambda$ then $V_t^{\Phi}(\lambda) \geq 0$, since otherwise (x, y) gives strong arbitrage in the submodel. This establishes (1). For (2), if in addition $V_{t+1}^{\Phi}(\mu) > 0$ for at least one $\mu \in \text{succ}\,\lambda$ then $V_t^{\Phi}(\lambda) > 0$, since otherwise (x, y) gives arbitrage in the submodel.

Now we will see that Φ is not an arbitrage opportunity. Suppose that $V_T^{\Phi} \geq 0$ and $V_T^{\Phi}(\omega) > 0$ for at least one $\omega \in \Omega$. By backward induction using (1) and (2) above, for each $t < T$ we have $V_t^{\Phi}(\lambda) \geq 0$ at every time-t node λ and $V_t^{\Phi}(\lambda) > 0$ for at least one such node. At time $t = 0$, since there is only one node this means that $V_0^{\Phi} > 0$. So Φ is not an arbitrage opportunity, and the model is viable.

For the converse, suppose that for some node λ at time $t < T$ the one-step submodel is not viable. This means that there is a portfolio (x, y) such that

$$xB_t + y \cdot S_t(\lambda) = 0$$

and

$$xB_{t+1} + y \cdot S_{t+1}(\mu) \geq 0$$

for all $\mu \in \text{succ}\,\lambda$ and

$$xB_{t+1} + y \cdot S_{t+1}(\mu') > 0$$

for at least one $\mu' \in \text{succ}\,\lambda$. Then an arbitrage opportunity for the multi-period model can be constructed as follows.

- Do nothing until time t.

- At time t, if we are at the node λ, obtain a free lunch by using the above portfolio (x, y) on the interval $(t, t+1]$. After this put all the wealth in bonds so as to ensure positive wealth at time T for all scenarios ω with $\omega \uparrow (t+1) = \mu'$.

- For any history that doesn't pass through λ continue to do nothing.

For the reader who wishes to see this in more detail, construct the strategy $\Phi = (x_t, y_t)_{t=0}^{T}$ with $x_0 = x_1$, $y_0 = y_1$ and zero initial cost as follows. First, for $s \le t$ set

$$x_s(\omega) := 0, \qquad y_s(\omega) := 0$$

for *all* ω. This means that $V_t^{\Phi}(\omega) = 0$ for *all* ω.

For all $\omega \in \Omega$ such that $\omega \uparrow t \neq \lambda$ set

$$x_s(\omega) := 0, \qquad y_s(\omega) := 0$$

for $s = t+1, t+2, \ldots, T$. So we do nothing until time t, and continue to do nothing for all time *unless the stock price history passes through the node* λ.

If the stock price history ω passes through the node λ, that is, if $\omega \uparrow t = \lambda$, put

$$x_{t+1}(\omega) := x, \qquad y_{t+1}(\omega) := y$$

which ensures that $V_{t+}^{\Phi}(\lambda) = V_t^{\Phi}(\lambda) = 0$. For each $\mu \in \text{succ}(\lambda)$ we have

$$V_{t+1}^{\Phi}(\mu) = x_{t+1}(\omega)B_{t+1} + y_{t+1}(\omega) \cdot S_{t+1}(\mu) \ge 0$$

and $V_{t+1}^{\Phi}(\mu') > 0$. If $t+1 = T$ there is nothing more to do. Otherwise, for $s > t+1$ and ω with $\omega \uparrow t = \lambda$ put

$$y_s(\omega) := 0,$$

so that all the wealth will be held in bonds. In order to maintain the self-financing property of Φ for $s = t+2$ and $\mu \in \text{succ}(\lambda)$ it is necessary to have

$$V_{t+1}^{\Phi}(\mu) = V_{(t+1)+}^{\Phi}(\mu) = x_{t+2}(\mu)B_{t+1}$$

so we set

$$x_{t+2}(\mu) = \frac{V_{t+1}^\Phi(\mu)}{B_{t+1}} \geq 0$$

and note that $x_{t+2}(\mu') > 0$. For subsequent times, for all ω with $\omega \uparrow t = \lambda$ make no changes in the portfolio; that is, set

$$x_s(\omega) = x_{t+2}(\omega) \geq 0$$

for $s > t+2$.

At time T this gives $V_T^\Phi(\omega) = 0$ if $\omega \uparrow t \neq \lambda$. If $\omega \uparrow t = \lambda$ then $V_T^\Phi(\omega) = (1+r)^{T-t-1} V_{t+1}^\Phi(\omega) \geq 0$, and if $\omega \uparrow (t+1) = \mu'$ then

$$V_T^\Phi(\omega) = (1+r)^{T-t-1} V_{t+1}^\Phi(\mu') > 0.$$

So Φ is an arbitrage opportunity, and the model is not viable. $\qquad\square$

Using Theorem 3.6, we can conclude from Theorem 4.29 that a model with a single stock S is viable if and only if at every node λ at time $t < T$ we have

$$\min\{S_{t+1}(\mu) : \mu \in \mathrm{succ}\,\lambda\} < (1+r)S_t(\lambda) < \max\{S_{t+1}(\mu) : \mu \in \mathrm{succ}\,\lambda\}, \quad (4.8)$$

or, equivalently, in discounted terms

$$\min\{\bar{S}_{t+1}(\mu)|\mu \in \mathrm{succ}\,\lambda\} < \bar{S}_t(\lambda) < \max\{\bar{S}_{t+1}(\mu)|\mu \in \mathrm{succ}\,\lambda\}. \quad (4.9)$$

Exercise 4.31 shows how Theorem 4.29 in conjunction with the results of Section 3.3 may be used to test the viability of multi-period models with more than one stock.

Exercise 4.31[†]

Determine whether Model 4.2 is viable if the interest rate is $r = 10\%$.

Hint. Use Proposition 3.37 and Lemma 3.42.

From the proof of Theorem 4.29 we can extract the following useful corollary.

Corollary 4.32

Suppose that Φ is a self-financing trading strategy in a viable model. The following holds for all t:

(1) If $V_t^\Phi \geq 0$, then $V_s^\Phi \geq 0$ for all $s \leq t$.

(2) If $V_t^\Phi \geq 0$ and $V_t^\Phi(\omega) > 0$ for some $\omega \in \Omega$, then $V_s^\Phi(\omega) > 0$ for all $s \leq t$.

Proof

This follows by backward induction using the properties (1) and (2) established at the beginning of the proof of Theorem 4.29.   □

Here is a useful result that can be deduced from Corollary 4.32 with the help of Exercise 4.24.

Corollary 4.33

Suppose that the model is viable.

(1) If Φ is a self-financing strategy with $V_T^\Phi = 0$, then $V_t^\Phi = 0$ for all t.

(2) If Φ and Φ' are two self-financing strategies with $V_T^\Phi = V_T^{\Phi'}$, then $V_t^\Phi = V_t^{\Phi'}$ for all t.

Proof

Apply Corollary 4.32 to Φ and $-\Phi$ to obtain (1). For (2), apply (1) to $\Phi - \Phi'$.   □

Exercise 4.34

Show that Corollary 4.33 is true in any single-stock model, even if it is not viable.

Hint. See Exercise 3.54.

4.4 Pricing Attainable Derivatives by Replication

4.4.1 Derivatives

The general notion of a derivative has a natural generalization to multi-period models: as the name suggests, it is an asset or financial instrument whose value is derived from, or dependent on, the history of the stock prices in the market.

The simplest derivatives are those that give a specified payoff at a fixed future date, such as the European call and put options discussed in earlier chapters. Such derivatives are called *European derivatives*, to distinguish them from *American derivatives* for example, where the specified payoff can be taken at any time. Since, in our models, each scenario ω is just a complete price history, a European derivative can be seen as a contract that specifies the value or *payoff to the owner* at the expiry date T for each scenario. In this chapter and the next the discussion is restricted to European derivatives, of which the precise mathematical definition reads as follows.

Definition 4.35 (European derivative)

A *European derivative* with expiry date T is a random variable; that is, a function $D : \Omega \to \mathbb{R}$. For each scenario $\omega \in \Omega$, the *payoff* of this derivative at time T is $D(\omega)$.

For now we will deal only with European derivatives, so until further notice the term *derivative* means *European derivative*.

Although a derivative only pays out at the expiry time, it is a contract that has value at any time from its date of issue up to the expiry date, and may be traded at any time. It is therefore relevant to ask whether there is a "fair" price of a derivative at any intermediate time $t < T$ as well as at time $t = 0$. This is the key question addressed by the theory to be developed in the rest of this chapter.

Here are some examples of derivatives, some of which have been seen earlier.

Examples 4.36

(a) A *European call option* on a stock S with strike price K and expiry date T is a derivative with payoff

$$C_T(\omega) = \max\{0, S_T(\omega) - K\} = [S_T(\omega) - K]_+$$

at time T for $\omega \in \Omega$.

(b) A *European put option* on a stock S with strike price K and expiry date T has payoff

$$P_T(\omega) = \max\{0, K - S_T(\omega)\} = [S_T(\omega) - K]_- = [K - S_T(\omega)]_+$$

at time T for $\omega \in \Omega$.

(c) A *long position in a forward contract* on a stock S with agreed forward price F and delivery date T has payoff

$$S_T(\omega) - F$$

for $\omega \in \Omega$. The payoff of the corresponding *short forward position* is

$$F - S_T(\omega)$$

for $\omega \in \Omega$. Section 4.5.2 gives more details on forward contracts.

The examples of European call and put options in earlier chapters illustrate clearly that these derivatives must have some value at their time of issue. Each is a contract that provides a payoff that is never negative, so at any time a trader would expect to have to pay something to own such a derivative. More generally, any derivative that gives a payoff (positive, negative or zero) at time T has *some* value at time t, which may be positive (meaning a trader would pay something to own this derivative), negative (meaning a trader would only accept this derivative upon receiving some money) or zero.

The theory of pricing in multi-period models both parallels the single step case, and builds on it. We continue to think of a multi-period model as a sequence of single-period models.

First, the notion of attainability is extended in a natural way.

Definition 4.37 (Attainability, replication)

A derivative D is *attainable* or *redundant* if there is a self-financing trading strategy Φ such that

$$V_T^\Phi(\omega) = D(\omega)$$

for $\omega \in \Omega$. The trading strategy Φ is then called a *replicating* or *hedging strategy* for D.

It follows from Exercise 4.39 below that if there is only one stock, then every attainable derivative has a unique replicating strategy. However, as with single-period models, if there is more than one stock there may be more than one way to replicate a given derivative. Bearing this in mind, we nevertheless have the following rephrasing of Corollary 4.33.

Theorem 4.38

In a viable model, if two self-financing strategies Φ, Φ' replicate a derivative D then their values are the same at all times; that is, $V_t^\Phi = V_t^{\Phi'}$ for all $t = 0, \ldots, T$.

The next exercise shows that in a model with a single stock we can deduce much more, without the need to assume viability.

Exercise 4.39

Show that if Φ and Φ' both replicate a derivative D in a multi-period model with a single stock, then $\Phi = \Phi'$.

Hint. See Exercises 3.54 and 4.34.

4.4.2 Fair Pricing

We begin by defining the notion of a fair price for a derivative D at time $t = 0$. For this it is necessary to define the notion of an extended strategy (involving buying or selling the derivative at any time) and the corresponding notion of an extended arbitrage opportunity.

Definition 4.40

(a) A *pricing structure* for a derivative D is an adapted stochastic process $\Pi = (\pi_t)_{t=0}^T$ such that $\pi_T = D$.

(b) An *extended (self-financing) trading strategy involving a derivative D* with pricing structure $\Pi = (\pi_t)_{t=0}^T$ for D is a process

$$\Psi = (x_t, y_t, z_t)_{t=0}^T$$

that is a self-financing trading strategy in the model obtained by thinking of the derivative D as a new risky asset priced at π_t at any time t; that is, z_t denotes the holding of D on the interval $(t-1, t]$ for $t > 0$ and $z_0 = z_1$.

(c) An *extended arbitrage opportunity* involving the pricing structure $\Pi = (\pi_t)_{t=0}^T$ is an extended trading strategy Ψ such that

$$V_0^\Psi := x_0 B_0 + y_0 \cdot S_0 + z_0 \pi_0 = 0,$$
$$V_T^\Psi := x_T B_T + y_T \cdot S_T + z_T D \geq 0$$

and $V_t^\Psi(\omega) > 0$ for at least one $\omega \in \Omega$.

(d) π is a *fair price* for D at time 0 if there is a pricing structure $\Pi = (\pi_t)_{t=0}^T$ for D with $\pi_0 = \pi$, such that there is no extended arbitrage opportunity involving Π.

Remark 4.41

A simpler notion of fair price would involve buying or selling the derivative at time $t = 0$ and holding it until the expiry time; however, in reality the derivative might be traded at intermediate times, so the above more elaborate definition is required to reflect that possibility. It turns out however that the two notions are equivalent—see Exercise 5.71 at the end of the next chapter.

The main theorem (and its proof) concerning pricing by replication in a multi-period model is the natural extension of the single-step result.

Theorem 4.42 (Law of One Price for multi-period models: $t = 0$)

Suppose that D is an attainable derivative with a replicating strategy Φ in a viable model. There is a unique fair price D_0 for D at time 0 given by

$$D_0 = V_0^\Phi.$$

Proof

First we show that if D is priced at π at time $t = 0$ and $\pi \neq V_0^\Phi$, then π is not a fair price. Suppose that $\pi > V_0^\Phi$. An extended arbitrage opportunity for any pricing structure for D with initial price π is constructed as follows. The idea is to adapt the replicating strategy Φ by selling the derivative and using the proceeds to increase the holding of bonds. To be precise, define

$$\Psi = (\Phi', -1)$$

where the holding of the derivative is constant ($z_t = -1$ for all t) and

$$\Phi' = \left(x_t + \frac{1}{B_0}(\pi - V_0^\Phi), y_t \right)_{t=0}^{T}.$$

This extended strategy is self-financing because z_t is constant and Φ is self-financing. Then

$$V_0^\Psi = V_0^{\Phi'} - \pi = x_0 B_0 + (\pi - V_0^\Phi) + y_0 \cdot S_0 - \pi = 0$$

and

$$V_T^\Psi = V_T^{\Phi'} - D = x_T B_T + \frac{1}{B_0}(\pi - V_0^\Phi) B_T + y_T \cdot S_T - D$$
$$= (1 + r)^T (\pi - V_0^\Phi) > 0.$$

So we have an extended arbitrage opportunity.

If $V_0 > \pi$ do the reverse of the above: that is, buy the derivative for π and finance the purchase by borrowing bonds. To be precise take $\Psi = (\Phi', 1)$ where

$$\Phi' = \left(x_t + \frac{1}{B_0}(V_0^\Phi - \pi), y_t \right)_{t=0}^{T}.$$

Thus any price other than V_0^Φ is not a fair price for D at time 0.

To complete the proof we must show that $\pi = V_0^\Phi$ is a fair price. Let $\Pi = (\pi_t)_{t=0}^{T}$ be the pricing structure for D given by the strategy Φ; that is, let

$$\pi_t := V_t^\Phi \tag{4.10}$$

for all t. We must show that there is no extended arbitrage for this Π. Suppose to the contrary that $\Psi = (\varphi'_t, z_t)_{t=0}^T$ is an extended arbitrage opportunity using Π; that is $V_0^\Psi = 0$ and $V_T^\Psi \geq 0$ with $V_T^\Psi(\omega) > 0$ for at least one $\omega \in \Omega$. We will see that this is impossible.

Let $\Phi' = (\varphi'_t)_{t=0}^T$ and $z = (z_t)_{t=0}^T$ and define a trading strategy in bonds and stock by

$$\Phi'' = \Phi' + z\Phi = \left(\varphi'_t + \varphi_t z_t\right)_{t=0}^T,$$

where $\Phi = (\varphi_t)_{t=0}^T$. Equation (4.10) gives

$$V_t^{\Phi''} = V_t^{\Phi'} + z_t V_t^{\Phi} = V_t^{\Phi'} + z_t \pi_t = V_t^{\Psi} \tag{4.11}$$

for all t. Also, the self-financing properties of Φ, Φ' and Ψ give

$$V_{t+}^{\Phi''} = V_{t+}^{\Phi'} + z_{t+1} V_{t+}^{\Phi} = V_{t+}^{\Phi'} + z_{t+1} V_t^{\Phi} = V_{t+}^{\Phi'} + z_t \pi_t = V_{t+}^{\Psi} = V_t^{\Psi} = V_t^{\Phi''},$$

for all t, so that Φ'' is self-financing. Thus, on account of equation (4.11) and the fact that Ψ is an extended arbitrage opportunity, Φ'' is an arbitrage opportunity in bonds and stock, which contradicts the viability of the model. $\square$

Note that if there is more than one replicating strategy for D, the uniqueness aspect of the above result shows that the fair price D_0 is the same whichever one is used to price D. This is consistent with the earlier result given in Theorem 4.38.

Example 4.43

Suppose that we want to calculate the fair price C_0 at time 0 of a European call option with expiry date 2 and strike $K = 115$ in the single-stock model in Model 4.1 with interest rate $r = 10\%$. In view of Theorem 4.42, this means we need to find a replicating strategy for this option.

Consider the single-step submodel at the node u. The replicating portfolio in this model satisfies the system

$$x_2(\mathrm{u})B_2 + y_2(\mathrm{u})S_2(\mathrm{uu}) = C(\mathrm{uu}),$$
$$x_2(\mathrm{u})B_2 + y_2(\mathrm{u})S_2(\mathrm{ud}) = C(\mathrm{ud}),$$

which translates as

$$1.21 x_2(\mathrm{u}) + 140 y_2(\mathrm{u}) = [140 - 115]_+ = 25,$$
$$1.21 x_2(\mathrm{u}) + 110 y_2(\mathrm{u}) = [110 - 115]_+ = 0.$$

The solution is

$$\left(x_2(\mathrm{u}), y_2(\mathrm{u})\right) = \left(-75\tfrac{25}{33}, \tfrac{5}{6}\right).$$

In exactly the same way, the replicating portfolio in the single-step submodel at d satisfies

$$1.21 x_2(\mathrm{d}) + 120 y_2(\mathrm{d}) = [120 - 115]_+ = 5,$$
$$1.21 x_2(\mathrm{d}) + 80 y_2(\mathrm{d}) = [80 - 115]_+ = 0,$$

which gives

$$\left(x_2(\mathrm{d}), y_2(\mathrm{d})\right) = \left(-8\tfrac{32}{121}, \tfrac{1}{8}\right).$$

To find the initial portfolio (x_1, y_1) for this strategy, we proceed by backward induction. Transactions at time 1 must be completed in a self-financing way; therefore (x_1, y_1) must satisfy the equations

$$x_1 B_1 + y_1 S_1(\mathrm{u}) = V_1(\mathrm{u}) = V_{1+}(\mathrm{u}) = x_2(\mathrm{u})B_1 + y_2(\mathrm{u})S_1(\mathrm{u}),$$
$$x_1 B_1 + y_1 S_1(\mathrm{d}) = V_1(\mathrm{d}) = V_{1+}(\mathrm{d}) = x_2(\mathrm{d})B_1 + y_2(\mathrm{d})S_1(\mathrm{d}),$$

or, in other words,

$$1.1 x_1 + 120 y_1 = -75\tfrac{25}{33} \times 1.1 + \tfrac{5}{6} \times 120 = 16\tfrac{2}{3},$$
$$1.1 x_1 + 90 y_1 = -8\tfrac{32}{121} \times 1.1 + \tfrac{1}{8} \times 90 = 2\tfrac{7}{44}.$$

The solution to this system is

$$(x_1, y_1) = \left(-37\tfrac{37}{121}, \tfrac{383}{792}\right).$$

The trading strategy Φ that we have constructed is self-financing and replicates the European call option. Its value process is given in Figure 4.6. The fair price C_0 at time 0 of the call is equal to the initial value of Φ, so it follows that

$$C_0 = V_0^\Phi = -37\tfrac{37}{121} + \tfrac{383}{792} \times 100 \approx 10.7553.$$

Exercise 4.44[†]

An *Asian call option* with strike K and exercise time T has payoff

$$\left[\tfrac{1}{T+1}\sum_{t=0}^{T} S_t - K\right]_+ ;$$

that is, it pays the difference between the average stock price over all $t \leq T$ and the strike price K, if this difference is positive. Find a replicating strategy in Model 4.7 for an Asian call option with exercise time 2 and strike 307, and compute its fair price at time 0.

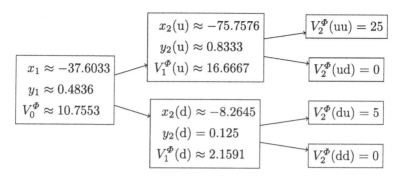

Figure 4.6   Value of the replicating strategy for a European call in a two-step single-stock binary model

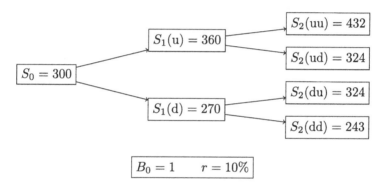

Model 4.7   Binary two-step model in Exercise 4.44

There are several equivalent ways to define the idea of a fair price at some intermediate time t with $0 < t < T$. Of course it will depend also on the scenario ω, or at least the history $\omega \uparrow t$, reflecting the fact that in a real situation we expect the current prices of any asset to depend on what is known about the evolution of the market to date. The most concise definition involves the notion of the *(full) submodel with root node* λ for any time-t node λ. The *(full) submodel with root node* λ for any time-t node λ is the model with trading dates $\{t, t+1, \ldots, T\}$ and with set of price histories

$$\mathcal{H}_\lambda := \big\{ \big(S_t(\omega), S_{t+1}(\omega), \ldots, S_T(\omega) \big) : \omega \uparrow t = \lambda \big\}$$

which is in one-to-one correspondence with the set

$$\Omega_\lambda := \{ \omega : \omega \uparrow t = \lambda \}$$

of scenarios in the full model that agree with λ up to time t.

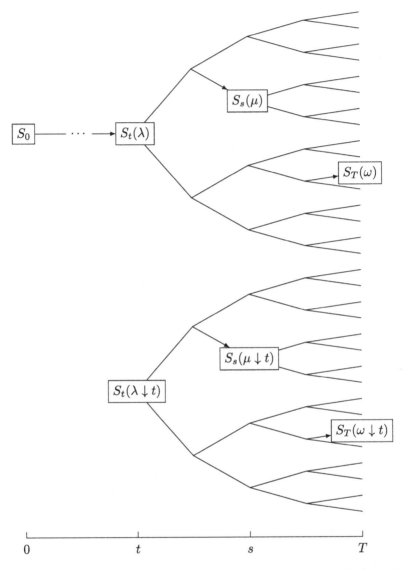

Figure 4.8 One-to-one correspondence between scenarios in Ω_λ (*above*) and price histories in the submodel at λ (*below*)

This one-to-one correspondence becomes abundantly clear if the full model is viewed as a tree; the following discussion should be compared with Figure 4.8. Every scenario $\omega \in \Omega_\lambda$ is a unique price history

$$\omega = \big(S_0, \dots, S_T(\omega)\big) = \big(S_0, \dots, S_t(\lambda), \dots, S_T(\omega)\big),$$

which corresponds to the truncated price history

$$\omega \downarrow t := \big(S_t(\omega), \dots, S_T(\omega)\big) = \big(S_t(\lambda), \dots, S_T(\omega)\big),$$

which is in $\mathcal{H}_\lambda$. Thus the node $\lambda = (S_0, \dots, S_t(\lambda))$ at time t in the full model corresponds to the root node $\lambda \downarrow t = (S_t(\lambda))$ in the submodel at λ, and any node $\mu = (S_0, \dots, S_s(\mu)) = (\lambda, S_{t+1}(\mu), \dots, S_s(\mu))$ at time $s > t$ in the full model whose price history includes λ (that is $\mu \uparrow t = \lambda$) corresponds to a unique node $\mu \downarrow t$ at time s in the submodel, where $\mu \downarrow t := (S_t(\mu), \dots, S_s(\mu))$.

This one-to-one correspondence between $\mathcal{H}_\lambda$ and $\varOmega_\lambda$ makes it convenient to identify $\mathcal{H}_\lambda$ with $\varOmega_\lambda$ and random variables on $\mathcal{H}_\lambda$ with functions defined on the subset $\varOmega_\lambda$ of $\varOmega$. We shall follow this convention below.

The idea of a submodel now gives a natural way to extend the notion of a fair price to any time $t < T$, as follows.

Definition 4.45 (Fair price)

Let D be a derivative and λ be a time-t node where $t < T$. A *fair price π for D at the node λ* is any price that is a fair price for D in the submodel with root node λ.

Here is the main pricing theorem for attainable derivatives.

Theorem 4.46 (Law of One Price for multi-period models)

Suppose that D is an attainable derivative with a replicating strategy $\varPhi = (\varphi_t)_{t=0}^T$ in a viable market. For any time $t < T$ and time-t node λ there is a unique fair price $D_t(\lambda)$ for D at λ given by

$$D_t(\lambda) = V_t^\varPhi(\lambda) = V_t^\varPhi(\omega) \quad \text{for any } \omega \text{ with } \omega \uparrow t = \lambda.$$

Proof

Theorem 4.42 says that for an attainable derivative in any model the unique fair price is given by the value of any replicating strategy at the root node.

It is possible to replicate D in the submodel at λ by restricting $\varPhi$ to $\varOmega_\lambda$. In detail, define the trading strategy $\varPhi' = (\varphi_s')_{s=t}^T$ in the submodel with root node λ by

$$\varphi_s'(\omega) := \varphi_s(\omega) \tag{4.12}$$

for all $s \geq t$ and all $\omega \in \varOmega_\lambda$; the definition (4.12) is equivalent to

$$\varphi_s'(\omega \downarrow t) := \varphi_s(\omega)$$

for all $\omega \downarrow t \in \mathcal{H}_\lambda$. The trading strategy Φ' is self-financing and replicates D in the submodel, since

$$V_T^{\Phi'}(\omega) = V_T^{\Phi}(\omega) = D(\omega)$$

for all $\omega \in \Omega_\lambda$. The fair price of D in the submodel is therefore equal to the value of Φ' at the root node $\lambda \downarrow t$ of the submodel, which is $V_t^{\Phi'}(\lambda \downarrow t) = V_t^{\Phi}(\lambda)$. □

Remark 4.47

The Law of One Price is sometimes stated as $D_t(\omega) = V_t^{\Phi}(\omega)$ or simply $D_t = V_t^{\Phi}$ for an attainable derivative, with the fact that both sides depend only on $\omega \uparrow t$ being understood.

To complete the notational picture, we write $D_T = D$; in fact the above theory applies trivially to the case $t = T$ to give $D_T = D$.

Example 4.48

Continuing with Example 4.43, we see that the fair price for the European call at the intermediate time $t = 1$ is

$$C_1(\mathrm{u}) = V_1^{\Phi}(\mathrm{u}) = 16\tfrac{2}{3} \approx 16.6667,$$
$$C_1(\mathrm{d}) = V_1^{\Phi}(\mathrm{d}) = 2\tfrac{7}{44} \approx 2.1591.$$

Note that the preceding pricing theory works only for attainable derivatives. At the end of the next chapter we will see how to extend the results of Chapter 3 on pricing non-attainable derivatives to multi-period models.

Before taking a first look at the characterization of the completeness of multi-period models, we examine some specific applications of the preceding theory.

4.5 Some General Results on Pricing

Here we will apply the above results on pricing to some specific kinds of derivative. However, the results are very general, in the sense that they apply to a wide range of models, not just the finite time step finite scenario models considered here. In that sense they are *model-independent*; the arguments use only the notion of extended arbitrage to define the notion of a fair price, which makes them very powerful.

4.5.1 European Puts and Calls

Consider a model with a single stock S (not paying dividends) and bond B, with interest rate r. As we have seen, the payoffs of European call and put options with strike price $K > 0$ and exercise date T are

$$C_T = [S_T - K]_+,$$
$$P_T = [S_T - K]_- = [K - S_T]_+.$$

Since European options may be traded at any time up to their expiry date, they must be priced for sale or purchase at any time $t < T$. We have the following result.

Theorem 4.49 (Put-call parity)

In a viable model, for any time t, the prices C_t of a European call and P_t of a European put, both with strike K and expiry T on the same stock S, should satisfy the identity

$$C_t - P_t = S_t - \frac{K}{(1+r)^{T-t}}. \tag{4.13}$$

Otherwise there would be an extended arbitrage opportunity.

Remark 4.50

It is clear from (4.13) that the price of a call and put option on the same stock should be the same if the strike price K is equal to the *forward price* $F_t = S_t(1+r)^{T-t}$ (see (4.15) below).

Proof (of Theorem 4.49)

Recall from Exercise 2.17(b) that $a = a_+ - a_-$ for any number a. Consider the derivative D that consists of buying one call and shorting one put. At the expiry time T this has the value

$$D = C_T - P_T = [S_T - K]_+ - [S_T - K]_- = S_T - K = S_T - \frac{K}{B_T}B_T.$$

This means that D is replicated by the constant strategy $\Phi = (x_t, y_t)_{t=0}^T$ with $(x_t, y_t) = (-\frac{K}{B_T}, 1)$ for all t. By the Law of One Price (Theorem 4.46) the unique fair price at time t for D is

$$D_t = V_t^{\Phi} = x_t B_t + y_t S_t = S_t - \frac{K}{B_T} B_t = S_t - \frac{K}{(1+r)^{T-t}}.$$

Therefore any prices C_t, P_t for the call and the put at time t that do not satisfy (4.13) allow extended arbitrage (involving trading in the stock, the bond and the options). □

Remark 4.51

There is a subtle issue in the above result. If the market is incomplete, it is quite possible that a given European call (or put) may not be attainable, and so there may not be a unique fair price for it. However, the *combination* $C_T - P_T$ (that is, one call and shorting one put) *is* attainable and for this combination there *is* a unique fair price. If the call happens to be attainable, then so is the put (and vice versa), and the fair prices C_t and P_t both exist and are related by put-call parity.

Example 4.52

Recall from Example 4.43 that the fair prices of a European call with expiry date 2 and strike $K = 115$ in the single-stock model in Model 4.1 with interest rate $r = 10\%$ are

$$C_0 \approx 10.7553,$$
$$C_1(\mathrm{u}) = 16\tfrac{2}{3} \approx 16.667,$$
$$C_1(\mathrm{d}) = 2\tfrac{7}{44} \approx 2.1591.$$

Put-call parity then gives

$$P_0 = C_0 + \frac{K}{(1+r)^2} - S_0 \approx 10.7553 + \tfrac{115}{1.21} - 100 \approx 5.7966,$$
$$P_0(\mathrm{u}) = C_1(\mathrm{u}) + \frac{K}{1+r} - S_1(\mathrm{u}) = 16\tfrac{2}{3} + \tfrac{115}{1.1} - 120 \approx 1.2121,$$
$$P_0(\mathrm{d}) = C_1(\mathrm{d}) + \frac{K}{1+r} - S_1(\mathrm{d}) = 2\tfrac{7}{44} + \tfrac{115}{1.1} - 90 \approx 16.705$$

for the fair prices of a put option with the same strike and expiry date.

Exercise 4.53[†]

Consider a binary model with two time steps, riskless interest rate $r = 5\%$, initial bond price $B_0 = 1$, and a single stock, whose price evolution is given in Model 4.9.

(a) Verify that this model is viable.

(b) Find a replicating strategy for a European call option with strike $K = 96$ and exercise date 2 on the stock. Compute a fair price for it at time 0.

(c) Find the fair price at time 0 of the corresponding put option.

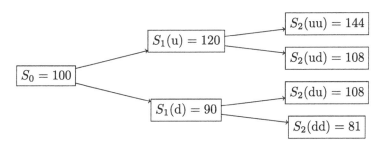

Model 4.9 Two-step binary model in Exercise 4.53

Exercise 4.54[‡]

Use put-call parity to show that

$$r = \sqrt[T-t]{\frac{K}{S_t + P_t - C_t}} - 1 \tag{4.14}$$

in a viable model where a European call is priced at C_t and a European put at P_t at time $t < T$. In a market where the prices of put and call options are determined solely by supply and demand, and are therefore quoted alongside stock prices, explain how a trader may use this relationship to identify an extended arbitrage opportunity if the interest rate is different from this.

4.5.2 Forwards

A *forward contract* (or just a *forward*) is a binding agreement to trade one share at the *delivery time* T, for a price F specified in advance, called the *forward price* (see Chapter 1). The trader who agrees to buy the share is said to be taking a *long forward position*; the trader who agrees to sell the share for F (worth S_T at time T) has a *short forward position*. In practice, forward contracts are usually settled by payment of the difference between the forward price and the actual stock price at time T. If $S_T > F$ then the trader in the short position pays $S_T - F$ to the trader in the long position, and if $S_T < F$ then the trader in the long position pays $F - S_T$ to the trader in the short position; if $S_T = F$ then there is no payment either way.

A forward differs from an option in that the initial agreement is made without cost to either party, whereas an option is normally bought or sold—with the above pricing theory indicating what is a fair price. Moreover, a forward contract obliges both parties to complete the trade at the final time, whereas the holder of an option may choose whether to exercise or not.

The payoff function of a position in a forward contract made at time 0, with forward price F and delivery time T is particularly simple: it is $S_T - F$ for the trader in the long forward position, and it is $F - S_T$ for the trader in the short forward position. This can of course be either positive or negative: one of the parties to a forward contract will always gain exactly what the other party is losing. As we will see below there is only one possible fair forward price for a forward contract made at time 0, namely

$$F_0 = (1+r)^T S_0. \tag{4.15}$$

We emphasize that a forward is a derivative that is exchanged without payment at time 0; that is to say it is priced at 0 at time 0. The issue therefore is not that of a fair price to pay at time 0, but what is a fair price to write into the contract. Thus we define a *fair forward price* to be any price F written into the contract that makes 0 a fair price at time 0. More generally, we can ask about the fair forward price for a forward contract made at time $t < T$ for delivery at time T.

Theorem 4.55

For any stock S there is a unique fair forward price for a forward contract made at time t on a stock S, with delivery time T, given by

$$F_t = (1+r)^{T-t} S_t.$$

We can rephrase this by saying that if the forward price F of a forward contract made at time t on a stock S, with delivery time T is not equal to $(1+r)^{T-t} S_t$, then an arbitrage opportunity may be created by trading in this forward; conversely, if the forward price written into the contract is $(1+r)^{T-t} S_t$ then no arbitrage is possible.

Proof

A forward that is agreed at time t for a price F is a derivative D with price $\pi_t = 0$ at the time of "purchase" (since a forward contract costs nothing), and with payoff $D = S_T - F$ for the trader in the long forward position. The derivative D is attainable, because it may be replicated by the self-financing strategy $\Phi = (x_s, y_s)_{s=0}^T$ with $(x_s, y_s) = (\frac{-F}{B_T}, 1)$ for all times $s \geq t$. Thus, according to the Law of One Price, arbitrage is avoided (in other words 0 is a fair price for the forward contract) if and only if

$$0 = D_t = x_t B_t + y_t S_t = \frac{-F}{B_T} B_t + S_t = -(1+r)^{t-T} F + S_t$$

which gives $F = F_t$ as the unique fair forward price. $\square$

Exercise 4.56[†]

Suppose that a forward contract is made at time 0 with the fair price F_0 written into the contract. Suppose that the trader in the long forward position wishes to sell the contract at some later time $t > 0$. What is the fair price at which to sell it?

Exercise 4.57

It is conceivable that a "forward contract" may be written with the price F not equal to the fair forward price. Suppose that a trader decides to take a short position at time t in such a contract with price F on the stock with delivery date T. What is the value of his position at time t?

4.6 Completeness

Just as with single-period models, it is useful to characterize those models in which all derivatives are attainable. In the next chapter there is an important result that generalizes the characterization of completeness in one-step models using risk-neutral probabilities (Theorem 5.57). Here we discuss the generalization in terms of notions from linear algebra, using the theory developed for one-step models and again viewing a multi-period model as a succession of these. Some notation is necessary.

In a model with m stocks, at a non-terminal time-t node λ, let n_λ be the number of successor nodes to λ, and consider for each stock S^k its values at these nodes. This is given by a vector of possible future prices

$$\left(S_{t+1}^k(\mu_1), \ldots, S_{t+1}^k(\mu_{n_k})\right) \tag{4.16}$$

where $\mu_1, \ldots, \mu_{n_k}$ are the successors to λ. The corresponding vector for the bond is $(B_{t+1}, \ldots, B_{t+1})$.

It is possible that for some k, the vector (4.16) may be written as a linear combination of the other assets (including the bond). For the purposes of hedging or replication, this means that S^k is redundant at this node, and could be disregarded: any derivative that is replicable using S^k at λ could be replicated using the other assets. This leads to the following definition: let m_λ be the size of the smallest collection of stocks that, together with the bond, allows the values of every other stock at the successors of λ to be written as a linear combination of them. It is clear that $m_\lambda \leq m$, and $m_\lambda > 0$ by Assumption 4.2.

In the terminology of linear algebra, $m_\lambda + 1$ is the dimension of the subspace of $\mathbb{R}^{n_\lambda}$ spanned (or generated by) the constant vector $(B_{t+1}, \ldots, B_{t+1})$ and the vectors $(S_{t+1}^k(\mu_1), \ldots, S_{t+1}^k(\mu_{n_k}))$ for $k = 1, 2 \ldots, n$. Note that $m_\lambda + 1 \leq n_\lambda$ always.

Here is the generalization of the relevant parts of Theorem 3.58.

Theorem 4.58

In a viable model the following statements are equivalent:

(1) The model is complete.

(2) Every submodel is complete.

(3) Every single-step submodel is complete.

(4) $n_\lambda = m_\lambda + 1$ for every non-terminal node λ.

Examples 4.59

(a) Model 4.1 is complete, because every non-terminal node λ has exactly two successors (that is, $n_\lambda = 2$) and there is one risky asset, so $m_\lambda = 1$. More generally, by the same reasoning, a viable single-stock model is complete if and only if there is binary branching at every node.

(b) Model 4.2 is not complete. The node u has three successors (that is, $n_u = 3$), but $m_u = 1$ because

$$S_2^2(\omega) = \tfrac{230}{B_2} B_2 - S_2^1(\omega)$$

for $\omega = \mathrm{uu}, \mathrm{um}, \mathrm{ud}$, so the model has only one independent risky asset at the successors of u.

Proof (of Theorem 4.58)

From Theorem 3.58 we know that (3) and (4) are equivalent.

To see that (3) implies (1), assume that every single-step submodel is complete and consider any derivative D. To find a hedging strategy $\Phi \equiv (x_t, y_t)_{t=1}^T$ for D, define (x_t, y_t) recursively beginning with $t = T$ and working backwards in time.

Let a λ be a node at time $T - 1$ with successors $\omega_1, \ldots, \omega_{n_\lambda} \in \Omega$. We need a portfolio $(x_T(\lambda), y_T(\lambda)) \in \mathbb{R}^{m+1}$ to satisfy the equations

$$x_T(\lambda)B_T + y_T(\lambda) \cdot S_T(\omega_1) = D(\omega_1),$$
$$x_T(\lambda)B_T + y_T(\lambda) \cdot S_T(\omega_2) = D(\omega_2),$$
$$\vdots$$
$$x_T(\lambda)B_T + y_T(\lambda) \cdot S_T(\omega_{n_\lambda}) = D(\omega_{n_\lambda}).$$

This system admits a solution $(x_T(\lambda), y_T(\lambda))$ because the one-step submodel with root node λ is complete. In this way we can define $(x_T(\lambda), y_T(\lambda))$ for all nodes λ at time $T - 1$.

Suppose now that if $t < T$, and we have found a replicating strategy $(x_s, y_s)_{s=t+1}^T$ for D between time t and time T; that is, $(x_s, y_s)_{s=t+1}^T$ is self-financing at the trading dates $t + 1, \ldots, T - 1$ and $x_T B_T + y_T \cdot S_T = D$. We need to define (x_t, y_t). Let λ be a node at time $t - 1$ with successor nodes $\mu_1, \ldots, \mu_{n_\lambda}$. We must choose (x_t, y_t) so that

$$x_t(\mu)B_t + y_t(\mu) \cdot S_t(\mu) = x_{t+1}(\mu)B_t + y_{t+1}(\mu) \cdot S_t(\mu)$$

for each $\mu \in \operatorname{succ} \lambda$ in order to maintain the self-financing property at the trading date t. Defining

$$D_{t+}(\mu) := x_{t+1}(\mu)B_t + y_{t+1}(\mu) \cdot S_t(\mu)$$

for each $\mu \in \operatorname{succ} \lambda$, this means we need to solve the system

$$x_t(\lambda)B_t + y_t(\lambda) \cdot S_t(\mu_1) = D_{t+}(\mu_1),$$
$$x_t(\lambda)B_t + y_t(\lambda) \cdot S_t(\mu_2) = D_{t+}(\mu_2),$$
$$\vdots$$
$$x_t(\lambda)B_t + y_t(\lambda) \cdot S_t(\mu_{n_\lambda}) = D_{t+}(\mu_{n_\lambda}).$$

Since each $D_{t+}(\mu_i)$ is known and the one-step model with root node λ is complete, these equations have a solution, as required.

In this way an entire strategy $\Phi \equiv (x_t, y_t)_{t=0}^T$ may be defined, and by construction $V_T^\Phi = D$, so we have established (1).

The fact that (3) implies (2) is proved in the same way, and (2) clearly implies (3).

To complete the proof we show that (1) implies (3). Assume that the full model is complete and let λ be a time-t node with $t < T$. Consider the one-step submodel with root node λ and a derivative D' in this model, which is specified by the values $\{D'(\mu) : \mu \in \text{succ }\lambda\}$. Consider the derivative

$$D(\omega) = \begin{cases} (1+r)^{T-t-1} D'(\mu) & \text{if } \omega \uparrow (t+1) = \mu, \\ 0 & \text{otherwise} \end{cases}$$

in the full model. Since the whole model is complete there is a replicating strategy Φ say for D. We will show that the portfolio $(x_{t+1}(\lambda), y_{t+1}(\lambda))$ replicates D' in the one-step submodel at λ. For this we need to show that

$$V_{t+1}^{\Phi}(\mu) = D'(\mu) \quad \text{for each } \mu \in \text{succ }\lambda.$$

Fix $\mu \in \text{succ }\lambda$ and consider the full submodel at μ; recall that this is the model with root node μ and trading dates $t+1, \ldots, T$. Its set of scenarios may be identified with the subset

$$\Omega_\mu = \{\omega : \omega \uparrow (t+1) = \mu\}$$

of Ω. Denote by D^μ the derivative with constant payoff $(1+r)^{T-t-1} D'(\mu)$ at time T on Ω_μ. Its fair price in this submodel is therefore simply its discounted value, namely $D'(\mu)$.

Now $D^\mu(\omega)$ is the same as D on the set Ω_μ, so the strategy Φ restricted to this submodel replicates D^μ and its value at the root node μ is $V_{t+1}^{\Phi}(\mu)$, which is the unique fair price of D^μ, which we have seen is $D'(\mu)$. Hence $V_{t+1}^{\Phi}(\mu) = D'(\mu)$ and this holds for each $\mu \in \text{succ }\lambda$ as required. $\qquad\square$

Remarks 4.60

(1) An easier proof of some of the equivalences of (1)–(3) in Theorem 4.58 is possible using the notion of equivalent martingale measure defined in the next chapter.

(2) Note that since we always have $m_\lambda + 1 \leq n_\lambda$, the condition Theorem 4.58(4) can also be written

$$n_\lambda \leq m_\lambda + 1,$$

or, in words, every non-terminal node has at most $m_\lambda + 1$ successors.

It follows that a necessary condition for completeness is that every node has at most $m + 1$ successors, but we have seen that even in a one-step model this is not sufficient.

As with single-period models, we will see that in an incomplete model any non-attainable derivative has a range of fair prices. This will be discussed in the next chapter using additional tools.

Multi-Period Models: Risk-Neutral Pricing

The goal of this chapter is first to extend the idea of a risk-neutral probability measure to multi-period market models. We will then find that, just as in the single-period case, this provides a powerful tool for pricing derivatives in viable markets. There are a few new concepts to be mastered, but in many respects the theory to be developed is again obtained by thinking of a multi-period model as a succession of single-step models.

We continue with the notation of Chapter 4, and Assumptions 4.1–4.2 that we made there. In particular, attention is restricted to models with n possible scenarios, and with m stocks. The initial stock prices $S_0^1, \ldots, S_0^m$ are assumed to be known at time 0, and every possible price history corresponds to a unique scenario. Together, these two assumptions mean that there is only one root node, and as in Chapter 4 we identify each scenario with a price history; that is, each scenario $\omega \in \Omega$ is a unique price history $(S_0, S_1(\omega), \ldots, S_T(\omega))$.

The new ideas centre around the concept of *information*. We have already encountered the basic idea in the definitions of predictable and adapted processes in Chapter 4. For any $t > 0$, the value $X_t(\omega)$ of a predictable process X in any scenario ω depends only on $\omega \uparrow (t-1)$. Putting this another way, the value $X_t(\omega)$ depends only on the *information* about the history of prices that has become available up to time $t-1$. Similarly, the value $Y_t(\omega)$ of an adapted process Y at time t in the scenario depends only on $\omega \uparrow t$, that is, the partial price history up to time t.

N.J. Cutland, A. Roux, *Derivative Pricing in Discrete Time*,
Springer Undergraduate Mathematics Series,
DOI 10.1007/978-1-4471-4408-3_5, © Springer-Verlag London 2012

Such ideas are best understood within the general framework of the *information structure* underlying the financial model. That is the topic of the first section.

5.1 Information and Measurability

While the set Ω represents the collection of all possible histories, in reality, the market will follow only one such history, say ω. At any given time $t < T$ however, a trader will not know all of ω, only the partial information

$$\omega \uparrow t = \big(S_0, S_1(\omega), \dots, S_t(\omega)\big) = \lambda.$$

Concerning future times, all the trader knows at time t is that the complete history ω will be a member of the set

$$\Omega_\lambda := \big\{\omega' \in \Omega : \omega' \uparrow t = \lambda\big\}.$$

As time progresses, more information about the market becomes available, so the trader's knowledge about the possible future scenarios increases, thereby reducing the possibilities as to what the future holds. This is expressed mathematically by the fact that if $t < s$, then

$$\Omega_{\omega \uparrow s} \subset \Omega_{\omega \uparrow t}.$$

Example 5.1

Consider the stock price evolution in the following table, represented in tree form as Model 5.1.

ω	S_0^1	S_0^2	$S_1^1(\omega)$	$S_1^2(\omega)$	$S_2^1(\omega)$	$S_2^2(\omega)$
ω_1	10	18	10	24	13	29
ω_2	10	18	10	24	10	23
ω_3	10	18	10	24	7	22
ω_4	10	18	9	19	8	23
ω_5	10	18	9	19	9	20
ω_6	10	18	9	19	13	18
ω_7	10	18	12	17	10	25
ω_8	10	18	12	17	14	17
ω_9	10	18	12	17	15	20

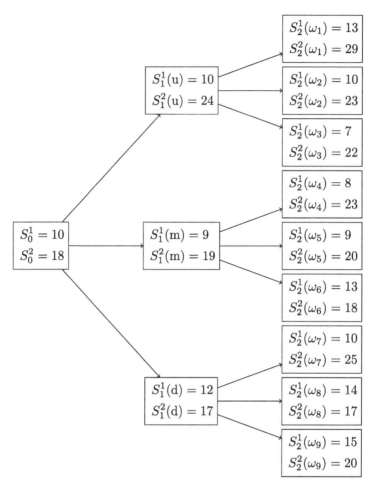

Model 5.1   Two-step ternary branching with two stocks in Examples 5.1, 5.10, 5.14, 5.21, 5.24, 5.31, 5.37, 5.50 and Exercise 5.51

This model has one node at time 0, three nodes at time 1, namely

$$\mathrm{u} := \omega_1 \uparrow 1 = \omega_2 \uparrow 1 = \omega_3 \uparrow 1,$$

$$\mathrm{m} := \omega_4 \uparrow 1 = \omega_5 \uparrow 1 = \omega_6 \uparrow 1,$$

$$\mathrm{d} := \omega_7 \uparrow 1 = \omega_8 \uparrow 1 = \omega_9 \uparrow 1,$$

and nine nodes at time 2.

As there is only one root node, we have $\omega \uparrow 0 = \emptyset$ for all $\omega \in \Omega$, and

$$\Omega_{\omega \uparrow 0} = \Omega_\emptyset = \Omega.$$

This reflects the fact that any trader in this model will not be able to distinguish between any of the scenarios in this model by using only the information available at time 0.

For the three nodes at time 1 we have

$$\Omega_{\mathrm{u}} = \{\omega_1, \omega_2, \omega_3\}, \qquad \Omega_{\mathrm{m}} = \{\omega_4, \omega_5, \omega_6\}, \qquad \Omega_{\mathrm{d}} = \{\omega_7, \omega_8, \omega_9\}.$$

At time 1, using the stock price history up to this time, any trader in this model will know at which of the nodes u, m or d the market has arrived, and from this fact the set of possible scenarios can be narrowed down to Ω_{u}, Ω_{m} or Ω_{d}. For example, if the stock price vector at time 1 is $(10, 24)$, then the market has clearly arrived at the node u, which means that the scenario being followed must be one of the elements of Ω_{u}, namely ω_1, ω_2 or ω_3.

At the terminal time 2 the full stock price history is known, and this is one of the terminal nodes $\omega_1, \dots, \omega_9$.

The general idea of *information available at time t* is captured mathematically by means of the collection $\mathcal{F}_t$ of subsets of Ω for which membership depends only on $\omega \uparrow t$. It is defined as follows.

Definition 5.2

For any t, let $\mathcal{F}_t$ be the collection of all sets $A \subseteq \Omega$ with the property that, if $\omega \in A$ and there is another $\omega' \in \Omega$ such that

$$\omega' \uparrow t = \omega \uparrow t, \tag{5.1}$$

then $\omega' \in A$.

Remarks 5.3

(1) We may equivalently define $\mathcal{F}_t$ as the collection of all sets $A \subseteq \Omega$ such that whenever $\omega' \uparrow t = \omega \uparrow t$, then $\omega \in A$ if and only if $\omega' \in A$.

(2) $\Omega \in \mathcal{F}_t$ for all t.

(3) For every time-t node λ we have $\Omega_\lambda \in \mathcal{F}_t$.

The following proposition helps to see that Definition 5.2 does encapsulate the idea of information about the scenario ω up to time t.

Proposition 5.4

Fix any time t.

(1) For any $\omega \in \Omega$

$$\Omega_{\omega\uparrow t} = \bigcap\{A \in \mathcal{F}_t : \omega \in A\}.$$

(2) For any ω, ω' the following are equivalent:

(i) ω and ω' belong to the same sets of $\mathcal{F}_t$.

(ii) $\omega \uparrow t = \omega' \uparrow t$.

Remark 5.5

Item (1) in the above proposition shows that if we know which sets of $\mathcal{F}_t$ a scenario ω belongs to, then we know $\Omega_{\omega\uparrow t}$ and hence $\omega \uparrow t$. Conversely, (2) shows that if we know only $\omega \uparrow t$, then we know which sets of $\mathcal{F}_t$ contain ω.

Proof (of Proposition 5.4)

Part (1) is given as Exercise 5.6 below. For (2), it follows from the definition of $\mathcal{F}_t$ that (2)(ii) implies (2)(i). For the converse, suppose that (2)(ii) does not hold; that is, $\omega \uparrow t \neq \omega' \uparrow t$. Then $\Omega_{\omega\uparrow t} \in \mathcal{F}_t$ and $\omega \in \Omega_{\omega\uparrow t}$ but $\omega' \notin \Omega_{\omega\uparrow t}$, so ω and ω' do not belong to the same sets of $\mathcal{F}_t$. Thus (2)(i) does not hold either. □

Exercise 5.6[‡]

Prove Proposition 5.4(1).

Next we note that $\mathcal{F}_t$ is a σ-algebra.

Theorem 5.7

For any t, the collection $\mathcal{F}_t$ has the following properties:

(1) $\Omega \in \mathcal{F}_t$.

(2) If $A \in \mathcal{F}_t$ then $\Omega \setminus A \in \mathcal{F}_t$.

(3) If $A_n \in \mathcal{F}_t$ for $n \in \mathbb{N}$, then $\bigcup_{n\in\mathbb{N}} A_n \in \mathcal{F}_t$.

Proof

The definition of $\mathcal{F}_t$ immediately gives (1).

Fix any $A \in \mathcal{F}_t$ and $\omega \in \Omega \setminus A$, and suppose there is another $\omega' \in \Omega$ such that $\omega \uparrow t = \omega' \uparrow t$. If $\omega' \in A$ then $\omega \in A$ (because $A \in \mathcal{F}_t$), which is a contradiction. So we must have $\omega' \in \Omega \setminus A$. This establishes (2).

For (3), let $A = \bigcup_{n \in \mathbb{N}} A_n$ with each $A_n \in \mathcal{F}_t$. Suppose that $\omega \in A$ and $\omega \uparrow t = \omega' \uparrow t$. Then $\omega \in A_n$ for some $n \in \mathbb{N}$, so $\omega' \in A_n$ also. Hence $\omega' \in A$. Thus $A \in \mathcal{F}_t$. $\square$

It is straightforward to verify that $\mathcal{F}_0 = \{\emptyset, \Omega\}$ and $\mathcal{F}_T$ is the collection of all subsets of Ω. More generally, the following result shows that for any time t the sets of the form Ω_λ, where λ is a time-t node, are the *atoms*, or the smallest non-empty members, of $\mathcal{F}_t$; and $\mathcal{F}_t$ is *generated* by these atoms. This is made precise by the following result.

Proposition 5.8

Fix any time t.

(1) If λ is a node at time t then Ω_λ is an atom of $\mathcal{F}_t$; that is, if $A \in \mathcal{F}_t$ and $A \subseteq \Omega_\lambda$, then either $A = \emptyset$ or $A = \Omega_\lambda$.

(2) If $A \in \mathcal{F}_t$, then

$$A = \bigcup_{\omega \in A} \Omega_{\omega \uparrow t}, \tag{5.2}$$

and this is a union of disjoint sets.

Proof

If $\emptyset \neq A \in \mathcal{F}_t$, take any $\omega \in A$. By definition of $\mathcal{F}_t$ this means that $\Omega_{\omega \uparrow t} \subseteq A$. Since $A \subseteq \Omega_\lambda$ then $\omega \in \Omega_\lambda$ so $\omega \uparrow t = \lambda$. Hence $\Omega_\lambda \subseteq A$, and (1) is established.

Since $\omega \in \Omega_{\omega \uparrow t}$ for any ω, we have $A \subseteq \bigcup_{\omega \in A} \Omega_{\omega \uparrow t}$. By definition of $\mathcal{F}_t$, if $\omega \in A$ then $\Omega_{\omega \uparrow t} \subseteq A$ and so $A \supseteq \bigcup_{\omega \in A} \Omega_{\omega \uparrow t}$. This gives (5.2). It is easy to check that $\omega \uparrow t \neq \omega' \uparrow t$ implies that $\Omega_{\omega \uparrow t} \cap \Omega_{\omega' \uparrow t} = \emptyset$, so the union in (5.2) is disjoint. $\square$

It is natural to expect that the amount of information available in the model should increase over time, as follows.

Proposition 5.9

The sequence $(\mathcal{F}_t)_{t=0}^{T}$ is a *filtration*; that is, $\mathcal{F}_t \subseteq \mathcal{F}_s$ whenever $t \leq s$.

Proof

Fix any $A \in \mathcal{F}_t$ and $\omega \in A$, and suppose that $\omega \uparrow s = \omega' \uparrow s$ for some $\omega' \in \Omega$. If $t \le s$ we must have $\omega \uparrow t = \omega' \uparrow t$, and, since $A \in \mathcal{F}_t$, we have $\omega' \in A$. Thus $A \in \mathcal{F}_s$. $\qquad\qquad\square$

The family $(\mathcal{F}_t)_{t=0}^T$ is called the *filtration generated by S*. It is also straightforward to check from Assumption 4.2 and Definition 5.2 that $\mathcal{F}_t \subset \mathcal{F}_s$ whenever $t < s$, so $\mathcal{F}_t = \mathcal{F}_s$ if and only if $t = s$.

Example 5.10

Consider Model 5.1. The only node at time 0 is $\emptyset$, so clearly $\mathcal{F}_0 = \{\emptyset, \Omega\}$. The collection $\mathcal{F}_1$ is generated by the sets Ω_u, Ω_m and Ω_d, each corresponding to a node at time 1. We therefore have

$$\mathcal{F}_1 = \{\emptyset, \Omega_u, \Omega_m, \Omega_d, \Omega_u \cup \Omega_m, \Omega_u \cup \Omega_d, \Omega_m \cup \Omega_d, \Omega_u \cup \Omega_m \cup \Omega_d\}$$

$$= \{\emptyset, \{\omega_1, \omega_2, \omega_3\}, \{\omega_4, \omega_5, \omega_6\}, \{\omega_7, \omega_8, \omega_9\}, \{\omega_1, \omega_2, \omega_3, \omega_4, \omega_5, \omega_6\},$$

$$\{\omega_1, \omega_2, \omega_3, \omega_7, \omega_8, \omega_9\}, \{\omega_4, \omega_5, \omega_6, \omega_7, \omega_8, \omega_9\}, \Omega\}.$$

Likewise, the collection $\mathcal{F}_2$ is generated by the sets corresponding to the time-2 nodes, namely $\{\omega_1\}, \ldots, \{\omega_9\}$. Clearly this means that $\mathcal{F}_2$ is the collection of *all* subsets of Ω. Note that $\mathcal{F}_0 \subset \mathcal{F}_1 \subset \mathcal{F}_2$.

Exercise 5.11

Give a tree representation of the stock price evolution in the following table and determine $\mathcal{F}_0$, $\mathcal{F}_1$ and $\mathcal{F}_2$.

ω	S_0	$S_1(\omega)$	$S_2(\omega)$
ω_1	10	12	15
ω_2	10	12	11
ω_3	10	11	13
ω_4	10	11	12
ω_5	10	11	9
ω_6	10	9	10
ω_7	10	9	8

The idea that the value $X(\omega)$ of a random variable depends only on the partial price history $\omega \uparrow t = (S_0, S_1(\omega), \ldots, S_t(\omega))$ has been encountered several times in Chapter 4. This idea is captured mathematically by the notion of $\mathcal{F}_t$-*measurability*, a notion that is also important in much wider contexts.

Definition 5.12 (Measurability)

For any t, a real-valued random variable X on Ω is said to be *measurable with respect to $\mathcal{F}_t$* if

$$\{X \le x\} := \{\omega \in \Omega | X(\omega) \le x\} \in \mathcal{F}_t$$

for every $x \in \mathbb{R}$. A k-dimensional random variable $X = (X^1, \ldots, X^k)$ is called $\mathcal{F}_t$-measurable if each of its components $X^1, \ldots, X^k$ is $\mathcal{F}_t$-measurable.

Note that Proposition 5.9 means that a random variable that is $\mathcal{F}_t$-measurable is also $\mathcal{F}_s$-measurable for all $s \ge t$.

The next result provides a simple test for measurability in our finite setting, namely that a random variable X is $\mathcal{F}_t$-measurable if and only if it takes a single value on the set Ω_λ for every node λ at time t.

Proposition 5.13

For any t, a random variable X is $\mathcal{F}_t$-measurable if and only if

$$X(\omega) = X(\omega') \tag{5.3}$$

whenever $\omega \uparrow t = \omega' \uparrow t$. Equivalently, X is $\mathcal{F}_t$-measurable if and only if X is constant on each set Ω_λ corresponding to a time-t node λ.

Proof

It is sufficient to show this in the case where X is real-valued. First suppose that X is $\mathcal{F}_t$-measurable and $\omega \uparrow t = \omega' \uparrow t$. For any $x \in \mathbb{R}$ it follows from Proposition 5.4 that $X(\omega) \le x$ if and only if $X(\omega') \le x$, so the only possibility is that $X(\omega) = X(\omega')$. For the converse, for any $\omega \in \Omega$ equation (5.3) implies that, if $X(\omega) \le x$ and $\omega \uparrow t = \omega' \uparrow t$, then $X(\omega') = X(\omega) \le x$. So $\{X \le x\} \in \mathcal{F}_t$ and therefore X is $\mathcal{F}_t$-measurable. $\qquad\square$

Note that X is $\mathcal{F}_0$-measurable if and only if X is a constant, since

$$\omega \uparrow 0 = (S_0) = \omega' \uparrow 0$$

for all $\omega, \omega' \in \Omega$. It is also easy to verify that *every* random variable on Ω is $\mathcal{F}_T$-measurable.

If X is $\mathcal{F}_t$-measurable we sometimes abuse notation by writing

$$X(\omega \uparrow t) = X(\omega) = X(\omega')$$

for any ω' with $\omega' \uparrow t = \omega \uparrow t$. In other words, if $\lambda = \omega \uparrow t$ then we can write $X(\lambda)$ unambiguously to mean $X(\omega')$ for any ω' with $\omega' \uparrow t = \lambda$.

Example 5.14

Consider again Model 5.1. The random variable

$$X(\omega) := \begin{cases} 6 & \text{if } \omega = \omega_1, \omega_2, \\ 7 & \text{if } \omega = \omega_3, \\ 8 & \text{if } \omega \uparrow 1 = m, \\ 10 & \text{if } \omega \uparrow 1 = d \end{cases}$$

is clearly $\mathcal{F}_2$-measurable. However, it is not $\mathcal{F}_1$-measurable, because it does not take a single value on $\Omega_u = \{\omega_1, \omega_2, \omega_3\}$, and therefore

$$\{\omega \in \Omega | X(\omega) \le 6.5\} = \{\omega_1, \omega_2\} \notin \mathcal{F}_1.$$

In contrast, the random variable

$$Z(\omega) := \begin{cases} 6 & \text{if } \omega \uparrow 1 = u, \\ 8 & \text{if } \omega \uparrow 1 = m, \\ 10 & \text{if } \omega \uparrow 1 = d \end{cases}$$

is measurable with respect to both $\mathcal{F}_1$ and $\mathcal{F}_2$.

Exercise 5.15

In Model 5.2, determine whether the following random variables X and Y are measurable with respect to $\mathcal{F}_0$, $\mathcal{F}_1$ and $\mathcal{F}_2$, where

$$X(\omega) := \begin{cases} 5 & \text{if } S_1(\omega) > 110, \\ 10 & \text{if } S_1(\omega) \le 110, \end{cases} \qquad Y(\omega) := \begin{cases} 7 & \text{if } S_2(\omega) > 110, \\ -3 & \text{if } S_2(\omega) \le 110. \end{cases}$$

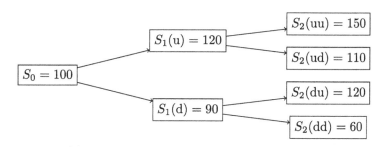

Model 5.2   Binary two-step model in Exercise 5.15

The next definitions extend the concept of dependence on available information from single random variables to *stochastic processes*; recall that a *stochastic* or *random process* $X = (X_t)_{t=0}^T$ or $X = (X_t)_{t=1}^T$ is a family of random variables indexed by time, which can be thought of alternatively as a random quantity that changes over time.

Definition 5.16 (Adaptedness, predictability)

(1) A stochastic process $X = (X_t)_{t=0}^T$ is *adapted* to the filtration $(\mathcal{F}_t)_{t=0}^T$ if for each $t = 0, \ldots, T$, the random variable X_t is $\mathcal{F}_t$-measurable.

(2) A stochastic process $Y = (Y_t)_{t=1}^T$ is *predictable* or *previsible* to the filtration $(\mathcal{F}_t)_{t=0}^T$ if for each $t > 0$ the random variable Y_t is $\mathcal{F}_{t-1}$-measurable.

Remark 5.17

It is straightforward to check that Definition 5.16 is consistent with Definitions 4.6 and 4.10 in the previous chapter.

In the discrete-time, discrete-space financial models that we consider here, the filtration $(\mathcal{F}_t)_{t=0}^T$ generated by the stock price S is the only filtration of real interest. In the sequel, we implicitly assume that the measurability properties of all random variables and stochastic processes are with respect to this filtration; thus *adapted* means *adapted to* $(\mathcal{F}_t)_{t=0}^T$, and similarly with *predictable*.

•

Remarks 5.18

(1) Both the stock price S and the bond price B are adapted processes. For the stock price, adaptedness follows by construction. Since B_t is a known real number, and therefore $\mathcal{F}_0$-measurable for all t the process B is trivially adapted (and it is also predictable).

(2) Recall that any trading strategy $\Phi \equiv (x_t, y_t)_{t=0}^T$ is predictable by definition. This means that the portfolio (x_t, y_t) that a trader chooses at any time $t-1$ must depend only on the price evolution of S and B up to time $t-1$.

(3) Recall that the value process V^Φ of a strategy Φ is defined as

$$V_t^\Phi = x_t B_t + y_t \cdot S_t$$

for all t. The quantities involved in the calculation of V_t^Φ are all $\mathcal{F}_t$-measurable, so V_t^Φ itself must be $\mathcal{F}_t$-measurable. Hence V^Φ is adapted.

5.2 Equivalent Martingale Measures

In Chapters 2 and 3 we saw that there is a close connection between risk-neutral probabilities and the viability and completeness of single-period models. This connection extends readily to models with multiple time steps, via *martingale theory* and *equivalent martingale measures*, which form the subject of this section.

Recall from Assumption 4.1(6) that the state space Ω admits a real-world probability $\mathbb{P}$. The probability of any $\omega \in \Omega$ occurring is given by $\mathbb{P}(\omega) \geq 0$; of course, $\mathbb{P}(\Omega) = 1$. We also assume that

$$\mathbb{P}(\omega) > 0 \quad \text{for all } \omega \in \Omega, \tag{5.4}$$

thus disregarding any scenarios that have zero probability of occurring.

5.2.1 Equivalent Probability Measures on Ω

The first notion we need is that of *equivalent probability measures* on Ω. In our finite setting, where every subset of Ω is measurable, the definition is as follows.

Definition 5.19 (Equivalence of probability measures)

Two probability measures $\mathbb{Q}$ and $\mathbb{Q}'$ on Ω are called *equivalent* if, for any set $A \subseteq \Omega$, we have $\mathbb{Q}(A) = 0$ if and only if $\mathbb{Q}'(A) = 0$.

Every probability measure is equivalent to itself. It follows from (5.4) that a probability $\mathbb{Q}$ is equivalent to the real-world probability $\mathbb{P}$ if and only if $\mathbb{Q}(\omega) > 0$ for all $\omega \in \Omega$. From here on, we will suppress mentioning the real-world measure, and simply refer to such a probability $\mathbb{Q}$ as an equivalent probability measure.

Results such as Theorems 2.30 and 3.36 together with Theorem 4.29 mean that viability is equivalent to the existence of a family of one-step probabilities, one at each node of the tree, corresponding to a risk-neutral probability in the single-period model at that node. In order to utilize this we must first show how an equivalent probability measure on Ω may be converted into such a family, and vice versa.

Definition 5.20 (Conditional probability)

Suppose that $\mathbb{Q}$ is an equivalent probability measure on Ω. For each $t > 0$, and node λ at time t, define the *conditional probability*

$$q_\lambda := \frac{\mathbb{Q}(\Omega_\lambda)}{\mathbb{Q}(\Omega_{\lambda\uparrow(t-1)})}.$$

In other words, for a time-t node λ, the probability q_λ is the probability that the stock price history reaches the node λ, conditional on the fact that it has reached the parent node $\lambda \uparrow (t-1)$ of λ.

Example 5.21

Suppose that the equivalent probability measure $\mathbb{Q}$ in Model 5.1 is given in the following table.

ω	ω_1	ω_2	ω_3	ω_4	ω_5	ω_6	ω_7	ω_8	ω_9
$\mathbb{Q}(\omega)$	$\frac{1}{18}$	$\frac{1}{18}$	$\frac{1}{18}$	$\frac{1}{9}$	$\frac{2}{9}$	$\frac{3}{40}$	$\frac{3}{10}$	$\frac{1}{12}$	$\frac{1}{24}$

The conditional probabilities of the nodes at time 1 are particularly simple, because the predecessor of each is the root node $\emptyset$ and $\Omega_\emptyset = \Omega$. Recalling the definitions of u, m and d from Example 5.1, we have

$$q_u = \frac{\mathbb{Q}(\Omega_u)}{\mathbb{Q}(\Omega_\emptyset)} = \frac{\mathbb{Q}(\Omega_u)}{\mathbb{Q}(\Omega)} = \mathbb{Q}(\Omega_u) = \mathbb{Q}(\omega_1) + \mathbb{Q}(\omega_2) + \mathbb{Q}(\omega_3) = \tfrac{1}{18} + \tfrac{1}{18} + \tfrac{1}{18} = \tfrac{1}{6},$$

and, in similar fashion,

$$q_m = \mathbb{Q}(\Omega_m) = \mathbb{Q}(\omega_4) + \mathbb{Q}(\omega_5) + \mathbb{Q}(\omega_6) = \tfrac{1}{9} + \tfrac{2}{9} + \tfrac{3}{40} = \tfrac{49}{120},$$

$$q_d = \mathbb{Q}(\Omega_d) = \mathbb{Q}(\omega_7) + \mathbb{Q}(\omega_8) + \mathbb{Q}(\omega_9) = \tfrac{3}{10} + \tfrac{1}{12} + \tfrac{1}{24} = \tfrac{17}{40}.$$

Now consider the conditional probabilities of the nodes at time 2. For $\omega \in \{\omega_1, \omega_2, \omega_3\}$ the parent node is u, so

$$q_\omega = \frac{\mathbb{Q}(\omega)}{\mathbb{Q}(\Omega_u)} = \tfrac{1}{3}.$$

For the other nodes at time 2, similar reasoning gives

$$q_{\omega_4} = \frac{\mathbb{Q}(\omega_4)}{\mathbb{Q}(\Omega_m)} = \tfrac{40}{147}, \qquad q_{\omega_7} = \frac{\mathbb{Q}(\omega_7)}{\mathbb{Q}(\Omega_d)} = \tfrac{12}{17},$$

$$q_{\omega_5} = \frac{\mathbb{Q}(\omega_5)}{\mathbb{Q}(\Omega_m)} = \tfrac{80}{147}, \qquad q_{\omega_8} = \frac{\mathbb{Q}(\omega_8)}{\mathbb{Q}(\Omega_d)} = \tfrac{10}{51},$$

$$q_{\omega_6} = \frac{\mathbb{Q}(\omega_6)}{\mathbb{Q}(\Omega_m)} = \frac{9}{49}, \qquad q_{\omega_9} = \frac{\mathbb{Q}(\omega_9)}{\mathbb{Q}(d)} = \frac{5}{51}.$$

It may be easily verified that adding these probabilities over the successors of any node yields a total of 1, for example $q_u + q_m + q_d = 1$.

Exercise 5.22[†]

Consider a model with 3 time steps and 8 scenarios, and suppose that the probability measure $\mathbb{Q}$ on Ω is given in the following table.

ω	ω_1	ω_2	ω_3	ω_4	ω_5	ω_6	ω_7	ω_8
$\mathbb{Q}(\omega)$	$\frac{1}{10}$	$\frac{3}{20}$	$\frac{1}{4}$	$\frac{1}{4}$	$\frac{1}{12}$	$\frac{1}{24}$	$\frac{1}{10}$	$\frac{1}{40}$

Assume that the underlying tree has two nodes u, d at time 1 with

$$u = \omega_1 \uparrow 1 = \omega_2 \uparrow 1 = \omega_3 \uparrow 1 = \omega_4 \uparrow 1,$$

$$d = \omega_5 \uparrow 1 = \omega_6 \uparrow 1 = \omega_7 \uparrow 1 = \omega_8 \uparrow 1,$$

and four nodes at time 2, namely uu, ud, du, dd with $\Omega_{uu} = \{\omega_1, \omega_2\}$, $\Omega_{ud} = \{\omega_3, \omega_4\}$, $\Omega_{du} = \{\omega_5, \omega_6\}$, $\Omega_{dd} = \{\omega_7, \omega_8\}$. Compute the conditional probabilities at every non-root node, and verify that at the successors of each non-terminal node they add up to 1.

The next result shows that in a finite market model (where the price tree can be given explicitly), any probability measure $\mathbb{Q}$ is characterized by its associated family of one-step conditional probabilities; that is, if we know the conditional probability at each node then we can find $\mathbb{Q}(\omega)$ for any ω.

Theorem 5.23

Suppose that $\mathbb{Q}$ is an equivalent probability measure on Ω, with conditional probabilities q_μ. For any non-terminal node λ, we have

$$\sum_{\mu \in \text{succ } \lambda} q_\mu = 1.$$

Moreover, $\mathbb{Q}$ can be reconstructed from the conditional probabilities through the relationship

$$\mathbb{Q}(\omega) = \prod_{t=1}^{T} q_{\omega \uparrow t}.$$

for any $\omega \in \Omega$.

Proof

Suppose that λ is any node at time $t < T$. Since

$$\Omega_\lambda = \bigcup_{\mu \in \mathrm{succ}\, \lambda} \Omega_\mu$$

is a union of disjoint sets, we have

$$\sum_{\mu \in \mathrm{succ}\, \lambda} q_\mu = \sum_{\mu \in \mathrm{succ}\, \lambda} \frac{\mathbb{Q}(\Omega_\mu)}{\mathbb{Q}(\Omega_\lambda)} = \frac{1}{\mathbb{Q}(\Omega_\lambda)} \sum_{\mu \in \mathrm{succ}\, \lambda} \mathbb{Q}(\Omega_\mu)$$

$$= \frac{1}{\mathbb{Q}(\Omega_\lambda)} \mathbb{Q}\left[\bigcup_{\mu \in \mathrm{succ}\, \lambda} \Omega_\mu \right] = \frac{\mathbb{Q}(\Omega_\lambda)}{\mathbb{Q}(\Omega_\lambda)} = 1.$$

For every $\omega \in \Omega$, we have

$$\prod_{t=1}^{T} q_{\omega \uparrow t} = \prod_{t=1}^{T} \frac{\mathbb{Q}(\Omega_{\omega \uparrow t})}{\mathbb{Q}(\Omega_{\omega \uparrow t-1})} = \frac{\mathbb{Q}(\Omega_{\omega \uparrow T})}{\mathbb{Q}(\Omega_{\omega \uparrow 0})} = \frac{\mathbb{Q}(\omega)}{\mathbb{Q}(\Omega)} = \mathbb{Q}(\omega).$$

$\square$

Example 5.24

Consider the equivalent probability $\mathbb{Q}$ for Model 5.1 in Example 5.21. Using Theorem 5.23, we can recover $\mathbb{Q}$ from the one-step conditional probabilities computed in Example 5.21. We have

$$\mathbb{Q}(\omega) = q_\mathrm{u} q_\omega = \tfrac{1}{6} \times \tfrac{1}{3} = \tfrac{1}{18}$$

for $\omega \in \{\omega_1, \omega_2, \omega_3\}$, together with

$$\mathbb{Q}(\omega_4) = q_\mathrm{m} q_{\omega_4} = \tfrac{49}{120} \times \tfrac{40}{147} = \tfrac{1}{9}, \qquad \mathbb{Q}(\omega_7) = q_\mathrm{d} q_{\omega_7} = \tfrac{17}{40} \times \tfrac{12}{17} = \tfrac{3}{10},$$

$$\mathbb{Q}(\omega_5) = q_\mathrm{m} q_{\omega_5} = \tfrac{49}{120} \times \tfrac{80}{147} = \tfrac{2}{9}, \qquad \mathbb{Q}(\omega_8) = q_\mathrm{d} q_{\omega_8} = \tfrac{17}{40} \times \tfrac{10}{51} = \tfrac{1}{12},$$

$$\mathbb{Q}(\omega_6) = q_\mathrm{m} q_{\omega_6} = \tfrac{49}{120} \times \tfrac{9}{49} = \tfrac{3}{40}, \qquad \mathbb{Q}(\omega_9) = q_\mathrm{d} q_{\omega_9} = \tfrac{17}{40} \times \tfrac{5}{51} = \tfrac{1}{24}.$$

The converse to Theorem 5.23 is also true: given a collection of probabilities (one for every non-root node of the price tree) with the property that the probabilities of successor nodes add to 1, there is a unique probability measure on Ω having these families as its one-step conditional probabilities. Here is the result.

Theorem 5.25

Suppose that for every non-root node μ a probability $p_\mu > 0$ is given, and that

$$\sum_{\mu \in \text{succ } \lambda} p_\mu = 1 \tag{5.5}$$

for every non-terminal node λ. Then there is a unique equivalent probability measure $\mathbb{Q}$ on Ω with conditional probability $q_\lambda = p_\lambda$ for each non-root node λ, given by

$$\mathbb{Q}(\omega) = \prod_{t=1}^{T} p_{\omega \uparrow t}$$

for $\omega \in \Omega$.

Proof

To see that $\mathbb{Q}$ as defined in the theorem is a probability measure (and that $q_\lambda = p_\lambda$ for each node λ), we first show that for any non-root node λ at time $u > 0$ we have

$$\mathbb{Q}(\Omega_\lambda) = \prod_{s=1}^{u} p_{\lambda \uparrow s}. \tag{5.6}$$

This is done by backward induction, as follows. At time T, each node λ is simply a scenario ω, so that

$$\mathbb{Q}(\Omega_\lambda) = \mathbb{Q}(\omega) = \prod_{s=1}^{T} p_{\omega \uparrow s} = \prod_{s=1}^{T} p_{\lambda \uparrow s}.$$

For $t < T$, suppose that (5.6) holds for all nodes at time $u = t + 1$; we deduce it for each node λ at time t. Using the disjoint union $\Omega_\lambda = \bigcup_{\mu \in \text{succ } \lambda} \Omega_\mu$, we have

$$\mathbb{Q}(\Omega_\lambda) = \sum_{\mu \in \text{succ } \lambda} \mathbb{Q}(\Omega_\mu) = \sum_{\mu \in \text{succ } \lambda} \prod_{s=1}^{t+1} p_{\mu \uparrow s}.$$

If $\mu \in \text{succ } \lambda$, then $\mu \uparrow (t+1) = \mu$ and $\mu \uparrow s = \lambda \uparrow s$ for all $s \leq t$. So (5.6) with $u = t + 1$ gives

$$\mathbb{Q}(\Omega_\lambda) = \sum_{\mu \in \text{succ } \lambda} \left[\prod_{s=1}^{t} p_{\lambda \uparrow s} \right] p_\mu = \prod_{s=1}^{t} p_{\lambda \uparrow s} \left[\sum_{\mu \in \text{succ } \lambda} p_\mu \right] = \prod_{s=1}^{t} p_{\lambda \uparrow s},$$

with the last equality following from (5.5). This concludes the inductive step and establishes (5.6) for all $u > 0$.

Now use (5.6) at time $u = 1$ together with the disjoint union

$$\Omega = \Omega_\emptyset = \bigcup_{\mu \in \text{succ}\, \emptyset} \Omega_\mu$$

and (5.5) to obtain

$$\mathbb{Q}(\Omega) = \mathbb{Q}(\Omega_\emptyset) = \sum_{\lambda \in \text{succ}\, \emptyset} \mathbb{Q}(\Omega_\lambda) = \sum_{\lambda \in \text{succ}\, \emptyset} p_{\lambda\uparrow 1} = \sum_{\lambda \in \text{succ}\, \emptyset} p_\lambda = 1.$$

Thus $\mathbb{Q}$ is an equivalent probability measure. Moreover, for any node λ at time $t > 0$, equation (5.6) gives

$$q_\lambda = \frac{\mathbb{Q}(\Omega_\lambda)}{\mathbb{Q}(\Omega_{\lambda\uparrow t-1})} = \frac{\prod_{s=1}^{t} p_{\lambda\uparrow s}}{\prod_{s=1}^{t-1} p_{\lambda\uparrow s}} = p_{\lambda\uparrow t} = p_\lambda.$$

The uniqueness of $\mathbb{Q}$ follows from Theorem 5.23, which states that a measure is defined uniquely from its conditional probabilities.   □

Exercise 5.26

Consider Model 5.3. Using the one-step probabilities in the figure, design a state space Ω and construct a probability measure $\mathbb{Q}$ on this set with these conditional probabilities.

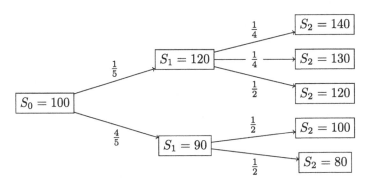

Model 5.3 Two-period model in Exercises 5.26 and 5.38

5.2.2 Conditional Expectation and Martingales

Before formally defining a martingale, a key notion in what follows, we need the definition of *conditional expectation*. In a general setting, for any t the conditional expectation of a random variable X on a set Ω (with respect to the

σ-algebra $\mathcal{F}_t$ and equivalent probability $\mathbb{Q}$) is defined to be the $\mathcal{F}_t$-measurable random variable Y satisfying

$$\sum_{\omega \in A} \mathbb{Q}(\omega) Y(\omega) = \sum_{\omega \in A} \mathbb{Q}(\omega) X(\omega) \tag{5.7}$$

for every $A \in \mathcal{F}_t$. Any such Y is unique and we write

$$\mathbb{E}_{\mathbb{Q}}(X|\mathcal{F}_t) := Y.$$

Intuitively, the conditional expectation of X with respect to $\mathcal{F}_t$ is simply a random variable that is $\mathcal{F}_t$-measurable and has the same average as X over the sets in $\mathcal{F}_t$; X itself may not be $\mathcal{F}_t$-measurable. *It is most important to note that the conditional expectation is a random variable, and not just a number.*

If the set Ω is finite, as in our case, the property (5.7) can be achieved by ensuring that it is true for the atoms of the σ-algebra under consideration. We are interested in the σ-algebras $\mathcal{F}_t$ generated by partial stock price histories, and as we saw in Proposition 5.8, the atoms of $\mathcal{F}_t$ are the sets Ω_λ for the nodes λ at time t. Taking $A = \Omega_\lambda$ in (5.7), we have

$$\sum_{\omega \in \Omega_\lambda} \mathbb{Q}(\omega) Y(\omega) = \sum_{\omega \in \Omega_\lambda} \mathbb{Q}(\omega) X(\omega). \tag{5.8}$$

We saw in Proposition 5.13 that any $\mathcal{F}_t$-measurable random variable Y must be constant on Ω_λ with value $Y(\lambda)$. The left hand side of equation (5.8) is therefore $\mathbb{Q}(\Omega_\lambda) Y(\lambda)$, and rearrangement gives

$$Y(\lambda) = \frac{1}{\mathbb{Q}(\Omega_\lambda)} \sum_{\omega \in \Omega_\lambda} \mathbb{Q}(\omega) X(\omega).$$

This leads to the following, which we shall use as our working definition.

Definition 5.27 (Conditional expectation)

Suppose that $\mathbb{Q}$ is an equivalent probability on Ω, and fix t. The conditional expectation of a random variable X on Ω with respect to $\mathcal{F}_t$ is the random variable $\mathbb{E}_{\mathbb{Q}}(X|\mathcal{F}_t)$ given by

$$\mathbb{E}_{\mathbb{Q}}(X|\mathcal{F}_t)(\omega') := \frac{1}{\mathbb{Q}(\Omega_{\omega'\uparrow t})} \sum_{\omega \in \Omega_{\omega'\uparrow t}} \mathbb{Q}(\omega) X(\omega)$$

for $\omega' \in \Omega$. Since this depends only on $\omega' \uparrow t$ we may also write

$$\mathbb{E}_{\mathbb{Q}}(X|\mathcal{F}_t)(\lambda) = \frac{1}{\mathbb{Q}(\Omega_\lambda)} \sum_{\omega \in \Omega_\lambda} \mathbb{Q}(\omega) X(\omega)$$

for any node λ at time t.

In other words, the number $\mathbb{E}_{\mathbb{Q}}(X|\mathcal{F}_t)(\lambda)$ is the average (weighted by $\mathbb{Q}$) of the values of $X(\omega)$ for those scenarios ω whose price histories up to time t are the same as λ. Another useful way to think of the conditional expectation $\mathbb{E}_{\mathbb{Q}}(X|\mathcal{F}_t)$ is that it is a simplification (by averaging) of the original random variable X to make it $\mathcal{F}_t$-measurable. The main use of this notion is when X is *not* $\mathcal{F}_t$-measurable, and thus more complex than $\mathbb{E}_{\mathbb{Q}}(X|\mathcal{F}_t)$.

Exercise 5.28[‡]

Show that Definition 5.27 implies that (5.7) holds for any $A \in \mathcal{F}_t$.

Hint. See Proposition 5.8(2).

Exercise 5.29[‡]

Verify the following for any random variable X.

(a) $\mathbb{E}_{\mathbb{Q}}(X|\mathcal{F}_0) = \mathbb{E}_{\mathbb{Q}}(X)$ (so in this case, and this case only, the conditional expectation is a number).

(b) If X is $\mathcal{F}_t$-measurable, then $\mathbb{E}_{\mathbb{Q}}(X|\mathcal{F}_t) = X$.

(c) If X is $\mathcal{F}_{t+1}$-measurable, then

$$\mathbb{E}_{\mathbb{Q}}(X|\mathcal{F}_t)(\lambda) = \sum_{\mu \in \text{succ } \lambda} q_\mu X(\mu) \qquad (5.9)$$

for any node λ at time t.

The *tower property* of conditional expectations, below, plays an important role in stochastic analysis. According to this property, taking the conditional expectation of a random variable X with respect to $\mathcal{F}_t$, and then taking the conditional expectation of this with respect to $\mathcal{F}_s$ where $s < t$ gives the same result as simply taking the conditional expectation of X with respect to $\mathcal{F}_s$.

Theorem 5.30 (Tower property of conditional expectation)

For any random variable X, if $s < t$ then

$$\mathbb{E}_{\mathbb{Q}}\big(\mathbb{E}_{\mathbb{Q}}(X|\mathcal{F}_t)|\mathcal{F}_s\big) = \mathbb{E}_{\mathbb{Q}}(X|\mathcal{F}_s).$$

Proof

Let $Y := \mathbb{E}_{\mathbb{Q}}(X|\mathcal{F}_t)$. We have to prove that

$$\mathbb{E}_{\mathbb{Q}}(Y|\mathcal{F}_s) = \mathbb{E}_{\mathbb{Q}}(X|\mathcal{F}_s).$$

For any set $A \in \mathcal{F}_s$, we have

$$\sum_{\omega \in A} \mathbb{Q}(\omega) \mathbb{E}_{\mathbb{Q}}(Y|\mathcal{F}_s)(\omega) = \sum_{\omega \in A} \mathbb{Q}(\omega) Y(\omega).$$

Since $\mathcal{F}_s \subseteq \mathcal{F}_t$, we also have $A \in \mathcal{F}_t$ and so, using the definition of Y above,

$$\sum_{\omega \in A} \mathbb{Q}(\omega) Y(\omega) = \sum_{\omega \in A} \mathbb{Q}(\omega) \mathbb{E}_{\mathbb{Q}}(X|\mathcal{F}_t)(\omega) = \sum_{\omega \in A} \mathbb{Q}(\omega) X(\omega).$$

Therefore

$$\sum_{\omega \in A} \mathbb{Q}(\omega) \mathbb{E}_{\mathbb{Q}}(Y|\mathcal{F}_s)(\omega) = \sum_{\omega \in A} \mathbb{Q}(\omega) X(\omega),$$

so $\mathbb{E}_{\mathbb{Q}}(Y|\mathcal{F}_s)$ satisfies the defining property (5.7) for $\mathbb{E}_{\mathbb{Q}}(X|\mathcal{F}_s)$. $\qquad\square$

Example 5.31

Let us calculate the conditional expectations of the random variables X and Z in Example 5.14 in Model 5.1, under the probability $\mathbb{Q}$ in Example 5.21 and the σ-algebras $\mathcal{F}_0$, $\mathcal{F}_1$ and $\mathcal{F}_2$.

The random variables X and Z are both $\mathcal{F}_2$-measurable, and therefore

$$\mathbb{E}_{\mathbb{Q}}(X|\mathcal{F}_2)(\omega) = X(\omega) = \begin{cases} 6 & \text{if } \omega = \omega_1, \omega_2, \\ 7 & \text{if } \omega = \omega_3, \\ 8 & \text{if } \omega \uparrow 1 = \mathrm{m}, \\ 10 & \text{if } \omega \uparrow 1 = \mathrm{d}, \end{cases}$$

$$\mathbb{E}_{\mathbb{Q}}(Z|\mathcal{F}_2)(\omega) = Z(\omega) = \begin{cases} 6 & \text{if } \omega \uparrow 1 = \mathrm{u}, \\ 8 & \text{if } \omega \uparrow 1 = \mathrm{m}, \\ 10 & \text{if } \omega \uparrow 1 = \mathrm{d}. \end{cases}$$

Using (5.9), the conditional expected value of X with respect to $\mathcal{F}_1$ is

$$\mathbb{E}_{\mathbb{Q}}(X|\mathcal{F}_1)(\omega) = \begin{cases} 6p_{\omega_1} + 6p_{\omega_2} + 7p_{\omega_3} & \text{if } \omega \uparrow 1 = \mathrm{u}, \\ 8p_{\omega_4} + 8p_{\omega_5} + 8p_{\omega_6} & \text{if } \omega \uparrow 1 = \mathrm{m}, \\ 10p_{\omega_7} + 10p_{\omega_8} + 10p_{\omega_9} & \text{if } \omega \uparrow 1 = \mathrm{d}, \end{cases}$$

$$= \begin{cases} 6(\frac{1}{3} + \frac{1}{3}) + 7 \times \frac{1}{3} = 6\frac{1}{3} & \text{if } \omega \uparrow 1 = \mathrm{u}, \\ 8 & \text{if } \omega \uparrow 1 = \mathrm{m}, \\ 10 & \text{if } \omega \uparrow 1 = \mathrm{d}. \end{cases}$$

Since Z is $\mathcal{F}_1$-measurable, we once again have

$$\mathbb{E}_{\mathbb{Q}}(Z|\mathcal{F}_1) = Z.$$

Finally, the conditional expectations of X and Z with respect to the trivial σ-algebra $\mathcal{F}_0$ are

$$\mathbb{E}_{\mathbb{Q}}(X|\mathcal{F}_0) = 6\mathbb{Q}(\{\omega_1, \omega_2\}) + 7\mathbb{Q}(\omega_3) + 8\mathbb{Q}(\Omega_{\mathrm{m}}) + 10\mathbb{Q}(\Omega_{\mathrm{d}})$$
$$= 6\left(\tfrac{1}{18} + \tfrac{1}{18}\right) + 7 \times \tfrac{1}{18} + 8 \times \tfrac{49}{120} + 10 \times \tfrac{17}{40} = 8\tfrac{103}{180} \approx 8.5722$$

and

$$\mathbb{E}_{\mathbb{Q}}(Z|\mathcal{F}_0) = 6p_{\mathrm{u}} + 8p_{\mathrm{m}} + 10p_{\mathrm{d}} = 6 \times \tfrac{1}{6} + 8 \times \tfrac{49}{120} + 10 \times \tfrac{17}{40} = 8\tfrac{31}{60} \approx 8.5167.$$

Exercise 5.32[‡]

Fix t and suppose that $\mathbb{Q}$ is an equivalent probability measure on Ω. Prove the following statements:

(a) If X and Y are random variables, then

$$\mathbb{E}_{\mathbb{Q}}(X + Y|\mathcal{F}_t) = \mathbb{E}_{\mathbb{Q}}(X|\mathcal{F}_t) + \mathbb{E}_{\mathbb{Q}}(Y|\mathcal{F}_t).$$

(b) If X is a random variable and Y is $\mathcal{F}_t$-measurable, then

$$\mathbb{E}_{\mathbb{Q}}(XY|\mathcal{F}_t) = Y\mathbb{E}_{\mathbb{Q}}(X|\mathcal{F}_t).$$

We are now in a position to define what we mean by a *martingale*, a central concept in the whole of mathematical finance.

Definition 5.33 (Martingale)

Suppose that $\mathbb{Q}$ is an equivalent probability measure in a discrete-time market model. A process M is a *martingale* with respect to $\mathbb{Q}$ if it is adapted and

$$\mathbb{E}_{\mathbb{Q}}(M_t|\mathcal{F}_s) = M_s$$

whenever $s \leq t$.

Intuitively, a martingale is the stochastic equivalent of a constant process. The following result gives a straightforward test for establishing whether an adapted process is a martingale in a discrete setting.

Proposition 5.34

An adapted process M is a martingale with respect to $\mathbb{Q}$ if and only if

$$\mathbb{E}_{\mathbb{Q}}(M_{t+1}|\mathcal{F}_t) = M_t \tag{5.10}$$

for every $t < T$.

Proof

If M is a martingale, then we clearly have (5.10). For the converse, fix s and assume (5.10) for all t. Then proceed by induction on $t > s$ to prove (5.10). For the induction step, if we know that $\mathbb{E}_{\mathbb{Q}}(M_t|\mathcal{F}_s) = M_s$ then the tower property combined with (5.10) gives

$$\mathbb{E}_{\mathbb{Q}}(M_{t+1}|\mathcal{F}_s) = \mathbb{E}_{\mathbb{Q}}\big(\mathbb{E}_{\mathbb{Q}}(M_{t+1}|\mathcal{F}_t)|\mathcal{F}_s\big) = \mathbb{E}_{\mathbb{Q}}(M_t|\mathcal{F}_s) = M_s.$$

$\square$

Exercise 5.35[†]

Consider a binary model with 2 time steps where the two nodes at time 1 are $\lambda_1 := \omega_1 \uparrow 1 = \omega_2 \uparrow 1$ and $\lambda_2 := \omega_3 \uparrow 1 = \omega_4 \uparrow 1$ at time 1. Determine whether the process M in the following table is a martingale with respect to the probability measure $\mathbb{Q}$.

ω	M_0	$M_1(\omega)$	$M_2(\omega)$	$\mathbb{Q}(\omega)$
ω_1	11	10	11	$\frac{2}{5}$
ω_2	11	10	9	$\frac{2}{5}$
ω_3	11	15	12	$\frac{1}{20}$
ω_4	11	15	16	$\frac{3}{20}$

Let M be an adapted process. Recalling that $\Delta M_{t+1} = M_{t+1} - M_t$ for any $t < T$, the martingale property (5.10) is equivalent to

$$\mathbb{E}_{\mathbb{Q}}(\Delta M_{t+1}|\mathcal{F}_t) = 0$$

for every $t < T$.

It follows from Exercise 5.29 that at any node λ at time t the current value $M_t(\lambda)$ of a martingale M is equal to the $\mathbb{Q}$-weighted average of the values $M_{t+1}(\mu)$ at each of the successors μ of λ. In other words, M is a martingale with respect to $\mathbb{Q}$ if and only if

$$M_t(\lambda) = \sum_{\mu \in \text{succ}\, \lambda} q_\mu M_{t+1}(\mu)$$

or, equivalently

$$\sum_{\mu \in \text{succ}\, \lambda} q_\mu \Delta M_{t+1}(\mu) = 0$$

for every node λ at time t.

5.2.3 Equivalent Martingale Measures

Recall that the *discounted process* $\bar{X}$ corresponding to a stochastic process X is defined by

$$\bar{X}_t := \frac{X_t}{(1+r)^t}$$

for all t.

An *equivalent martingale measure* for our market model is now be defined to be any equivalent measure under which the *discounted* stock price process is a martingale.

Definition 5.36 (Equivalent martingale measure)

A probability measure $\mathbb{Q}$ on the set Ω is called an *equivalent martingale measure* if it is equivalent to the real-world probability $\mathbb{P}$, and the discounted stock price process $\bar{S}$ is a martingale under $\mathbb{Q}$.

In the case of a single-period model, an equivalent martingale measure is nothing other than a risk-neutral probability (see Definition 3.32).

Example 5.37

If Model 5.1 is coupled with a riskless interest rate $r = 0.1$, then an equivalent martingale measure $\mathbb{Q}$ may be constructed by first calculating the conditional probabilities at each non-root node, as follows.

The martingale condition at time 0 tells us that we need $\mathbb{Q}$ such that

$$\mathbb{E}_\mathbb{Q}(\bar{S}_1) = \bar{S}_0.$$

That is, we need conditional probabilities q_u, q_m and q_d satisfying the system

$$q_u \bar{S}_1^1(u) + q_m \bar{S}_1^1(m) + q_d \bar{S}_1^1(d) = \bar{S}_0^1,$$
$$q_u \bar{S}_1^2(u) + q_m \bar{S}_1^2(m) + q_d \bar{S}_1^2(d) = \bar{S}_0^2,$$
$$q_u + q_m + q_d = 1.$$

That is,

$$10q_u + 9q_m + 12q_d = 11,$$
$$24q_u + 19q_m + 17q_d = 19.8,$$
$$q_u + q_m + q_d = 1.$$

It is routine to see that there is a unique solution to this system, namely

$$(q_u, q_m, q_d) = \left(\tfrac{32}{85}, \tfrac{7}{85}, \tfrac{46}{85}\right).$$

At time 1, the martingale condition means that $\mathbb{Q}$ should satisfy

$$\mathbb{E}_{\mathbb{Q}}(\bar{S}_2 | \mathcal{F}_1) = \bar{S}_1.$$

Finding the one-step conditional probabilities at time 2, involves solving a system similar to the one above for each node at time 1. At the node u, we need a solution $(q_{\omega_1}, q_{\omega_2}, q_{\omega_3})$ to the system

$$q_{\omega_1} \bar{S}_2^1(\omega_1) + q_{\omega_2} \bar{S}_2^1(\omega_2) + q_{\omega_3} \bar{S}_2^1(\omega_3) = \bar{S}_1^1(u),$$
$$q_{\omega_1} \bar{S}_2^2(\omega_1) + q_{\omega_2} \bar{S}_2^2(\omega_2) + q_{\omega_3} \bar{S}_2^2(\omega_3) = \bar{S}_1^2(u),$$
$$q_{\omega_1} + q_{\omega_2} + q_{\omega_3} = 1.$$

That is,

$$13q_{\omega_1} + 10q_{\omega_2} + 7q_{\omega_3} = 11,$$
$$29q_{\omega_1} + 23q_{\omega_2} + 22q_{\omega_3} = 26.4,$$
$$q_{\omega_1} + q_{\omega_2} + q_{\omega_3} = 1.$$

Note that this is just the system of equations for the single-step submodel with root node u. It has a unique solution

$$(q_{\omega_1}, q_{\omega_2}, q_{\omega_3}) = \left(\tfrac{46}{75}, \tfrac{8}{75}, \tfrac{7}{25}\right).$$

For the node m, the system to be solved is

$$8q_{\omega_4} + 9q_{\omega_5} + 13q_{\omega_6} = 9.9,$$
$$23q_{\omega_4} + 20q_{\omega_5} + 18q_{\omega_6} = 20.9,$$
$$q_{\omega_4} + q_{\omega_5} + q_{\omega_6} = 1$$

yielding

$$(q_{\omega_4}, q_{\omega_5}, q_{\omega_6}) = \left(\tfrac{27}{50}, \tfrac{1}{10}, \tfrac{9}{25}\right).$$

Finally, the system for the node d is

$$10q_{\omega_7} + 14q_{\omega_8} + 15q_{\omega_9} = 13.2,$$
$$25q_{\omega_7} + 17q_{\omega_8} + 20q_{\omega_9} = 18.7,$$
$$q_{\omega_7} + q_{\omega_8} + q_{\omega_9} = 1$$

which has the solution

$$(q_{\omega_7}, q_{\omega_8}, q_{\omega_9}) = \left(\tfrac{41}{200}, \tfrac{31}{40}, \tfrac{1}{50}\right).$$

These conditional probabilities give an equivalent martingale measure $\mathbb{Q}$ for this model using Theorem 5.25, as follows.

$$\mathbb{Q}(\omega_1) = q_u q_{\omega_1} = \tfrac{32}{85} \times \tfrac{46}{75} = \tfrac{1472}{6375} \approx 0.2309,$$
$$\mathbb{Q}(\omega_2) = q_u q_{\omega_2} = \tfrac{32}{85} \times \tfrac{8}{75} = \tfrac{256}{6375} \approx 0.0402,$$
$$\mathbb{Q}(\omega_3) = q_u q_{\omega_3} = \tfrac{32}{85} \times \tfrac{7}{25} = \tfrac{224}{2125} \approx 0.1054,$$
$$\mathbb{Q}(\omega_4) = q_m q_{\omega_4} = \tfrac{7}{85} \times \tfrac{27}{50} = \tfrac{189}{4250} \approx 0.0445,$$
$$\mathbb{Q}(\omega_5) = q_m q_{\omega_5} = \tfrac{7}{85} \times \tfrac{1}{10} = \tfrac{7}{850} \approx 0.0082,$$
$$\mathbb{Q}(\omega_6) = q_m q_{\omega_6} = \tfrac{7}{85} \times \tfrac{9}{25} = \tfrac{63}{2125} \approx 0.0296,$$
$$\mathbb{Q}(\omega_7) = q_d q_{\omega_7} = \tfrac{46}{85} \times \tfrac{41}{200} = \tfrac{943}{8500} \approx 0.1109,$$
$$\mathbb{Q}(\omega_8) = q_d q_{\omega_8} = \tfrac{46}{85} \times \tfrac{31}{40} = \tfrac{713}{1700} \approx 0.4194,$$
$$\mathbb{Q}(\omega_9) = q_d q_{\omega_9} = \tfrac{46}{85} \times \tfrac{1}{50} = \tfrac{23}{2125} \approx 0.0108.$$

Exercise 5.38[†]

Assume that Model 5.3 has an initial bond price of 1 and an interest rate of 10%. Find an equivalent martingale measure for this model.

Exercise 5.39[‡]

(a) Show that the set of equivalent martingale measures is convex.

(b) If D is any random variable, show that the set

$$\left\{\mathbb{E}_{\mathbb{Q}}(\bar{D}) : \mathbb{Q} \text{ an equivalent martingale measure}\right\}$$

is convex.

Hint. See Exercise 3.34(b).

An important fact for the further development of the theory is that under an equivalent martingale measure $\mathbb{Q}$ the value process of any self-financing trading strategy is again a martingale.

Theorem 5.40

If $\mathbb{Q}$ is an equivalent martingale measure for a discrete-time model, then the discounted value process $\bar{V}^{\Phi}$ of any self-financing trading strategy Φ is a martingale with respect to $\mathbb{Q}$.

Proof

Using the martingale property of $\bar{S}$ and the predictability of Φ, together with $\Delta \bar{V}_t^{\Phi} = y_t \cdot \Delta \bar{S}_t$ (from Theorem 4.20) we obtain

$$\mathbb{E}_{\mathbb{Q}}(\Delta \bar{V}_{t+1}^{\Phi}|\mathcal{F}_t) = \mathbb{E}_{\mathbb{Q}}(y_{t+1} \cdot \Delta \bar{S}_{t+1}|\mathcal{F}_t) = y_{t+1} \cdot \mathbb{E}_{\mathbb{Q}}(\Delta \bar{S}_{t+1}|\mathcal{F}_t) = 0$$

for $t < T$. Therefore $\bar{V}^{\Phi}$ is a martingale under $\mathbb{Q}$. $\qquad\square$

The simple idea in this proof actually gives one half of the following characterization of an equivalent martingale measure, which will be useful later.

Theorem 5.41

The following statements are equivalent for any equivalent measure $\mathbb{Q}$:

(1) $\mathbb{Q}$ is an equivalent martingale measure.

(2) For any predictable process $(y_t)_{t=1}^T$,

$$\mathbb{E}_{\mathbb{Q}}\left(\sum_{t=1}^{T} y_t \cdot \Delta \bar{S}_t\right) = 0.$$

Proof

If $\mathbb{Q}$ is an equivalent martingale measure then for any predictable $(y_t)_{t=1}^T$, the tower property gives

$$\mathbb{E}_{\mathbb{Q}}(y_t \cdot \Delta \bar{S}_t) = \mathbb{E}_{\mathbb{Q}}(\mathbb{E}_{\mathbb{Q}}(y_t \cdot \Delta \bar{S}_t|\mathcal{F}_{t-1})) = \mathbb{E}_{\mathbb{Q}}(y_t \cdot \mathbb{E}_{\mathbb{Q}}(\Delta \bar{S}_t|\mathcal{F}_{t-1})) = 0$$

for each $t > 0$. Thus (1) implies (2).

For the converse, fix a node λ at any time $t < T$ and define

$$
y_s = \begin{cases} 1 & \text{if } s = t+1 \text{ and } \omega \in \Omega_\lambda, \\ 0 & \text{otherwise.} \end{cases}
$$

Then assuming (2) we have

$$
0 = \mathbb{E}_\mathbb{Q}\left(\sum_{s=1}^T y_s \cdot \Delta \bar{S}_s\right) = \sum_{\omega \in \Omega_\lambda} \mathbb{Q}(\omega)\Delta \bar{S}_{t+1}(\omega) = \mathbb{Q}(\Omega_\lambda)\mathbb{E}_\mathbb{Q}(\Delta \bar{S}_{t+1}|\mathcal{F}_t)(\lambda)
$$

by the definition of conditional expectation. Thus $\mathbb{E}_\mathbb{Q}(\Delta \bar{S}_{t+1}|\mathcal{F}_t) = 0$ for all $t < T$ and $\mathbb{Q}$ is an equivalent martingale measure. $\qquad\square$

5.3 Viability

It is an immediate consequence of Theorem 5.40 that the existence of an equivalent martingale measure in the model precludes arbitrage.

Theorem 5.42

If a multi-period model with an arbitrary number of stocks admits an equivalent martingale measure, then it is viable.

Proof

Suppose that Φ is any self-financing strategy satisfying $V_T^\Phi \geq 0$, together with $V_T^\Phi(\omega) > 0$ for some $\omega \in \Omega$. If $\mathbb{Q}$ is an equivalent martingale measure, then $\mathbb{E}_\mathbb{Q}(V_T^\Phi) > 0$, so that

$$
\mathbb{E}_\mathbb{Q}(\bar{V}_T^\Phi) > 0.
$$

It then follows from Theorem 5.40 that

$$
V_0^\Phi = \bar{V}_0^\Phi = \mathbb{E}_\mathbb{Q}(\bar{V}_T^\Phi|\mathcal{F}_0) = \mathbb{E}_\mathbb{Q}(\bar{V}_T^\Phi) > 0,
$$

so Φ cannot be an arbitrage opportunity. We conclude that the model is viable. $\qquad\square$

Example 5.43

We constructed an equivalent martingale measure for Model 5.1 in Example 5.37. It follows from Theorem 5.42 that the model is viable.

We now come to the *First Fundamental Theorem of Asset Pricing* for a general multi-period model.

Theorem 5.44 (First Fundamental Theorem of Asset Pricing)

A multi-period model with a finite number of scenarios and an arbitrary number of stocks is viable if and only if it admits an equivalent martingale measure.

Proof

We showed in Theorem 5.42 that a model with an equivalent martingale measure must be viable.

For the converse, suppose that the model is viable. Theorem 4.29 showed that each single-step submodel is viable, so by Theorem 3.36 each single-step submodel has a risk-neutral probability; that is for every node λ at any time $t < T$ there is a probability assignment $\{q_\mu | \mu \in \mathrm{succ}\,\lambda\}$ satisfying $q_\mu > 0$ for $\mu \in \mathrm{succ}\,\lambda$, together with

$$\sum_{\mu \in \mathrm{succ}\,\lambda} q_\mu = 1, \qquad \sum_{\mu \in \mathrm{succ}\,\lambda} q_\mu S_{t+1}(\mu) = (1+r)S_t(\lambda).$$

The second equation means that

$$\sum_{\mu \in \mathrm{succ}\,\lambda} q_\mu \bar{S}_{t+1}(\mu) = \bar{S}_t(\lambda). \tag{5.11}$$

Now take $\mathbb{Q}$ to be the unique equivalent probability measure given by Theorem 5.25. Since the conditional probability at any node λ is given by q_λ, the equality (5.11) together with Exercise 5.29(c) shows that the discounted price process $\bar{S}$ is a martingale under $\mathbb{Q}$. Thus $\mathbb{Q}$ is an equivalent martingale measure. $\qquad\square$

Remark 5.45

An alternative and more sophisticated proof of the existence of an equivalent martingale measure in a viable model can be given by generalizing the second proof of the single step case (Proposition 3.39). This is outlined in Exercise 5.46. It has the advantage of brevity. However, the proof above, which exploits the fine structure of discrete time models, is useful to have in mind to guide the intuition even for more complex models.

Exercise 5.46[‡]

For enthusiasts only. Fill in the details in the following outline of an alternative proof of the existence of an equivalent martingale measure in a viable model. This closely resembles the second proof of Proposition 3.39 in the single-period case, which should be consulted.

(a) Writing $\Omega = \{\omega_1, \ldots, \omega_n\}$, define $\mathcal{A} \subseteq \mathbb{R}^n$ by

$$\mathcal{A} := \left\{ \left(\sum_{t=1}^{T} y_t(\omega_1) \cdot \Delta \bar{S}_t(\omega_1), \ldots, \sum_{t=1}^{T} y_t(\omega_n) \cdot \Delta \bar{S}_t(\omega_n) \right) \right.$$

$$\left. : y \text{ a predictable process} \right\}.$$

Show that $\mathcal{A}$ is a *subspace* of $\mathbb{R}^n$.

(b) Show that the viability of the model implies that $\mathcal{A} \cap \mathcal{R} = \varnothing$ where

$$\mathcal{R} := \left\{ z \in \mathbb{R}^n_+ \mid \sum_{j=1}^{n} z_j = 1 \right\}.$$

Hint. Use the Implicit Arbitrage Lemma (Lemma 4.28).

(c) Show that there is a vector $q \in \mathbb{R}^n$ with

$$q \cdot z = 0 \quad \text{for all } z \in \mathcal{A}, \tag{5.12}$$

$$q \cdot z > 0 \quad \text{for all } z \in \mathcal{R} \tag{5.13}$$

and $q_j > 0$ for each i, and $\sum_{j=1}^{n} q_j = 1$.

(d) Let $\mathbb{Q}$ be the probability with $\mathbb{Q}(\omega_j) = q_j$ for all j. Show that $\mathbb{Q}$ is an equivalent martingale measure.

Hint. Use Theorem 5.41.

5.4 Derivative Pricing in a Viable Model

The major drawback of the arbitrage pricing methodology developed in Chapter 4 is the practical problem of actually finding a replicating strategy. This is not difficult, but may become tedious when pricing a large number of derivatives. Now that we know that a market model with an equivalent martingale measure is viable, we can use equivalent martingale measures as an alternative, computationally more efficient method of pricing derivatives, in the same way that risk-neutral probabilities were used in one-step models. Here is the key result for an attainable derivative.

Theorem 5.47

Suppose that D is an attainable derivative in a viable multi-period model with equivalent martingale measure $\mathbb{Q}$. At any time t the unique fair price D_t of this derivative is the random variable

$$D_t = (1+r)^{t-T} \mathbb{E}_{\mathbb{Q}}(D|\mathcal{F}_t).$$

In particular, the fair price at time 0 is

$$D_0 = (1+r)^{-T} \mathbb{E}_{\mathbb{Q}}(D) = \mathbb{E}_{\mathbb{Q}}(\bar{D}),$$

where $\bar{D} := (1+r)^{-T} D$.

Proof

Since D is attainable, there is a self-financing trading strategy Φ with final value V_T^{Φ} equal to D. According to Theorem 4.46, the unique fair price D_t of D at any time t is the value V_t^{Φ} of this strategy at time t. Theorem 5.40 gives

$$V_t^{\Phi} = (1+r)^t \bar{V}_t^{\Phi} = (1+r)^t \mathbb{E}_{\mathbb{Q}}\left(\bar{V}_T^{\Phi}|\mathcal{F}_t\right)$$
$$= (1+r)^{t-T} \mathbb{E}_{\mathbb{Q}}\left(V_T^{\Phi}|\mathcal{F}_t\right) = (1+r)^{t-T} \mathbb{E}_{\mathbb{Q}}(D|\mathcal{F}_t).$$

$\square$

Note that the fair price $(D_t)_{t=0}^T$ of any attainable derivative D is an adapted stochastic process. This means that each D_t is an $\mathcal{F}_t$-measurable random variable. In particular, D_0 is a constant and $D_T = D$.

Theorem 5.47 shows incidentally that the risk-neutral price of any attainable derivative is the same for any choice of the equivalent martingale measure. It is not unusual for viable models to have more than one equivalent martingale measure, so the following result is worth noting.

Corollary 5.48

If $\mathbb{Q}$ and $\mathbb{Q}'$ are two different equivalent martingale measures and D is the payoff of an attainable derivative at time T, then

$$\mathbb{E}_{\mathbb{Q}}(D|\mathcal{F}_t) = \mathbb{E}_{\mathbb{Q}'}(D|\mathcal{F}_t)$$

for all t.

Proof

Both of these quantities are equal to $(1 + r)^{T-t} D_t$, where D_t is the fair price of D at time t. $\square$

The following corollary provides us with a convenient way to price derivatives by backward induction.

Corollary 5.49

The discounted value process $(\bar{D}_t)_{t=0}^T$ of an attainable derivative with payoff D at time T is a martingale under any equivalent martingale measure $\mathbb{Q}$; that is, $\bar{D}_t = \mathbb{E}_{\mathbb{Q}}(\bar{D}_{t+1}|\mathcal{F}_t)$ for all $t < T$.

Proof

If Φ is any replicating strategy for D, then $\bar{D}_t = \bar{V}_t^\Phi$ for all t. The result follows from the fact that $\bar{V}^\Phi$ is a martingale under any equivalent martingale measure. $\square$

Example 5.50

We know from Theorem 4.58 that Model 5.1 is complete, since it has three independent assets and ternary branching at every node. Finding a replicating strategy for any but the most trivial contingent claim is rather tedious. Let us see how the equivalent martingale measure $\mathbb{Q}$ from Example 5.37 provides a more convenient way of pricing derivatives in this model.

Consider the problem of pricing a European call option with strike 10 and exercise time 2 on S^1. The payoff of this derivative is

$$C_2 = \left[S_2^1 - 10\right]_+.$$

In detail, note that

$$C_2(\omega_1) = 3, \qquad C_2(\omega_6) = 3, \qquad C_2(\omega_8) = 4, \qquad C_2(\omega_9) = 5$$

and $C_2(\omega) = 0$ for all other ω. The unique fair price of this option at time 0 may be calculated directly as

$$C_0 = \frac{1}{(1+r)^2} \mathbb{E}_{\mathbb{Q}}(C_2) = \sum_{k=1}^9 \mathbb{Q}(\omega_k) \bar{C}_2(\omega_k)$$

$$= \frac{1}{1.21}\left(\frac{1472}{6375} \times 3 + \frac{63}{2125} \times 3 + \frac{713}{1700} \times 4 + \frac{23}{2125} \times 5\right) = 2\frac{43}{557} \approx 2.0772.$$

In general, it is easiest to price an option at all intermediate times by backward induction on the stock price tree. Using the conditional probabilities in Example 5.37, the price of the option at time 1 is

$$C_1(\omega) = \frac{1}{1+r} \mathbb{E}_{\mathbb{Q}}(C_2|\mathcal{F}_1)(\omega)$$

$$= \begin{cases} \frac{1}{1+r}\left(q_{\omega_1}C_2(\omega_1) + q_{\omega_2}C_2(\omega_2) + q_{\omega_3}C_2(\omega_3)\right) & \text{if } \omega\uparrow 1 = \mathrm{u}, \\ \frac{1}{1+r}\left(q_{\omega_4}C_2(\omega_4) + q_{\omega_5}C_2(\omega_5) + q_{\omega_6}C_2(\omega_6)\right) & \text{if } \omega\uparrow 1 = \mathrm{m}, \\ \frac{1}{1+r}\left(q_{\omega_7}C_2(\omega_7) + q_{\omega_8}C_2(\omega_8) + q_{\omega_9}C_2(\omega_9)\right) & \text{if } \omega\uparrow 1 = \mathrm{d} \end{cases}$$

$$= \begin{cases} \frac{1}{1.1} \times \frac{46}{75} \times 3 = 1\frac{37}{55} & \text{if } \omega\uparrow 1 = \mathrm{u}, \\ \frac{1}{1.1} \times \frac{9}{25} \times 3 = \frac{54}{55} & \text{if } \omega\uparrow 1 = \mathrm{m}, \\ \frac{1}{1.1}\left(\frac{31}{40} \times 4 + \frac{1}{50} \times 5\right) = 2\frac{10}{11} & \text{if } \omega\uparrow 1 = \mathrm{d}. \end{cases}$$

If we had not already found C_0 directly, the final step at time 0 gives

$$C_0 = \frac{1}{1+r}\mathbb{E}_{\mathbb{Q}}(C_1) = \frac{1}{1+r}\left(q_{\mathrm{u}}C_1(\mathrm{u}) + q_{\mathrm{m}}C_1(\mathrm{m}) + q_{\mathrm{d}}C_1(\mathrm{d})\right)$$

$$= \frac{1}{1.1}\left(\frac{32}{85} \times 1\frac{37}{55} + \frac{7}{85} \times \frac{54}{55} + \frac{46}{85} \times 2\frac{10}{11}\right) = 2\frac{43}{557} \approx 2.0772.$$

Exercise 5.51[†]

Using the equivalent martingale measure $\mathbb{Q}$ from Example 5.37, compute the fair price at times 0 and 1 of a European put option P_2 with strike 24 and exercise time 2 on S^2 in Model 5.1.

Exercise 5.52

Consider a European call option with expiry date 2 and strike $K = 115$ in Model 4.1 with interest rate $r = 10\%$.

(a) Find an equivalent martingale measure in this complete model.

(b) Compute the fair price at time 0 of the call option, and verify that it coincides with the price computed in Example 4.43.

Exercise 5.53

Consider a European call option C^E and a European put option P^E with strike K and expiry T in a viable model with single stock S. Show that, if $\mathbb{Q}$ is any equivalent martingale measure, then

$$\mathbb{E}_{\mathbb{Q}}\left(C_T^E|\mathcal{F}_t\right) - \mathbb{E}_{\mathbb{Q}}\left(P_T^E|\mathcal{F}_t\right) = (1+r)^T S_t - K \tag{5.14}$$

for all t. Deduce put-call parity from Theorem 5.47.

Exercise 5.54[‡]

Suppose that D and E are the payoffs of two attainable derivatives at time T with equivalent martingale measure $\mathbb{Q}$. Show that for any constants α and β the derivative G with payoff $\alpha D + \beta E$ is attainable, and show that its unique fair price at any time t is given by $\alpha D_t + \beta E_t$.

Hint. See Exercise 4.24.

Exercise 5.55[†]

Suppose that fair prices at time 0 of three call options with exercise date T in a viable model are as in the following table. You may assume that $S_0 = 100$ and $r = 0\%$.

Strike	Price
95	12.34
100	10.56
105	9.86

Give fair prices at time 0 for the following derivatives with expiry date T:

(a) A *bull spread* with payoff

$$D = \begin{cases} 0 & \text{if } S_T < 95, \\ S_T - 95 & \text{if } 95 \leq S_T < 105, \\ 10 & \text{if } S_T \geq 105. \end{cases}$$

(b) A *butterfly spread* with payoff

$$G = \begin{cases} 0 & \text{if } S_T < 95, \\ S_T - 95 & \text{if } 95 \leq S_T < 100, \\ 105 - S_T & \text{if } 100 \leq S_T < 105, \\ 0 & \text{if } S_T \geq 105. \end{cases}$$

5.5 Completeness

Suppose that a model is viable and complete. This means that every derivative in this model is attainable, in particular the derivative

$$D^A(\omega) := \begin{cases} (1+r)^T & \text{if } \omega \in A, \\ 0 & \text{if } \omega \notin A \end{cases}$$

for any set $A \subseteq \Omega$. The viability of the model means that it contains an equivalent martingale measure $\mathbb{Q}$, and the unique fair price of D^A at time 0 is

$$(1+r)^{-T} \mathbb{E}_{\mathbb{Q}}(D^A) = \mathbb{Q}(A).$$

If $\mathbb{Q}'$ is also an equivalent martingale measure in the model, by the same argument the fair price of D^A at time 0 is

$$(1+r)^{-T} \mathbb{E}_{\mathbb{Q}'}(D^A) = \mathbb{Q}'(A).$$

Since the fair price is unique, it follows that $\mathbb{Q}(A) = \mathbb{Q}'(A)$ for any $A \subseteq \Omega$, and therefore $\mathbb{Q} = \mathbb{Q}'$. In conclusion, if a viable model is complete, then it has a unique equivalent martingale measure.

Remark 5.56

In view of Theorems 3.58 and Theorem 5.25 the above argument shows that if a viable model is complete, then every single-step submodel that it contains is also complete. This gives a straightforward proof of the implication of (2) by (1) in Theorem 4.58.

As to be expected, the generalization of Theorem 3.58 to multi-period models is the following.

Theorem 5.57 (Second Fundamental Theorem of Asset Pricing)

A viable finite-state discrete-time model with an arbitrary number of stocks is complete if and only if the equivalent martingale measure is unique.

Proof

Recall from Theorem 4.58 that the model is complete if and only if each single-step submodel is complete. Using Theorem 3.58, the model is therefore complete if and only if the conditional probabilities at each node are given uniquely. It follows from Theorem 5.25 that the model is complete if and only if the equivalent martingale measure is unique.   □

Remark 5.58

The discussion at the beginning of this section gave an alternative (and more sophisticated) proof of the uniqueness of the equivalent martingale measure in a complete model (but not the converse).

For a single-stock model we can specialize the above result as follows.

Theorem 5.59

In a viable discrete-time single-stock model with finitely many outcomes the following statements are equivalent:

(1) The model is complete.

(2) There is binary branching at every node.

(3) The equivalent martingale measure is unique.

Remark 5.60

The equivalence of (1) and (2) was already noted in the previous chapter—see Theorem 4.58 and Example 4.59(a).

Exercise 5.61[†]

Verify that the single-stock model with $B_0 = 1$ and $r = 10\%$ in the following table is viable and complete. Then find the fair prices at time $t = 1$ and $t = 0$ of the derivatives with payoffs $D = (S_2 - 100)^2$ and $D' = |S_2 - 110|$ at time 2.

ω	S_0	$S_1(\omega)$	$S_2(\omega)$
ω_1	100	120	140
ω_2	100	120	120
ω_3	100	90	105
ω_4	100	90	85

5.6 Pricing Non-attainable Derivatives

We are now in a position to obtain a complete picture of the set of possible fair prices for a derivative that is not attainable. Not surprisingly we find that, as with one-step models, the super- and sub-replicating approach gives the same set of fair prices as that given by equivalent martingale measures. This result is established quite easily using the theory for one-step models. We assume for the remainder of this section that the model is viable.

The notions of super- and sub-replication extend in a natural way from the single-period case.

Definition 5.62 (Super- and sub-replication)

Let D be a derivative and Φ a self-financing trading strategy.

(a) Φ is said to *super-replicate* D if $V_T^\Phi \geq D$.

(b) Φ is said to *sub-replicate* D if $V_T^\Phi \leq D$.

Let D be any derivative. The bid and ask prices of D are defined as

$$\pi_D^a := \inf\{V_0^\Phi : \Phi \text{ super-replicates } D\},$$
$$\pi_D^b := \sup\{V_0^\Phi : \Phi \text{ sub-replicates } D\}.$$

Also define

$$\pi_D^+ := \sup\{\mathbb{E}_\mathbb{Q}(\bar{D}) : \mathbb{Q} \text{ an equivalent martingale measure}\},$$
$$\pi_D^- := \inf\{\mathbb{E}_\mathbb{Q}(\bar{D}) : \mathbb{Q} \text{ an equivalent martingale measure}\}.$$

We already know that if D is attainable with unique fair price D_0 at time 0, then

$$\pi_D^a = \pi_D^b = \pi_D^+ = \pi_D^- = D_0.$$

Thus the main concern in the rest of this section will be with non-attainable derivatives, though some results are stated in generality. The main result to be established is the natural generalization of the single period result, as follows. Let F_D be the set of fair prices for D at time 0.

Theorem 5.63

Let D be any derivative.

(1) If D is attainable then

$$F_D = \{\pi_D^a\} = \{\pi_D^b\} = \{\pi_D^+\} = \{\pi_D^-\} = \{D_0\}.$$

(2) If D is not attainable then $\pi_D^b < \pi_D^a$ and

$$F_D = (\pi_D^b, \pi_D^a) = (\pi_D^-, \pi_D^+).$$

The path to this result is somewhat similar to that for single-period models in Section 3.6. We begin with the characterization of F_D using equivalent martingale measures.

Proposition 5.64

For any derivative D in a viable multi-period model

$$F_D = \left\{ \mathbb{E}_{\mathbb{Q}}(\bar{D}) : \mathbb{Q} \text{ an equivalent martingale measure} \right\} \qquad (5.15)$$

and F_D is non-empty and is an interval. Hence

$$\left(\pi_D^-, \pi_D^+ \right) \subseteq F_D \subseteq \left[\pi_D^-, \pi_D^+ \right].$$

Proof

This proof is similar to that of Theorem 3.68 in the single-period case. Suppose first that $\pi = \mathbb{E}_{\mathbb{Q}}(\bar{D}) = (1 + r)^{-T} \mathbb{E}_{\mathbb{Q}}(D)$ for some equivalent martingale measure $\mathbb{Q}$ is a price for D at time 0. Take the pricing structure $(\pi_t)_{t=0}^T$ for D given by

$$\pi_t = (1 + r)^{t-T} \mathbb{E}_{\mathbb{Q}}(D | \mathcal{F}_t)$$

for all t. We clearly have $\pi_0 = \pi$ and $\pi_T = D$. If we now regard D as an additional asset in the model, with the pricing structure $(\pi_t)_{t=0}^T$, then $\mathbb{Q}$ is an equivalent martingale measure for this extended model, so that it is viable. Thus there can be no extended arbitrage opportunity Ψ involving $(\pi_t)_{t=0}^T$, and so π is a fair price for D at time 0.

For the converse, suppose that π is a fair price for D. This means that there is a pricing structure $(\pi_t)_{t=0}^T$ with $\pi_0 = \pi$ for which there is no extended arbitrage opportunity. Thus the extended model with the new asset D with pricing structure $(\pi_t)_{t=0}^T$ is viable, and admits an equivalent martingale measure $\mathbb{Q}$. The discounted stock and bond price processes are martingales with respect to $\mathbb{Q}$, so it is also an equivalent martingale measure for the basic model with bond and stock. At the same time we have

$$\mathbb{E}_{\mathbb{Q}}(\bar{D}) = \mathbb{E}_{\mathbb{Q}}(\bar{\pi}_T) = \bar{\pi}_0 = \pi_0 = \pi,$$

which concludes the proof of (5.15).

The set of all equivalent martingale measures is convex (Exercise 5.39) and the expectation $\mathbb{E}_{\mathbb{Q}}$ depends linearly on $\mathbb{Q}$. Thus F_D is a convex subset of $\mathbb{R}$, and must be an interval with end points $\pi_D^- = \inf F_D$ and $\pi_D^+ = \sup F_D$. $\qquad \square$

Turning to the sub- and super-replicating approach, an easy first observation is given in the following exercise.

Exercise 5.65[‡]

If Φ sub-replicates a derivative D and Φ' super-replicates D, show that

$$V_0^\Phi \leq \mathbb{E}_\mathbb{Q}(\bar{D}) \leq V_0^{\Phi'} \tag{5.16}$$

for any equivalent martingale measure $\mathbb{Q}$. Show that these inequalities are strict if D is not attainable.

From Exercise 5.65 we can deduce that $\pi_D^b \leq \pi_D^a$ for any derivative. Continuing with the super- and sub-replicating approach, we see that if D is not attainable, then any prices given by sub- or super-replication are not fair. (This was given as Exercise 3.17 for single-period single-stock models.)

Theorem 5.66

Suppose that D is a non-attainable derivative in a multi-step model. If Φ either sub- or super-replicates D, then V_0^Φ is not a fair price for D.

Proof

Suppose that $\Phi = (x_t, y_t)_{t=0}^T$ super-replicates D and $\Pi = (\pi_t)_{t=0}^T$ is any pricing structure for D with $\pi_0 = V_0^\Phi$. Consider the extended trading strategy $\Psi = (x_t, y_t, -1)_{t=0}^T$. Then

$$V_0^\Psi = V_0^\Phi - \pi_0 = 0$$

and by super-replication

$$V_T^\Psi = V_T^\Phi - D \geq 0$$

and, since Φ does not replicate D, there exists at least one $\omega \in \Omega$ such that

$$V_T^\Psi(\omega) = V_T^\Phi(\omega) - D(\omega) > 0.$$

It follows that Ψ is an extended arbitrage opportunity with D and the pricing structure Π, so $\pi_0 = V_0^\Phi$ is unfair.

The proof in the case of Φ sub-replicating D is similar (or apply the above to $-D$). $\qquad\square$

The following proposition is the key to establishing the main result. The proof utilizes the single-period result.

Proposition 5.67

Let D be a derivative in a viable multi-period model. The following statements are equivalent:

(1) $\pi \geq \pi_D^+$.

(2) $\pi = V_0^\Phi$ for some super-replicating trading strategy Φ.

Similarly, we have $\pi \leq \pi_D^-$ if and only if $\pi = V_0^\Phi$ for some sub-replicating strategy Φ.

Proof

Exercise 5.65 together with Proposition 5.64 shows that (2) implies (1).

The converse is proved by induction on the number of time steps in the model. For a single-period model, it follows from Theorems 3.66 and 3.70 that there exists a super-replicating portfolio $\varphi^a = (x^a, y^a)$ for D such that

$$V_0^{\varphi^a} = \pi_D^a = \pi_D^+.$$

The portfolio

$$\varphi := \left(x^a + \tfrac{1}{B_0}\left(\pi - V_0^{\varphi^a}\right), y^a \right)$$

then satisfies

$$V_0^\varphi = V_0^{\varphi^a} + \left(\pi - V_0^{\varphi^a}\right) = \pi,$$
$$V_1^\varphi = V_1^{\varphi^a} + (1+r)\left(\pi - V_0^{\varphi^a}\right) \geq V_1^{\varphi^a} \geq D$$

as required.

Suppose then that $\pi \geq \pi_D^+$ for a derivative D in a model with $T > 1$ time steps, and denote by $\lambda_1, \ldots, \lambda_k$ the nodes at time 1. Each of these nodes corresponds to a submodel with $T - 1$ steps that is viable due to Theorem 4.29 and the viability of the full model. For each i by the ith *submodel* we mean the submodel with root node λ_i. Its price histories may be identified with the set $\Omega_{\lambda_i} = \{\omega : \omega \uparrow 1 = \lambda_i\}$. Let D_i be D restricted to Ω_{λ_i}; then D_i is a derivative in the ith submodel. Write

$$\pi_i^+ = \sup F_{D_i} = \sup_{\mathbb{Q} \in \mathcal{Q}_i} \left[(1+r)^{1-T} \mathbb{E}_{\mathbb{Q}_i}(D_i) \right]$$

where $\mathcal{Q}_i$ is the collection of equivalent martingale measures in the ith submodel.

Using Theorems 5.25 and 5.23, each equivalent martingale measure $\mathbb{Q}$ for the full model can be viewed as a risk-neutral probability $q = (q_1, \ldots, q_k)$ on the single-step submodel at the root node, together with a collection $\mathbb{Q}_1, \ldots, \mathbb{Q}_k$ of equivalent martingale measures in the submodels at $\lambda_1, \ldots, \lambda_k$, and then

$$\mathbb{E}_\mathbb{Q}(D) = \sum_{i=1}^{k} q_i \mathbb{E}_{\mathbb{Q}_i}(D_i).$$

Let $\mathcal{Q}$ be the collection of risk-neutral probabilities in the single-step submodel at the root note. Then

$$\pi_D^+ = \sup\{(1+r)^{-T}\mathbb{E}_\mathbb{Q}(D) : \mathbb{Q} \text{ an equivalent martingale measure}\}$$

$$= \frac{1}{1+r} \sup_{q \in \mathcal{Q}} \sum_{i=1}^{k} q_i \sup_{\mathbb{Q} \in \mathcal{Q}_i} \left[(1+r)^{1-T}\mathbb{E}_{\mathbb{Q}_i}(D_i)\right]$$

$$= \frac{1}{1+r} \sup_{q \in \mathcal{Q}} \sum_{i=1}^{k} q_i \pi_i^+ = \frac{1}{1+r} \sup_{q \in \mathcal{Q}} \mathbb{E}_q(D')$$

$$= \pi_{D'}^+,$$

where D' is the derivative in the one-step submodel at the root node defined by $D'(\lambda_i) = \pi_i^+$ for $1 \le i \le k$.

Since $\pi \ge \pi_{D'}^+$, it follows from Theorems 3.66 and 3.70 that there is a portfolio φ with $V_0^\varphi = \pi$ that super-replicates D'; that is

$$V_1^\varphi(\lambda_i) \ge D'(\lambda_i) = \pi_i^+$$

for each i. Using the induction hypothesis, for each i there is a self-financing trading strategy Φ_i in the ith submodel that super-replicates D_i, with initial value $V_1^\varphi(\lambda_i)$. Combine the strategies Φ_i with the portfolio φ over the first time interval to give a self-financing trading strategy Φ with $V_0^\Phi = V_0^\varphi = \pi$ and $V_T^\Phi \ge D$, since on each Ω_{λ_i} we have $V_T^\Phi = V_T^{\Phi_i} \ge D_i = D$.

The proof for $\pi \le \pi_D^-$ is similar (or just apply the above to $-D$). $\qquad\square$

The following corollary to Proposition 5.67 is an immediate consequence of the definitions of the bid and ask prices π_D^b and π_D^a.

Corollary 5.68

For any derivative D, we have $\pi_D^a = \pi_D^+$ and $\pi_D^b = \pi_D^-$.

The proof of Theorem 5.63 now follows easily.

Proof (of Theorem 5.63)

The claim (1) was established in Section 5.4.

For (2), Theorems 5.66 and 5.67 show that for a non-attainable derivative D the prices π_D^+ and π_D^- are not fair. It follows from Proposition 5.64 that $F_D = (\pi_D^-, \pi_D^+) = (\pi_D^b, \pi_D^a)$, and $\pi_D^b < \pi_D^a$ because $F_D \neq \emptyset$. $\square$

We illustrate the above theory with an example.

Example 5.69

Consider the following two-step *trinomial model* with one stock defined as follows. The initial bond price is $B_0 = 1$. At each non-terminal node there is three way branching of the stock price, with each branch designated by u, m or d. Thus there are 9 scenarios; that is,

$$\Omega = \{uu, um, ud, mu, mm, md, du, dm, dd\},$$

and there are three nodes at time 1, namely

$$u := uu \uparrow 1 = um \uparrow 1 = ud \uparrow 1,$$

$$m := mu \uparrow 1 = mm \uparrow 1 = md \uparrow 1,$$

$$d := du \uparrow 1 = dm \uparrow 1 = dd \uparrow 1.$$

The model has three parameters, namely u, m, d with $d < m < u$. If we write $\omega = \alpha_1 \alpha_2$ where each $\alpha_i \in \{u, m, d\}$, then the stock price at time 1 is

$$S_1(\omega) = \begin{cases} S_0 u & \text{if } \alpha_1 = u, \\ S_0 m & \text{if } \alpha_1 = m, \\ S_0 d & \text{if } \alpha_1 = d, \end{cases}$$

and the stock price at time 2 satisfies

$$S_2(\omega) = \begin{cases} S_1 u & \text{if } \alpha_2 = u, \\ S_1 m & \text{if } \alpha_2 = m, \\ S_1 d & \text{if } \alpha_2 = d. \end{cases}$$

If $d < 1 + r < u$, then we know from earlier theory that this model is viable, but incomplete.

Consider the particular trinomial model with $S_0 = 100$, $r = 0.1$, $u = 1.2$, $m = 1.1$ and $d = 0.9$ depicted in Model 5.4. The conditional risk-neutral probabilities at the root node satisfy the system

$$\tfrac{120}{1.1}q_{\mathrm{u}} + \tfrac{110}{1.1}q_{\mathrm{d}} + \tfrac{90}{1.1}q_{\mathrm{m}} = 100, \qquad q_{\mathrm{u}} + q_{\mathrm{m}} + q_{\mathrm{d}} = 1.$$

It is straightforward to check that the solution to this system is

$$(q_{\mathrm{u}}, q_{\mathrm{m}}, q_{\mathrm{d}}) = \left(\tfrac{2}{3}(1 - \lambda), \lambda, \tfrac{1}{3}(1 - \lambda)\right)$$

for any $\lambda \in (0,1)$. Similar calculations at the nodes u, m and d show that the other conditional risk-neutral probabilities follow the same pattern, with respective parameters λ_{u}, λ_{m} and λ_{d} in $(0,1)$. Full details of the conditional probabilities are shown in Model 5.4. Thus a general equivalent martingale measure $\mathbb{Q}$ in this model is characterized by four parameters, namely λ, λ_{u}, λ_{m} and λ_{d}.

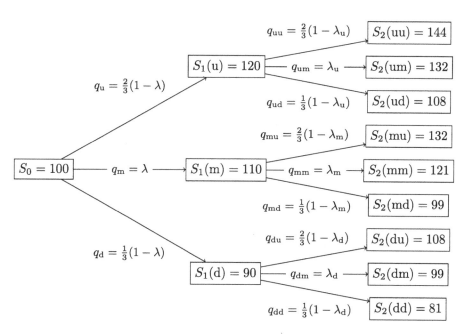

Model 5.4 Two-step trinomial model with parameters $S_0 = 100$, $r = 0.1$, $u = 1.2$, $m = 1.1$ and $d = 0.9$ in Example 5.69

Consider now a call option with exercise date 2 and strike $K = 110$ in Model 5.4. Is payoff $C_2 = [S_2 - K]_+$ satisfies

$$C_2(\mathrm{uu}) = 34, \qquad C_2(\mathrm{um}) = 22, \qquad C_2(\mathrm{mu}) = 22, \qquad C_2(\mathrm{mm}) = 11,$$

with $C_2(\omega) = 0$ for all other ω. Thus for any equivalent martingale measure $\mathbb{Q}$ we have

$$\mathbb{E}_{\mathbb{Q}}(\bar{C}_2) = \tfrac{1}{1.21}(34q_u q_{uu} + 22q_u q_{um} + 22q_m q_{mu} + 11q_m q_{mm})$$

$$= \tfrac{1}{1.21}\left(34 \times \tfrac{2}{3}(1-\lambda) \times \tfrac{2}{3}(1-\lambda_u) + 22 \times \tfrac{2}{3}(1-\lambda)\lambda_u\right.$$

$$\left. + 22\lambda \times \tfrac{2}{3}(1-\lambda_m) + 11\lambda\lambda_m\right)$$

$$= \tfrac{1}{1.21}\left(\tfrac{4}{9}(1-\lambda)(34-\lambda_u) + \tfrac{11}{3}\lambda(4-\lambda_m)\right).$$

Writing

$$f(\lambda, \lambda_u, \lambda_m) := \tfrac{1}{1.21}\left(\tfrac{4}{9}(1-\lambda)(34-\lambda_u) + \tfrac{11}{3}\lambda(4-\lambda_m)\right),$$

it follows from Theorem 5.63 that

$$\pi^a_{C_2} = \pi^+_{C_2} = \sup\{f(\lambda, \lambda_u, \lambda_m) : \lambda, \lambda_u, \lambda_m \in (0,1)\},$$

$$\pi^b_{C_2} = \pi^-_{C_2} = \inf\{f(\lambda, \lambda_u, \lambda_m) : \lambda, \lambda_u, \lambda_m \in (0,1)\}.$$

The supremum and infimum of f can be found at extreme values of the parameters λ, λ_u, λ_m. This leads to the following table of values for f.

λ	λ_u	λ_m	$f(\lambda, \lambda_u, \lambda_m)$
0	0	$[0,1]$	$\tfrac{1}{1.21} \times \tfrac{4}{9} \times 34 \approx 12.4885$
0	1	$[0,1]$	$\tfrac{1}{1.21} \times \tfrac{4}{9} \times 33 \approx 12.1212$
1	$[0,1]$	0	$\tfrac{1}{1.21} \times \tfrac{11}{3} \times 4 \approx 12.1212$
1	$[0,1]$	1	$\tfrac{1}{1.21} \times \tfrac{11}{3} \times 3 \approx 9.0909$

It is clear from the table that $\pi^+_{C_2} = \pi^a_{C_2} = 12.4885$ and $\pi^-_{C_2} = \pi^b_{C_2} = 9.0909$, so the set of fair prices for the call option is $F_{C_2} = (9.0909, 12.485)$.

Exercise 5.70[†]

Consider a two-period single-stock trinomial model such as in Example 5.69 with parameters $S_0 = 100$, $r = 0.25$, $u = 1.3$, $m = 1.25$ and $d = 1.1$.

(a) Show that the risk-neutral conditional probabilities at the root node take the form

$$(q_u, q_m, q_d) = \left(\tfrac{3}{4}(1-\lambda), \lambda, \tfrac{1}{4}(1-\lambda)\right)$$

for $\lambda \in (0,1)$.

(b) Give a general formula for an equivalent martingale measure $\mathbb{Q}$ in this model.

(c) Find the set of fair prices for a put option P_2 with exercise date 2 and strike 150 in this model.

We end this chapter with an exercise designed to explore an alternative approach to option pricing and hedging.

Exercise 5.71[‡]

For enthusiasts only. Define a *simple extended (self-financing) trading strategy* in D to be an extended trading strategy of the form

$$\Psi = (\varphi_t, z)_{t=0}^T,$$

where $\Phi = (\varphi_t)_{t=0}^T$ is a self-financing trading strategy in bonds and stock, and the holding z of the derivative D is constant, i.e. it is traded only at time $t = 0$. Define π to be a *simple fair price* for D if there no simple extended arbitrage opportunity based on this price; that is, there is no simple extended trading strategy Ψ as above with

$$V_0^\Psi = V_0^\Phi + z\pi = 0,$$
$$V_T^\Psi = V_T^\Phi + zD \geq 0$$

and $V_T^\Psi(\omega) > 0$ for some $\omega \in \Omega$.

Show that the following statements are equivalent for any derivative D:

(1) π is not a simple fair price for D.

(2) $\pi = V_0^\Phi$ for some trading strategy Φ that super- or sub-replicates D, but is not a replicating strategy for D.

(3) π is not a fair price for D.

Thus the notions of *fair* and *simply fair* prices are equivalent.

Hint. It is routine to see from the definition that a fair price is simply fair, since a fair price involves consideration of a more extensive set of extended trading strategies. For attainable D the pricing theory of earlier sections shows quite easily that any price other than D_0 is simply unfair. So what remains is to show that for non-attainable D if π is unfair then it is simply unfair. In this case either $\pi \geq \pi_D^+$ or $\pi \leq \pi_D^-$. Then use Proposition 5.67.

6

The Cox-Ross-Rubinstein Model

In this chapter we examine in detail a simple multi-period binary model with one stock and one bond, known as the *Cox-Ross-Rubinstein model* (after the authors who first introduced it). It has been widely studied and used. The key feature of the Cox-Ross-Rubinstein model is that all nodes behave in the same way. This allows substantial simplification in pricing and replicating the large class of derivatives that are *path-independent*; see Section 6.3. The Cox-Ross-Rubinstein model is also important because in a well-defined sense the famous Black-Scholes continuous-time market model is a limit of suitably scaled versions of the Cox-Ross-Rubinstein model.

6.1 Description and Basic Properties

The Cox-Ross-Rubinstein model is a special case of the multi-period models studied in Chapters 4 and 5. Thus Assumptions 4.1–4.2 remain in force, with the following modifications.

Assumption 6.1 (Bond evolution in the Cox-Ross-Rubinstein model)

The model contains a single bond with initial value $B_0 = 1$.

N.J. Cutland, A. Roux, *Derivative Pricing in Discrete Time*,
Springer Undergraduate Mathematics Series,
DOI 10.1007/978-1-4471-4408-3_6, © Springer-Verlag London 2012

In view of Assumption 4.1(2), this means that the value of the bond at any time t is

$$B_t = (1+r)^t,$$

where the interest rate $r > -1$ is fixed and known at time 0.

The following should be compared with Assumptions 4.1(3) and 4.2.

Assumptions 6.2 (Stock price evolution in the Cox-Ross-Rubinstein model)

(1) Every non-terminal node has exactly two successors.

(2) The model has a single stock with initial price $S_0 > 0$. If the stock price at a node λ at time $t < T$ is $S_t(\lambda)$, then the stock prices at the two successors of λ at time $t+1$ are $S_t(\lambda)u$ and $S_t(\lambda)d$, where the parameters u and d are fixed and known at time 0, with $0 < d < u$.

Thus branching on the price tree is binary, and is completely determined by the two parameters u ("up") and d ("down").

Exercise 6.1

Show that $S_t > 0$ for all t in the Cox-Ross-Rubinstein model.

Hint. Use induction.

The Cox-Ross-Rubinstein model has parameters T, r, S_0, u and d so it encompasses a class of particular models. In addition the Cox-Ross-Rubinstein model has a real-world probability measure $\mathbb{P}$, but as we have seen more generally, this plays no role in the theory of derivative pricing, so it is usually not mentioned. When we speak of *a* Cox-Ross-Rubinstein model we mean the Cox-Ross-Rubinstein model given by a particular choice of the parameters, which completely determines the model.

Example 6.2

Model 6.1 illustrates how the first few steps of the stock price evolution in the Cox-Ross-Rubinstein model can be written out explicitly once the parameters S_0, u and d are given.

Note that at each time step there are several nodes with the same current price. For example, at time 3 there are three nodes with the same price $S_0 u^2 d$; these nodes are the price histories

$$\left(S_0, S_0 u, S_0 u^2, S_0 u^2 d\right),$$

$$\left(S_0, S_0u, S_0ud, S_0u^2d\right),$$

$$\left(S_0, S_0d, S_0ud, S_0u^2d\right).$$

Since a node at any time t is the whole price history up to that time (not just the final price), these nodes are clearly different.

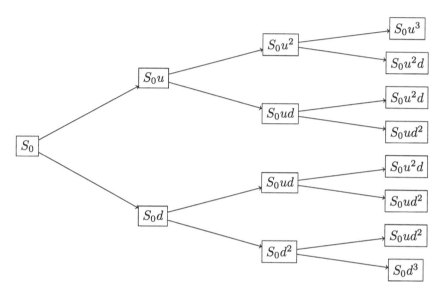

Model 6.1   Three-step Cox-Ross-Rubinstein model

Cox-Ross-Rubinstein models such as this are often represented graphically as in Figure 6.2 below. Care should be taken when using this type of representation, as it puts multiple nodes at the same place; the shaded boxes in Figure 6.2 all represent more than one node.

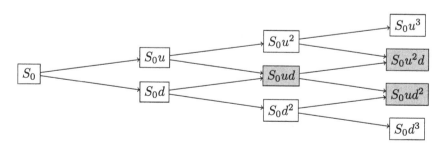

Figure 6.2   Three-step Cox-Ross-Rubinstein model (recombinant representation)

It is convenient to describe scenarios and nodes by means of the sequence of "ups" and "downs" in their price histories. Thus any time-t node in a Cox-Ross-Rubinstein model can be represented as a sequence of length t consisting of the symbols u (for "up") and d (for "down"). Figure 6.3 illustrates the relationship between any node λ at time t and its two successors λu and λd. It is clear that the model has 2^t nodes at any time t, and 2^T scenarios in total.

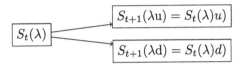

Figure 6.3   One-step submodel of Cox-Ross-Rubinstein model

The 2^t nodes at any time t result in only $t + 1$ different stock prices at time t. In ascending order, these prices are

$$S_0 d^t, S_0 u d^{t-1}, S_0 u^2 d^{t-2}, \ldots, S_0 u^t.$$

For any $s \leq t$, simple combinatorics shows that the number of different nodes with s "up" jumps and $t - s$ "down" jumps is

$$\binom{t}{s} = \frac{t!}{s!(t-s)!},$$

(where $0! := 1$), each giving the same time-t price

$$S_t(\lambda) = S_0 u^s d^{t-s}.$$

Stock price models with this property (that different time-t nodes may have the same stock price at time t) are called *recombinant*.

Example 6.3

Consider Model 6.1. At the final time $T = 3$ there are $8 = 2^T$ different nodes, namely uuu, uud, udu, udd, duu, dud, ddu and ddd, which are the different scenarios of the model. However, the time-3 stock prices for these nodes are not all different: there are only $4 = T + 1$ possibilities, namely $S_0 d^3$, $S_0 d^2 u$, $S_0 d u^2$ and $S_0 u^3$.

Note that a term such as udu is shorthand for the price history

$$\text{udu} = \left(S_0, S_0 u, S_0 u d, S_0 u^2 d \right)$$

as in Chapters 4 and 5. It is *not* a product, and so it is not the number $u^2 d$.

Exercise 6.4[†]

Consider a Cox-Ross-Rubinstein model in which $T = 3$, $S_0 = 100$, $u = 1.1$ and $d = 0.9$.

(a) Give a tree representation of this model, and list the nodes at times 0, 1, 2 and 3.

(b) Determine the filtration $(\mathcal{F}_t)_{t=0}^3$ generated by the stock price in this model.

6.2 Viability and Completeness

The theory developed in Chapters 4 and 5 for multi-period models is directly applicable to Cox-Ross-Rubinstein models. First consider the question of viability.

Theorem 6.5

The following statements are equivalent for a Cox-Ross-Rubinstein model with parameters T, r, S_0, u and d.

(1) The model is viable.

(2) $d < 1 + r < u$.

Proof

Theorem 4.29 shows that the Cox-Ross-Rubinstein model is viable if and only if each single-period submodel is viable. The single-period submodel at any node λ at time $t < T$ is binary as depicted in Figure 6.3, and it is viable if and only if

$$S_t(\lambda)d < S_t(\lambda)(1 + r) < S_t(\lambda)u.$$

Since $S_t(\lambda) > 0$, this is equivalent to $d < 1 + r < u$.   □

The previous result reflects the fact that a Cox-Ross-Rubinstein model is simply a sequence of single-step binary models that all behave in the same way; the only differences are in the initial prices of these models. Thus, if any one single-period submodel is viable then they all are, and so is the whole model. Again drawing on the general theory, we now obtain the following description of the equivalent martingale measure in a Cox-Ross-Rubinstein model.

Theorem 6.6

If a Cox-Ross-Rubinstein model is viable, then it admits a unique equivalent martingale measure $\mathbb{Q}$. The one-step conditional risk-neutral probabilities at the successors of a non-terminal node λ are all the same, and given by

$$(q_{\lambda u}, q_{\lambda d}) \equiv (q, 1 - q) := \left(\frac{1 + r - d}{u - d}, \frac{u - (1 + r)}{u - d} \right).$$

For any $\omega \in \Omega$, if

$$S_T(\omega) = S_0 u^s d^{T-s}$$

for some $s \leq T$, then

$$\mathbb{Q}(\omega) = q^s (1 - q)^{T-s}, \qquad (6.1)$$

which depends only on $S_T(\omega)$ and not the full price history of ω.

Proof

The (one-step) risk-neutral probabilities $(q_{\lambda u}, q_{\lambda d})$ at the successors of a node λ at time $t < T$ satisfy

$$q_{\lambda u} S_{t+1}(\lambda u) + q_{\lambda d} S_{t+1}(\lambda d) = (1 + r) S_t(\lambda), \qquad q_{\lambda u} + q_{\lambda d} = 1.$$

Since $S_{t+1}(\lambda u) = S_t(\lambda) u$ and $S_{t+1}(\lambda d) = S_t(\lambda) d$, this gives

$$(q_{\lambda u}, q_{\lambda d}) = \left(\frac{1 + r - d}{u - d}, \frac{u - (1 + r)}{u - d} \right);$$

note that this is independent of λ, reflecting the fact that all one-step submodels behave in the same way.

   If the final price $S_T(\omega)$ is given for some scenario $\omega \in \Omega$, then we can deduce from it the number of "up" jumps in the price history. To be precise, rearranging the equality $S_T(\omega) = S_0 u^s d^{T-s}$ gives

$$\left(\frac{u}{d} \right)^s = \frac{S_T(\omega)}{S_0 d^T}$$

and taking logarithms shows that the number s of "up" jumps in ω is

$$s = \ln \frac{S_T(\omega)}{S_0 d^T} \bigg/ \ln \frac{u}{d} . \qquad (6.2)$$

This is well-defined because $u > d$, and so $\ln \frac{u}{d} \neq 0$. The formula (6.1) for $\mathbb{Q}$ follows from Theorem 5.25. $\square$

Example 6.7

Model 6.4 is a Cox-Ross-Rubinstein model with parameters $T = 2$, $r = 0.1$, $S_0 = 100$, $u = 1.2$ and $d = 0.9$. The one-step conditional risk-neutral probabilities are

$$(q, 1 - q) = \left(\frac{1.1 - 0.9}{1.2 - 0.9}, \frac{1.2 - 1.1}{1.2 - 0.9} \right) = \left(\frac{2}{3}, \frac{1}{3} \right).$$

Thus the equivalent martingale measure $\mathbb{Q}$ for this model is

$$\mathbb{Q}(\text{uu}) = q^2 = \tfrac{4}{9}, \qquad \mathbb{Q}(\text{ud}) = q(1 - q) = \tfrac{2}{9},$$
$$\mathbb{Q}(\text{du}) = q(1 - q) = \tfrac{2}{9}, \qquad \mathbb{Q}(\text{dd}) = (1 - q)^2 = \tfrac{1}{9}.$$

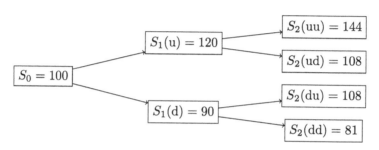

Model 6.4 Cox-Ross-Rubinstein model with $T = 2$, $r = 0.1$, $S_0 = 100$, $u = 1.2$, $d = 0.9$

A Cox-Ross-Rubinstein model has binary branching at each node, so it is complete whenever it is viable. Therefore every derivative in this model may be priced using the theory developed in Chapters 4 and 5, as follows.

Theorem 6.8

If a Cox-Ross-Rubinstein model is viable, then it is complete. The unique fair price D_t at time t of any derivative with payoff D at time T is

$$D_t = V_t^{\Phi} = (1 + r)^{t - T} \mathbb{E}_{\mathbb{Q}}(D | \mathcal{F}_t),$$

where Φ is the unique replicating strategy for D and $\mathbb{Q}$ is the unique equivalent martingale measure given in Theorem 6.6.

Proof

This follows directly from Theorems 4.46, 5.47 and 5.59. □

Exercise 6.9[†]

Consider the Cox-Ross-Rubinstein model in Exercise 6.4 with $r = 0$.

(a) Show that the model is viable and find the equivalent martingale measure.

(b) Compute the fair price at time 0 of an *up-and-in call option* with barrier $B = 110$ and strike $K = 100$ in this model, whose payoff at time 3 is

$$D(\omega) = \begin{cases} [S_3(\omega) - K]_+ & \text{if } S_t(\omega) \geq B \text{ for some } t \leq 3, \\ 0 & \text{if } S_t(\omega) < B \text{ for all } t. \end{cases}$$

An up-and-in call option can be thought of as a call option with strike K that is only activated if the stock price crosses the barrier B.

Hint. $S_3(\text{uud}) = S_3(\text{duu})$ but $D(\text{uud}) = 8.9 \neq 0 = D(\text{duu})$.

6.3 Pricing Path-Independent Derivatives

Recall that for any time s there are $\binom{T}{s}$ scenarios ω with the same final stock price

$$S_T(\omega) = S_0 u^s d^{T-s},$$

and that each of these scenarios has risk-neutral probability

$$\mathbb{Q}(\omega) = q^s (1 - q)^{T-s}.$$

Thus the $\mathbb{Q}$-probability that the final stock price is equal to $S_0 u^s d^{T-s}$ can be written explicitly as

$$\mathbb{Q}(\{\omega \in \Omega : S_T(\omega) = S_0 u^s d^{T-s}\}) = \binom{T}{s} q^s (1 - q)^{T-s}. \qquad (6.3)$$

This formula simplifies the computation of the fair price of any derivative whose payoff depends only on the final stock price.

Definition 6.10 (Path-independent derivative)

A derivative D is called *path-independent* if there exists a *payoff function* $\hat{D}$ such that

$$D = \hat{D}(S_T).$$

Remark 6.11

As we have seen in the proof of Theorem 6.6 the final stock price $S_T(\omega)$ is uniquely determined by the number of "up" jumps in ω. Thus an equivalent definition for a path-independent derivative would be that it depends on the *number* of "up" jumps in each scenario but not on their *order*.

Examples 6.12

(1) A call option with strike K and expiry T is a path-independent derivative with payoff function $\hat{C}(z) := [z - K]_+$.

(2) A put option with strike K and expiry T is a path-independent derivative with payoff function $\hat{P}(z) := [z - K]_-$.

(3) The up-and-in call option in Exercise 6.9 is not path-independent because the payoff depends on whether or not the barrier was crossed in the stock price history, which in turn depends on the order of the "up" jumps in each scenario.

Suppose that D is a path-independent derivative. From Theorem 6.8, the unique fair price of D at time 0 is

$$D_0 = (1+r)^{-T} \sum_{\omega \in \Omega} \mathbb{Q}(\omega) D(\omega) = (1+r)^{-T} \sum_{\omega \in \Omega} \mathbb{Q}(\omega) \hat{D}(S_T(\omega)). \qquad (6.4)$$

Since S_T must take one of the values $S_0 d^T, S_0 d^{T-1}u, \ldots, S_0 u^T$, this can be written as

$$D_0 = (1+r)^{-T} \sum_{s=0}^{T} \mathbb{Q}(\{\omega \in \Omega : S_T(\omega) = S_0 u^s d^{T-s}\}) \hat{D}(S_0 u^s d^{T-s}).$$

It follows from (6.3) that

$$D_0 = (1+r)^{-T} \sum_{s=0}^{T} \binom{T}{s} q^s (1-q)^{T-s} \hat{D}(S_0 u^s d^{T-s}). \qquad (6.5)$$

This is known as the (general) *Cox-Ross-Rubinstein formula* for the fair price of a path-independent derivative at time 0. Note that the sum in (6.5) contains $T+1$ terms, while the sum in (6.4) has 2^T terms. Thus the Cox-Ross-Rubinstein formula gives a substantial computational advantage in the pricing of path-independent derivatives, especially in models where T is large.

Exercise 6.13[†]

Use the Cox-Ross-Rubinstein formula (6.5) to compute the fair price at time 0 of a *straddle* with payoff $|S_3 - 100|$ at time 3 in the Cox-Ross-Rubinstein model with parameters $T = 3$, $r = 0.2$, $S_0 = 100$, $u = 1.5$ and $d = 1$.

In many practical applications, the Cox-Ross-Rubinstein formula can be simplified using the *binomial distribution function*, which for $n \in \mathbb{N}$, $k \in \mathbb{Z}$ and $p \in [0,1]$ is defined as

$$B(k;n,p) := \begin{cases} \sum_{l=0}^{k} \binom{n}{l} p^l (1-p)^{n-l} & \text{if } 0 \le k \le n, \\ 1 & \text{if } k > n. \end{cases}$$

This well-known function from probability theory gives the probability of *at most k* successes in n independent trials of an experiment with probability p of success. Also of importance is the *complementary binomial distribution function*

$$B'(k;n,p) := 1 - B(k-1;n,p) \tag{6.6}$$

which is the probability of *at least k* successes in n trials, each with probability p of success. It is well known that

$$B(n;n,p) = \sum_{l=0}^{n} \binom{n}{l} p^l (1-p)^{n-l} = 1,$$

from which it follows that

$$B'(k;n,p) = \begin{cases} \sum_{l=k}^{n} \binom{n}{l} p^l (1-p)^{n-l} & \text{if } 0 \le k \le n, \\ 0 & \text{if } k > n. \end{cases}$$

Most spreadsheet, statistical and mathematical software packages have built-in routines for computing both $B(k;n,p)$ and $B'(k;n,p)$, which greatly simplifies computation of option prices where these functions are involved, as in the following examples.

Example 6.14

A *binary call option* with strike K is a derivative with payoff

$$C_T^B(\omega) := \begin{cases} 1 & \text{if } S_T(\omega) > K, \\ 0 & \text{if } S_T(\omega) \le K. \end{cases}$$

The binary call option is path-independent since C_T^B depends only on S_T through the payoff function

$$\hat{C}^B(z) := \begin{cases} 1 & \text{if } z > K, \\ 0 & \text{if } z \leq K. \end{cases}$$

The complementary binomial function can be used to calculate its fair price C_0^B at time 0 as follows. Define

$$A := \min\{s \in \mathbb{N}_0 : S_0 u^s d^{T-s} > K\}, \tag{6.7}$$

where $\mathbb{N}_0 = \{0, 1, 2, \ldots\}$ is the set of *counting numbers*. Then the payoff of C_T^B is 1 in any scenario with A or more "up" jumps, and 0 in every other scenario. If $K > S_0 u^T$, then $A > T$ and $C_T^B = 0$, so that

$$C_0^B = 0 = (1+r)^{-T} B'(A; T, q).$$

If $A \leq T$, then the Cox-Ross-Rubinstein formula (6.5) gives the fair price of the binary call at time 0 as

$$C_0^B = (1+r)^{-T} \sum_{s=0}^{T} \binom{T}{s} q^s (1-q)^{T-s} \hat{D}\left(S_0 u^s d^{T-s}\right)$$

$$= (1+r)^{-T} \sum_{s=A}^{T} \binom{T}{s} q^s (1-q)^{T-s} \tag{6.8}$$

$$= (1+r)^{-T} B'(A; T, q). \tag{6.9}$$

Exercise 6.15[†]

In a viable Cox-Ross-Rubinstein model, consider a *binary put option* with strike K whose payoff at time T is

$$P_T^B(\omega) = \begin{cases} 1 & \text{if } S_T(\omega) \leq K, \\ 0 & \text{if } S_T(\omega) > K. \end{cases}$$

Derive a formula for its unique fair price P_0^B at time 0 in two different ways:

(a) Use an argument similar to Example 6.14.

(b) Show that if C_0^B is the fair price at time 0 of a binary call option, then

$$P_0^B + C_0^B = (1+r)^{-T},$$

and use (6.9) to derive a formula for P_0^B.

The best-known application of the Cox-Ross-Rubinstein formula is the pricing of call and put options.

Theorem 6.16 (Cox-Ross-Rubinstein formula for a call option)

The unique fair price at time 0 of a call option with strike K and expiry T in a Cox-Ross-Rubinstein model with parameters T, r, S_0, u and d is

$$C_0 = S_0 B'(A;T,q') - K(1+r)^{-T} B'(A;T,q), \qquad (6.10)$$

where $q' := q\frac{u}{1+r}$ and, as above,

$$A := \min\{s \in \mathbb{N}_0 : S_0 u^s d^{T-s} > K\}. \qquad (6.11)$$

Proof

The Cox-Ross-Rubinstein formula (6.5) gives the fair price at time 0 as

$$C_0 = (1+r)^{-T} \sum_{s=0}^{T} \binom{T}{s} q^s (1-q)^{T-s} \hat{C}(S_0 u^s d^{T-s}),$$

where $\hat{C}$ is the payoff function of the call option. Recall that $\hat{C}(z) = [z - K]_+$ for $z > 0$; this means that

$$\hat{C}(S_0 u^s d^{T-s}) = \begin{cases} 0 & \text{if } s < A, \\ S_0 u^s d^{T-s} - K & \text{if } s \geq A. \end{cases}$$

If $A > T$, then $C_0 = 0$ and (6.10) follows from $B'(A;T,q) = B'(A;T,q') = 0$. If on the other hand $A \leq T$, then

$$C_0 = (1+r)^{-T} \sum_{s=A}^{T} \binom{T}{s} q^s (1-q)^{T-s} (S_0 u^s d^{T-s} - K)$$

$$= \frac{S_0}{(1+r)^T} \sum_{s=A}^{T} \binom{T}{s} q^s (1-q)^{T-s} u^s d^{T-s} - \frac{K}{(1+r)^T} \sum_{s=A}^{T} \binom{T}{s} q^s (1-q)^{T-s}$$

$$= S_0 \sum_{s=A}^{T} \binom{T}{s} \left(q\frac{u}{1+r}\right)^s \left((1-q)\frac{d}{1+r}\right)^{T-s} - \frac{K}{(1+r)^T} B'(A;T,q).$$

$$(6.12)$$

Now let

$$q' := q\frac{u}{1+r} > 0,$$

and notice that

$$q' + (1-q)\frac{d}{1+r} = \frac{qu + (1-q)d}{1+r} = \frac{1+r}{1+r} = 1,$$

so

$$1 - q' = (1-q)\frac{d}{1+r} > 0$$

and $(q', 1-q')$ is a new artificial probability. Then (6.12) becomes

$$C_0 = S_0 \sum_{s=A}^{T} \binom{T}{s} q'^s (1-q')^{T-s} - K(1+r)^{-T} B'(A;T,q)$$

$$= S_0 B'(A;T,q') - K(1+r)^{-T} B'(A;T,q),$$

which completes the proof.   □

Exercise 6.17

Show that the fair price at time 0 of a European put option with expiry T and strike K in a viable Cox-Ross-Rubinstein model is

$$P_0 = K(1+r)^{-T} B(A-1;T,q) - S_0 B(A-1;T,q')$$

where $q' = q\frac{u}{1+r}$ and A is given by (6.11).

There is a convenient method for computing the critical number A, which simplifies the application of the Cox-Ross-Rubinstein formulae obtained above. By (6.11),

$$S_0 u^{A-1} d^{T-A+1} \leq K < S_0 u^A d^{T-A},$$

and rearrangement gives

$$\left(\frac{u}{d}\right)^{A-1} \leq \frac{K}{S_0 d^T} < \left(\frac{u}{d}\right)^A.$$

Taking natural logarithms, it follows that

$$A - 1 \leq \ln\left(\frac{K}{S_0 d^T}\right) \Big/ \ln\left(\frac{u}{d}\right) < A.$$

In other words, to find A, it is only necessary to calculate $\ln(\frac{K}{S_0 d^T})/\ln(\frac{u}{d})$ and then A is the smallest integer strictly greater than this number.

Example 6.18

Let us price a European call option with strike $K = 150$ in a Cox-Ross-Rubinstein model with parameters $T = 10$, $r = 0.02$, $S_0 = 100$, $u = 0.1$ and $d = -0.1$. Since

$$\ln\left(\frac{K}{S_0 d^T}\right) \bigg/ \ln\left(\frac{u}{d}\right) = \ln\frac{150}{100 \times 0.9^{100}} \bigg/ \ln\frac{1.1}{0.9} \approx 7.2710,$$

it follows that $A = 8$. Moreover,

$$q = \frac{1 + r - d}{u - d} = \frac{1.02 - 0.9}{1.1 - 0.9} = \frac{3}{5}, \qquad q' = q\frac{u}{1 + r} = \frac{3}{5} \times \frac{1.1}{1.02} = \frac{11}{17},$$

so that

$$C_0 = S_0 B'(A; T, q') - \frac{K}{(1 + r)^T} B'(A; T, q)$$

$$= 100 B'\left(8; 10, \tfrac{11}{17}\right) - \frac{150}{(1.02)^{10}} B'\left(8; 10, \tfrac{3}{5}\right)$$

$$= 100 \sum_{s=8}^{10} \binom{10}{s}\left(\tfrac{11}{17}\right)^s\left(\tfrac{6}{17}\right)^{10-s} - \frac{150}{(1.02)^{10}} \sum_{s=8}^{10} \binom{10}{s}\left(\tfrac{3}{5}\right)^s\left(\tfrac{2}{5}\right)^{10-s} \approx 4.9442.$$

By put-call parity (Theorem 4.49), the price of a put option with the same expiry and strike is then

$$P_0 = C_0 - S_0 + \frac{K}{(1 + r)^T} \approx 4.9442 - 100 + \frac{150}{(1.02)^{10}} \approx 27.9965.$$

Exercise 6.19[†]

Compute the fair price at time 0 of a put option with strike $K = 90$ in a Cox-Ross-Rubinstein model with parameters $T = 100$, $r = 1\%$, $S_0 = 10$, $u = 1.05$ and $d = 0.95$.

Exercise 6.20[†]

Consider a viable Cox-Ross-Rubinstein formula with parameters T, r, S_0, u and d.

(a) Derive a formula involving B' for the fair price C_0^{AN} at time 0 of an *asset-or-nothing call option* with strike K whose payoff at time T is

$$C_T^{AN}(\omega) = \begin{cases} S_T(\omega) & \text{if } S_T(\omega) > K, \\ 0 & \text{if } S_T(\omega) \leq K. \end{cases}$$

(b) The payoff of an *asset-or-nothing put option* with strike K and exercise date T is

$$P_T^{AN}(\omega) = \begin{cases} 0 & \text{if } S_T(\omega) > K, \\ S_T(\omega) & \text{if } S_T(\omega) \le K. \end{cases}$$

Show that if P_0^{AN} is the fair price at time 0 of the asset-or-nothing put option, then

$$P_0^{AN} + C_0^{AN} = S_0,$$

and use this to derive a formula for P_0^{AN}.

Exercise 6.21

Derive a formula for the fair price at time 0 of a straddle with payoff function $\hat{D}(z) = |z - K|$ in a viable Cox-Ross-Rubinstein model. Use this formula to check your answer to Exercise 6.13.

Hint. Combine Exercise 5.54 with the identity $|z| = z_+ + z_-$.

The following result gives a Cox-Ross-Rubinstein formula similar to that above for the price at any time t of a path-independent derivative D. It also gives formulae for the stock and bond holdings of the unique replicating strategy of D.

Theorem 6.22

Suppose that D is a path-independent derivative with payoff function $\hat{D}$ in a viable Cox-Ross-Rubinstein model.

(1) Its unique fair price at any time t is $D_t = \hat{D}_t(S_t)$ where

$$\hat{D}_t(z) := (1+r)^{t-T} \sum_{s=0}^{T-t} \binom{T-t}{s} q^s (1-q)^{T-t-s} \hat{D}(zu^s d^{T-t-s})$$

for $z > 0$.

(2) Its unique replicating strategy is $\Phi = (x_t, y_t)_{t=0}^T$ where $x_t = \hat{x}_t(S_{t-1})$ and $y_t = \hat{y}_t(S_{t-1})$ for all $t > 0$, and $\hat{x}_t$, $\hat{y}_t$ are given by

$$\hat{x}_t(z) := \frac{u\hat{D}_t(zd) - d\hat{D}_t(zu)}{(1+r)^t(u-d)},$$

$$\hat{y}_t(z) := \frac{\hat{D}_t(zu) - \hat{D}_t(zd)}{z(u-d)}$$

for $z > 0$.

The function $\hat{D}_t$ is called the *pricing function* at time t of the derivative D. Among practitioners, the function $\hat{y}_t$ is known as the *delta* of D at time t.

Proof

Theorem 6.8 states that the fair price of D at any time t is

$$D_t = (1+r)^{t-T}\mathbb{E}_{\mathbb{Q}}(D|\mathcal{F}_t) = V_t^{\Phi}.$$

Observe that $\hat{D}_T(z) = \hat{D}(z)$ so (1) is clearly true if $t = T$. For $t < T$ and any $\omega \in \Omega$, observe that the submodel at the time-t node $\omega \uparrow t$ behaves almost exactly like a Cox-Ross-Rubinstein model with parameters $T - t$, r, $S_t(\omega)$, u and d (the only difference is that the initial bond price is $B_t = (1+r)^t$ and not 1). A similar argument to the one that produced the Cox-Ross-Rubinstein formula (6.5) gives

$$D_t(\omega) = (1+r)^{t-T}\sum_{s=0}^{T-t}\binom{T-t}{s}q^s(1-q)^{T-t-s}\hat{D}\big(S_t(\omega)u^s d^{T-t-s}\big) = \hat{D}_t\big(S_t(\omega)\big),$$

which establishes (1).

Theorem 6.8 states that

$$x_t B_t + y_t S_t = V_t^{\Phi} = D_t$$

for all $t > 0$. For any $\omega \in \Omega$, let $\lambda := \omega \uparrow (t-1)$; then the successors λu and λd of λ are nodes at time t and

$$x_t(\lambda)B_t + y_t(\lambda)S_t(\lambda u) = D_t(\lambda u),$$
$$x_t(\lambda)B_t + y_t(\lambda)S_t(\lambda d) = D_t(\lambda d).$$

Using (1) and Assumptions 6.1–6.2, this can be rewritten as

$$x_t(\lambda)(1+r)^t + y_t(\lambda)S_{t-1}(\lambda)u = \hat{D}_t\big(S_t(\lambda u)\big) = \hat{D}_t\big(S_{t-1}(\lambda)u\big),$$
$$x_t(\lambda)(1+r)^t + y_t(\lambda)S_{t-1}(\lambda)d = \hat{D}_t\big(S_t(\lambda d)\big) = \hat{D}_t\big(S_{t-1}(\lambda)d\big).$$

The solution to this system is

$$x_t(\lambda) = \frac{u\hat{D}_t(S_{t-1}(\lambda)d) - d\hat{D}_t(S_{t-1}(\lambda)u)}{(1+r)^t(u-d)} = \hat{x}_t\big(S_{t-1}(\lambda)\big),$$

$$y_t(\lambda) = \frac{\hat{D}_t(S_{t-1}(\lambda)u) - \hat{D}_t(S_{t-1}(\lambda)d)}{S_{t-1}(\lambda)(u-d)} = \hat{y}_t\big(S_{t-1}(\lambda)\big).$$

This gives (2). $\square$

This section concludes with examples showing how the pricing functions of some well-known derivatives can be obtained. Define

$$A_t(z) := \min\{s \in \mathbb{N}_0 : zu^s d^{T-t-s} > K\} \tag{6.13}$$

for $z > 0$. Note that $A = A_0(S_0)$, where A is defined in (6.11) above. Generalizing the arguments involving A gives

$$A_t(z) - 1 \le \ln\left(\frac{K}{zd^{T-t}}\right) \bigg/ \ln\left(\frac{u}{d}\right) < A_t(z),$$

that is, $A_t(z)$ is the smallest integer that exceeds $\ln(\frac{K}{zd^{T-t}})/\ln(\frac{u}{d})$.

Example 6.23

Let us derive the pricing function $\hat{D}_t$ of a binary call option with strike K at some time t. The payoff function of the binary call is

$$\hat{D}(z) = \begin{cases} 1 & \text{if } z > K, \\ 0 & \text{if } z \le K. \end{cases}$$

This means that

$$\hat{D}(zu^s d^{T-t-s}) = \begin{cases} 1 & \text{if } s \ge A_t(z), \\ 0 & \text{if } s < A_t(z). \end{cases}$$

For any $z > 0$, if $A_t(z) \le T - t$, then

$$\hat{D}_t(z) = (1+r)^{t-T} \sum_{s=0}^{T-t} \binom{T-t}{s} q^s (1-q)^{T-t-s} \hat{D}(zu^s d^{T-t-s})$$

$$= (1+r)^{t-T} \sum_{s=A_t(z)}^{T-t} \binom{T-t}{s} q^s (1-q)^{T-t-s}$$

$$= (1+r)^{t-T} B'(A_t(z); T-t, q).$$

If $A_t(z) > T - t$, then

$$\hat{D}_t(z) = 0 = (1+r)^{t-T} B'(A_t(z); T-t, q).$$

Comparison with the formula for D_0 derived in Example 6.14 verifies that

$$D_0 = (1+r)^{-T} B'(A_0(S_0); T, q) = \hat{D}_0(S_0).$$

Exercise 6.24[†]

Suppose given a viable Cox-Ross-Rubinstein model with parameters T, r, S_0, u, r and d.

(a) Show that the pricing function at any time t of a call option with strike K is

$$\hat{C}_t(z) = zB'\left(A_t(z); T - t, q'\right) - K(1 + r)^{t-T}B'\left(A_t(z); T - t, q\right)$$

for $z > 0$, where $q' = q\frac{u}{1+r}$.

(b) Derive a formula for the pricing function at any time t of a put option with strike K.

6.4 Convergence of the Cox-Ross-Rubinstein Formula

In this section we demonstrate that the fair prices of options in a Cox-Ross-Rubinstein model converge as the number of time steps increases and the other parameters are suitably adjusted. This limit is given by the famous *Black-Scholes formula*. There is also a sense in which the Cox-Ross-Rubinstein models themselves converge to the *Black-Scholes model*.

The Black-Scholes model is a continuous-time market model with one bond and one stock over the interval $[0, T]$. In this section alone we allow $T \in (0, \infty)$ to reflect the fact that the time horizon T of a Black-Scholes model is a real date or time, which may not be an integer. The bond price at any time s is

$$B_s = e^{rs}, \tag{6.14}$$

where $r \in \mathbb{R}$ is the rate of *continuous compounding*. The stock price at any time $s > 0$ is

$$S_s = S_0 e^{(r - \frac{1}{2}\sigma^2)s + \sigma W_s}, \tag{6.15}$$

where $\sigma > 0$ is the *volatility* of the stock, the initial stock price $S_0 > 0$ is deterministic, and the process $(W_s)_{s \in [0,T]}$ is a *Brownian motion* under a probability measure $\mathbb{Q}$.

Brownian motion is defined as follows. For readers unfamiliar with the notions involved, they are explained below.

Definition 6.25 (Brownian motion)

A *Brownian motion* $W = (W_s)_{s \in [0,T]}$ is a stochastic process in continuous time with the following properties:

(1) $W_0 = 0$.

(2) W has *continuous but not differentiable trajectories*; that is, for almost all scenarios ω the mapping $s \mapsto W_s(\omega)$ is a function on $[0, T]$ that is continuous but not differentiable.

(3) W has *independent increments*; that is, if $0 \leq s_0 < \cdots < s_n \leq T$, then the random variables $W_{s_1} - W_{s_0}, \ldots, W_{s_n} - W_{s_{n-1}}$ are independent.

(4) W has *Gaussian increments*; that is, if $0 \leq s_1 < s_2 \leq T$, then the random variable $W_{s_2} - W_{s_1}$ is *Gaussian* under $\mathbb{Q}$ with mean 0 and variance $s_2 - s_1$.

Brownian motion is well studied in the literature. Proving that it exists as a mathematical entity is beyond the scope of this book, so we assume its existence without proof. Property (4) means that the Black-Scholes model has infinitely many scenarios, unlike the models that are the subject of this book.

Here are the notions from probability involved in the definition of Brownian motion. A collection $X_1, \ldots, X_n$ of real-valued random variables is called *independent* with respect to a probability measure $\mathbb{P}$ if

$$\mathbb{P}\left(\bigcap_{k=1}^{n} \{X_k \leq x_k\} \right) = \prod_{k=1}^{n} \mathbb{P}(X_k \leq x_k)$$

for all $x_1, \ldots, x_n \in \mathbb{R}$.

The *variance* $\mathrm{var}_\mathbb{P}(X)$ of a random variable X with respect to a probability measure $\mathbb{P}$ is defined as

$$\mathrm{var}_\mathbb{P}(X) := \mathbb{E}_\mathbb{P}\left((X - \mathbb{E}_\mathbb{P}(X))^2 \right) = \mathbb{E}_\mathbb{P}(X^2) - (\mathbb{E}_\mathbb{P}(X))^2.$$

It is well known from probability theory (and easy to check) that if $X_1, \ldots, X_n$ are independent random variables, then

$$\mathrm{var}_\mathbb{P}\left(\sum_{k=1}^{n} X_k \right) = \sum_{k=1}^{n} \mathrm{var}_\mathbb{P}(X_k). \tag{6.16}$$

A random variable X is called *Gaussian* or *normal* with mean μ and variance $\nu > 0$ under a probability measure $\mathbb{P}$ if

$$\mathbb{E}_\mathbb{P}(f(X)) = \int_{-\infty}^{\infty} \frac{1}{\sqrt{2\pi\nu}} e^{-\frac{(z-\mu)^2}{2\nu}} f(z)\, dz$$

for any function $f : \mathbb{R} \to \mathbb{R}$ for which the integral on the right hand side is defined. The Gaussian distribution is well known in statistics and probability theory.

Exercise 6.26[‡]

Suppose that X is a Gaussian random variable with mean μ and variance ν under a measure $\mathbb{P}$. Show that $Y := \frac{X-\mu}{\sqrt{\nu}}$ is a Gaussian random variable with mean 0 and variance 1 under $\mathbb{P}$.

Below we will define a sequence of viable Cox-Ross-Rubinstein models, all defined over the fixed time interval $[0, T]$ with a common initial stock price S_0, but with an increasing number of steps within this interval, so that the time between trading dates decreases. The convergence of the Cox-Ross-Rubinstein formula is a consequence of the *weak convergence* of the sequence of final stock prices of the Cox-Ross-Rubinstein models to the Black-Scholes final stock price S_T.

Definition 6.27 (Weak convergence)

Suppose that $(X_n)_{n=N}^{\infty}$ is a sequence of random variables, with each X_n defined on a set Ω_n endowed with a probability measure $\mathbb{P}_n$, and suppose that X is a random variable defined on a set Ω endowed with a probability $\mathbb{P}$. The sequence $(X_n)_{n=N}^{\infty}$ *converges weakly* to X if

$$\lim_{n \to \infty} \mathbb{E}_{\mathbb{P}_n}\big(f(X_n)\big) = \mathbb{E}_{\mathbb{P}}\big(f(X)\big)$$

for every bounded continuous function f.

The nth Cox-Ross-Rubinstein model will be a discrete time model with n time steps that approximates the Black-Scholes model. These n time steps will correspond to the division of the real time interval $[0, T]$ into n equal time steps of length $\Delta_n := T/n$. Thus any trading date $t \in \{0, 1, \ldots, n\}$ in the nth Cox-Ross-Rubinstein model corresponds to the real time $t\Delta_n \in [0, T]$ in the Black-Scholes model.

We would like the bond price $B_t^{(n)}$ at time step t in the nth Cox-Ross-Rubinstein model to be the same as the bond price $B_{t\Delta_n}$ at the real time $s = t\Delta_n$ in the Black-Scholes model. Equation (6.14) gives

$$B_s = B_{t\Delta_n} = e^{r\Delta_n t} = \big(e^{r\Delta_n}\big)^t,$$

so defining the interest rate

$$r_n := e^{r\Delta_n} - 1 \tag{6.17}$$

in the nth Cox-Ross-Rubinstein model gives

$$B_t^{(n)} = (1 + r_n)^t = e^{rt\Delta_n} = B_{t\Delta_n}$$

as desired.

Turning to stock prices, in the Black-Scholes model equation (6.15) gives

$$S_{t\Delta_n} = S_0 e^{(r-\frac{1}{2}\sigma^2)t\Delta_n + \sigma W_{t\Delta_n}} \tag{6.18}$$

$$= S_0 e^{(r-\frac{1}{2}\sigma^2)(t-1)\Delta_n + \sigma W_{(t-1)\Delta_n}} e^{(r-\frac{1}{2}\sigma^2)\Delta_n + \sigma(W_{t\Delta_n} - W_{(t-1)\Delta_n})}$$

$$= S_{(t-1)\Delta_n} e^{(r-\frac{1}{2}\sigma^2)\Delta_n + \sigma(W_{t\Delta_n} - W_{(t-1)\Delta_n})}. \tag{6.19}$$

The key idea for the stock price $S_t^{(n)}$ at time step t in the nth Cox-Ross-Rubinstein model is to approximate the Brownian increment $W_{t\Delta_n} - W_{(t-1)\Delta_n}$ over the real time interval $[(t-1)\Delta_n, t\Delta_n]$ by a random variable taking just two values: $\sqrt{\Delta_n}$ and $-\sqrt{\Delta_n}$. Then the approximation in the nth Cox-Ross-Rubinstein model is achieved by defining

$$u_n := e^{(r-\frac{1}{2}\sigma^2)\Delta_n + \sigma\sqrt{\Delta_n}}, \tag{6.20}$$

$$d_n := e^{(r-\frac{1}{2}\sigma^2)\Delta_n - \sigma\sqrt{\Delta_n}}. \tag{6.21}$$

The above can be summarized as follows.

Definition 6.28

The nth Cox-Ross-Rubinstein model has parameters n, r_n, S_0, u_n, d_n, where n is the number of time steps into which $[0, T]$ is divided, r_n is defined in (6.17), S_0 is the initial stock price in the Black-Scholes model, and u_n, d_n are defined in (6.20)–(6.21).

Denote the set of scenarios in the nth Cox-Ross-Rubinstein model by Ω_n. Let N be the smallest integer such that $N > \frac{1}{4}\sigma^2 T$.

Proposition 6.29

(1) The nth Cox-Ross-Rubinstein model is viable if and only if $n \geq N$.

(2) For $n \geq N$ the unique one-step conditional risk-neutral probabilities for the nth Cox-Ross-Rubinstein model are $(q_n, 1 - q_n)$, where

$$q_n = \frac{1 + r_n - d_n}{u_n - d_n} = \frac{e^{\frac{1}{2}\sigma^2\Delta_n} - e^{-\sigma\sqrt{\Delta_n}}}{e^{\sigma\sqrt{\Delta_n}} - e^{-\sigma\sqrt{\Delta_n}}}. \tag{6.22}$$

(3) For $n \geq N$ the unique equivalent martingale measure $\mathbb{Q}_n$ in the nth Cox-Ross-Rubinstein model satisfies

$$\mathbb{Q}_n\left(\{\omega \in \Omega_n : S_n^{(n)} = S_0 u_n^t d_n^{n-t}\}\right) = \binom{n}{t} q_n^t (1 - q_n)^{n-t}$$

for $t \leq n$.

Proof

By Theorem 6.5, the nth Cox-Ross-Rubinstein model is viable if and only if

$$d_n < 1 + r_n < u_n,$$

which by (6.17) and (6.20)–(6.21) is equivalent to

$$e^{-\sigma\sqrt{\Delta_n}} < e^{\frac{1}{2}\sigma^2\Delta_n} < e^{\sigma\sqrt{\Delta_n}}.$$

Taking logarithms and observing that $\sigma^2\Delta_n > 0$, this is equivalent to

$$\tfrac{1}{2}\sigma^2\Delta_n < \sigma\sqrt{\Delta_n},$$

which holds true if and only if $n > \tfrac{1}{4}\sigma^2 T$. This gives (1) since N is the smallest integer larger than $\tfrac{1}{4}\sigma^2 T$.

Theorem 6.6 gives (2) and the uniqueness of $\mathbb{Q}_n$. Claim (3) then follows directly from (6.3). □

Consider the nth Cox-Ross-Rubinstein model. It is helpful to define for each $t \in \{1,\dots,n\}$ the random variable $C_t^{(n)}$ with

$$C_t^{(n)}(\omega) := \begin{cases} \sqrt{\Delta_n} & \text{if } S_t^{(n)}(\omega) = S_{t-1}^{(n)}(\omega)u_n, \\ -\sqrt{\Delta_n} & \text{if } S_t^{(n)}(\omega) = S_{t-1}^{(n)}(\omega)d_n \end{cases} \tag{6.23}$$

for $\omega \in \Omega_n$. Note from (6.20)–(6.21) that

$$S_t^{(n)} = S_{t-1}^{(n)} e^{(r-\frac{1}{2}\sigma^2)\Delta_n + \sigma C_t^{(n)}}, \tag{6.24}$$

which should be compared with (6.19). Bearing in mind that the time steps $t-1$ and t in the nth Cox-Ross-Rubinstein model correspond to the real times $(t-1)\Delta_n$ and $t\Delta_n$ in the Black-Scholes model, we can think of $C_t^{(n)}$ as an approximation to the Brownian increment $W_{t\Delta_n} - W_{(t-1)\Delta_n}$ over the real time interval $[(t-1)\Delta_n, t\Delta_n]$.

Exercise 6.30[‡]

Show that $C_1^{(n)},\dots,C_n^{(n)}$ are independent random variables with respect to $\mathbb{Q}_n$ for any $n \geq N$, and that

$$\mathbb{E}_{\mathbb{Q}_n}\big(C_t^{(n)}\big) = 2\big(q_n - \tfrac{1}{2}\big)\sqrt{\Delta_n}, \tag{6.25}$$

$$\text{var}_{\mathbb{Q}_n}\big(C_t^{(n)}\big) = 4q_n(1 - q_n)\Delta_n \tag{6.26}$$

for $t = 1,\dots,n$.

Recall from Definition 6.25 that the increments $W_{\Delta_n} - W_0, \ldots, W_{n\Delta_n} - W_{(n-1)\Delta_n}$ are independent Gaussian random variables with mean 0 and variance Δ_n. We will see later (in Exercise 6.35(b)) that $q_n \to \frac{1}{2}$ as $n \to \infty$. Thus the mean and variance of $C_t^{(n)}$ approximate the mean and variance of $W_{t\Delta_n} - W_{(t-1)\Delta_n}$.

The stochastic process $W^{(n)} = (W_t^{(n)})_{t=0}^T$ defined by $W_0^{(n)} := 0$ and

$$W_t^{(n)} := \sum_{s=1}^t C_t^{(n)} \quad \text{for } t = 1, \ldots, n \tag{6.27}$$

is known as a *random walk* with final value $W_n^{(n)}$. It is clear from (6.24) that

$$
\begin{aligned}
S_t^{(n)} &= S_0 \prod_{s=1}^t e^{(r - \frac{1}{2}\sigma^2)\Delta_n + \sigma C_s^{(n)}} \\
&= S_0 e^{(r - \frac{1}{2}\sigma^2)t\Delta_n + \sigma \sum_{s=1}^t C_s^{(n)}} \\
&= S_0 e^{(r - \frac{1}{2}\sigma^2)t\Delta_n + \sigma W_t^{(n)}}.
\end{aligned}
\tag{6.28}
$$

Comparison with (6.18) shows that we can think of $W_t^{(n)}$ as an approximation to $W_{t\Delta_n}$, and in particular the final value $W_n^{(n)}$ is an approximation to $W_{n\Delta_n} = W_T$, the final value of the Brownian motion W. To make this precise we will prove at the end of this section that $(W_n^{(n)})_{n \geq N}$ converges weakly to W_T.

Example 6.31

Consider a Black-Scholes model with parameters $S_0 = 100$, $r = 0.1$, $\sigma = 0.15$ and $T = 1$. Figure 6.5 contains a sample path of the Brownian motion W and the corresponding Black-Scholes stock price S.

Note that

$$\tfrac{1}{4}\sigma^2 T = \tfrac{1}{4} \times (0.15)^2 \times 1 = 0.005625,$$

so $N = 1$ and therefore the nth Cox-Ross-Rubinstein model is viable for all $n \in \mathbb{N}$. The table below gives Δ_n, u_n, r_n, d_n and q_n for some values of n.

n	Δ_n	u_n	r_n	d_n	q_n
1	1.000	1.2697	0.1052	0.9406	0.5001408
5	0.200	1.0885	0.0202	0.9519	0.5000126
10	0.100	1.0579	0.0101	0.9622	0.5000044
20	0.050	1.0387	0.0050	0.9713	0.5000016
40	0.025	1.0263	0.0025	0.9787	0.5000006
100	0.010	1.0160	0.0010	0.9860	0.5000001

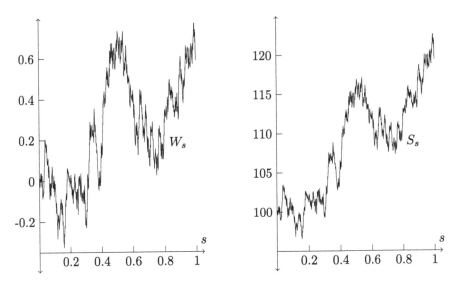

Figure 6.5 Sample paths of Brownian motion and Black-Scholes stock price in Example 6.31

The table suggests that $(q_n)_{n\in\mathbb{N}}$ converges to $\frac{1}{2}$; Exercise 6.35 at the end of this section gives a rigorous proof of this.

Figure 6.6 demonstrates how the sample path of Brownian motion shown in Figure 6.5 may be approximated by a path of the random walk $W^{(n)}$ in Cox-Ross-Rubinstein models with $n = 10$ and $n = 40$ steps. The processes $W^{(n)}$ and $S^{(n)}$ are discrete in time, and the straight lines connecting the points have only been added to emphasize the shapes of the paths. The approximation of the Black-Scholes stock price process S by Cox-Ross-Rubinstein stock prices in this figure uses the formula (6.28) for $n = 10, 40$.

Consider the approximation of W by $W^{(n)}$ in more detail. The approximation for $n = 10$ divides the real time interval $[0, 1]$ into 10 equal steps of length $\Delta_{10} = 0.1$, and at each step the random walk $W^{(10)}$ moves up or down by $\sqrt{\Delta_{10}} \approx 0.3162$. The approximation for $n = 40$ divides the interval $[0, 1]$ into 40 equal steps, each of length $\Delta_{40} = 0.025$, and the jump size at each step is $\sqrt{\Delta_{40}} \approx 0.1589$. Thus the approximation for $n = 40$ allows more time steps that are closer together in real time, and the jump size at each step is smaller than the approximation with 10 steps. This explains why the 40-step approximation appears visually to be much closer to the path of Brownian motion than the 10-step approximation.

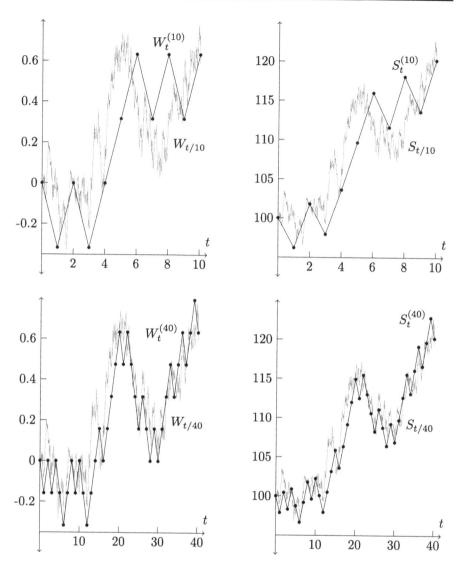

Figure 6.6 Cox-Ross-Rubinstein approximations with 10 (above) and 40 (below) steps in Example 6.31

Note that from (6.28)

$$S_n^{(n)} = \hat{S}\big(\tfrac{1}{\sqrt{T}} W_n^{(n)}\big),$$

where $\hat{S}$ is the continuous function

$$\hat{S}(z) := S_0 e^{(r - \frac{1}{2}\sigma^2)T + \sigma\sqrt{T}z} \quad \text{for } z \in \mathbb{R}.$$

By comparison, (6.15) gives the final Black-Scholes stock price as

$$S_T = \hat{S}\left(\tfrac{1}{\sqrt{T}}W_T\right).$$

We are now in a position to state and prove the main result of this section.

Theorem 6.32

Let $\hat{D}$ be a bounded continuous function. For each $n \geq N$, define the path-independent derivative $D^{(n)}$ in the nth Cox-Ross-Rubinstein model by

$$D^{(n)} := \hat{D}\big(S_n^{(n)}\big)$$

and let $D_0^{(n)}$ be its unique fair price at time 0. Then

$$\lim_{n \to \infty} D_0^{(n)} = e^{-rT}\mathbb{E}_{\mathbb{Q}}\big(\hat{D}(S_T)\big),$$

that is,

$$\lim_{n \to \infty} (1 + r_n)^{-n} \sum_{s=0}^{n} \binom{n}{s} q_n^s (1 - q_n)^{n-s} \hat{D}\big(S_0 u_n^s d_n^{n-s}\big)$$

$$= e^{-rT} \int_{-\infty}^{\infty} \frac{1}{\sqrt{2\pi}} e^{-\frac{1}{2}z^2} \hat{D}\big(\hat{S}(z)\big)\, dz. \tag{6.29}$$

Proof

This result depends on the weak convergence of $(W_n^{(n)})_{n=N}^{\infty}$ to W_T, which is proven in Exercise 6.35 at the end of this section.

From elementary calculus the function g defined by

$$g(z) := \hat{D}\big(\hat{S}(\tfrac{1}{\sqrt{T}}z)\big) \quad \text{for } z \in \mathbb{R}$$

is continuous, and is bounded because $\hat{D}$ is. The weak convergence of $(W_n^{(n)})_{n=N}^{\infty}$ to W_T then means that

$$\lim_{n \to \infty} \mathbb{E}_{\mathbb{Q}_n}\big(\hat{D}(S_n^{(n)})\big) = \lim_{n \to \infty} \mathbb{E}_{\mathbb{Q}_n}\big(\hat{D}\big(\hat{S}(\tfrac{1}{\sqrt{T}}W_n^{(n)})\big)\big)$$

$$= \lim_{n \to \infty} \mathbb{E}_{\mathbb{Q}_n}\big(g(W_n^{(n)})\big)$$

$$= \mathbb{E}_{\mathbb{Q}}\big(g(W_T)\big)$$

$$= \mathbb{E}_{\mathbb{Q}}\big(\hat{D}\big(\hat{S}(\tfrac{1}{\sqrt{T}}W_T)\big)\big) = \mathbb{E}_{\mathbb{Q}}\big(\hat{D}(S_T)\big).$$

Theorem 6.8 and the identity $(1 + r_n)^n = e^{rT}$ give

$$\lim_{n \to \infty} D_0^{(n)} = \lim_{n \to \infty} (1 + r_n)^{-n} \mathbb{E}_{\mathbb{Q}_n} \left(\hat{D}(S_n^{(n)}) \right)$$
$$= e^{-rT} \mathbb{E}_{\mathbb{Q}} \left(\hat{D}(S_T) \right).$$

Equation (6.29) then follows immediately from the Cox-Ross-Rubinstein formula (6.5) and the fact that $\frac{1}{\sqrt{T}} W_T$ is a Gaussian random variable with mean 0 and variance 1 under $\mathbb{Q}$ (see Exercise 6.26). □

The left hand side of (6.29) is the limit of the Cox-Ross-Rubinstein formulae for each of the viable Cox-Ross-Rubinstein models. The right hand side of (6.29) is known as the (general) *Black-Scholes formula*.

In similar fashion to the discrete-time case that we have studied in this book, it is possible to show that the Black-Scholes model is complete, that the probability measure $\mathbb{Q}$ is an equivalent martingale measure for this model, and that the unique fair price at time 0 of the derivative D with payoff function $\hat{D}$ and exercise date T is given by the Black-Scholes formula (the right hand side of (6.29)). A proof of this is outside the scope of this book. We can however show how this formula can be used to compute the fair price of put and call options in a Black-Scholes model.

Example 6.33

Recall that the payoff function of a put option with strike K is

$$\hat{P}(z) = [z - K]_- = \max\{K - z, 0\} \quad \text{for } z > 0.$$

This function becomes continuous and bounded for $z \in \mathbb{R}$ if we also define

$$\hat{P}(z) := K \quad \text{for } z \le 0;$$

note that in the calculations below the function $\hat{P}$ will only be applied to $z > 0$ so defining $\hat{P}$ for $z \le 0$ is for technical reasons only (to ensure continuity).

Assume that $n \ge N$. Exercise 6.17 shows that the fair price at time 0 of the put option in the nth Cox-Ross-Rubinstein model is

$$P_0^{(n)} = K(1 + r_n)^{-n} B(A_n - 1; n, q_n) - S_0 B(A_n - 1; n, q'_n), \tag{6.30}$$

where $q'_n = q_n \frac{u_n}{1 + r_n}$ and

$$A_n = \min\left\{ s \in \mathbb{N}_0 : S_0 u_n^s d_n^{n-s} > K \right\}.$$

According to Theorem 6.32, this converges to the price given by the Black-Scholes formula, namely

$$P_0 := e^{-rT} \mathbb{E}_{\mathbb{Q}}\big([S_T - K]_-\big)$$

$$= e^{-rT} \int_{-\infty}^{\infty} \frac{1}{\sqrt{2\pi}} e^{-\frac{1}{2}z^2} \big[S_0 e^{(r-\frac{1}{2}\sigma^2)T+\sigma\sqrt{T}z} - K\big]_- \, dz. \qquad (6.31)$$

This can be simplified using the well-known *cumulative standard Gaussian distribution function* Φ. It is defined as

$$\Phi(x) := \int_{-\infty}^{x} \frac{1}{\sqrt{2\pi}} e^{-\frac{1}{2}z^2} \, dz \quad \text{for } x \in \mathbb{R},$$

and it has the property that

$$\Phi(x) = 1 - \Phi(-x) \quad \text{for } x \in \mathbb{R}. \qquad (6.32)$$

Most spreadsheet and programming packages have built-in routines for computing Φ.

The integrand in (6.31) is zero unless

$$S_0 e^{(r-\frac{1}{2}\sigma^2)T+\sigma\sqrt{T}z} - K \leq 0.$$

Taking logs and rearranging, we deduce that the integrand in (6.31) is non-zero only when

$$z \leq \gamma := \frac{1}{\sigma\sqrt{T}}\left(\ln\frac{K}{S_0} - (r - \tfrac{1}{2}\sigma^2)T\right).$$

Thus

$$P_0 = e^{-rT} \int_{-\infty}^{\gamma} \frac{1}{\sqrt{2\pi}} e^{-\frac{1}{2}z^2} \left(K - S_0 e^{(r-\frac{1}{2}\sigma^2)T+\sigma\sqrt{T}z}\right) dz$$

$$= e^{-rT} K \Phi(\gamma) - S_0 \int_{-\infty}^{\gamma} \frac{1}{\sqrt{2\pi}} e^{-\frac{1}{2}z^2+\sigma\sqrt{T}z-\frac{1}{2}\sigma^2 T} \, dz. \qquad (6.33)$$

Since

$$-\tfrac{1}{2}z^2 + \sigma\sqrt{T}z - \tfrac{1}{2}\sigma^2 T = -\tfrac{1}{2}\big(z^2 - 2\sigma\sqrt{T}z + \sigma^2 T\big) = -\tfrac{1}{2}(z - \sigma\sqrt{T})^2,$$

the integral on the right hand side of (6.33) can be written as

$$\int_{-\infty}^{\gamma} \frac{1}{\sqrt{2\pi}} e^{-\frac{1}{2}z^2+\sigma\sqrt{T}z-\frac{1}{2}\sigma^2 T} \, dz = \int_{-\infty}^{\gamma} \frac{1}{\sqrt{2\pi}} e^{-\frac{1}{2}(z-\sigma\sqrt{T})^2} \, dz$$

$$= \int_{-\infty}^{\gamma-\sigma\sqrt{T}} \frac{1}{\sqrt{2\pi}} e^{-\frac{1}{2}y^2} \, dy$$

$$= \Phi(\gamma - \sigma\sqrt{T}).$$

To summarize, we have shown that

$$P_0 = e^{-rT} K \Phi(\gamma) - S_0 \Phi(\gamma - \sigma\sqrt{T}).$$

This is the well-known Black-Scholes formula for the price of a put option. It is customary to write this in the form

$$P_0 = e^{-rT} K \Phi(-d_2) - S_0 \Phi(-d_1) \tag{6.34}$$

where

$$d_1 = -\gamma + \sigma\sqrt{T} = \frac{1}{\sigma\sqrt{T}} \left(\ln \frac{S_0}{K} + \left(r + \tfrac{1}{2}\sigma^2 \right) T \right),$$

$$d_2 = -\gamma = \frac{1}{\sigma\sqrt{T}} \left(\ln \frac{S_0}{K} + \left(r - \tfrac{1}{2}\sigma^2 \right) T \right).$$

We have also shown that $P_0^{(n)} \to P_0$, where $P_0^{(n)}$ is given by (6.30). Note the similarity in form of $P_0^{(n)}$ and P_0.

The Black-Scholes price of a call option is

$$C_0 := e^{-rT} \mathbb{E}_{\mathbb{Q}} \big([S_T - K]_+ \big)$$
$$= e^{-rT} \mathbb{E}_{\mathbb{Q}} \big([S_T - K]_- \big) + e^{-rT} \mathbb{E}_{\mathbb{Q}}(S_T - K)$$
$$= P_0 + S_0 - e^{-rT} K,$$

where we have used $a_+ = a_- + a$ and the fact that $S_0 = e^{-rT} \mathbb{E}_{\mathbb{Q}}(S_T)$ in the Black-Scholes model. Equations (6.32) and (6.34) then give

$$C_0 = e^{-rT} K \Phi(-d_2) - S_0 \Phi(-d_1) + S_0 - e^{-rT} K$$
$$= S_0 \big(1 - \Phi(-d_1) \big) - e^{-rT} K \big(1 - \Phi(-d_2) \big)$$
$$= S_0 \Phi(d_1) - e^{-rT} K \Phi(d_2).$$

Now that the convergence of $(P_0^{(n)})_{n=N}^{\infty}$ to the Black-Scholes put price P_0 has been established, it is natural to consider the corresponding result for call options. Theorem 6.32 cannot be applied directly to find the limit of the Cox-Ross-Rubinstein formula for a call option, because the payoff function of a call option is not bounded. To be precise,

$$\lim_{z \to \infty} \hat{C}(z) = \lim_{z \to \infty} [z - K]_+ = \lim_{z \to \infty} (z - K) = \infty.$$

Nevertheless the convergence of the put option prices can be used, in combination with put-call parity, to show that the Cox-Ross-Rubinstein call option prices converge to the Black-Scholes price. Theorem 4.49 shows that for any $n \geq N$ the fair price $C_0^{(n)}$ at time 0 of the call option in the nth Cox-Ross-

Rubinstein model satisfies

$$C_0^{(n)} = P_0^{(n)} + S_0 - (1 + r_n)^{-n} K = P_0^{(n)} + S_0 - e^{-rT} K.$$

The convergence of $(P_0^{(n)})_{n=N}^{\infty}$ to P_0 then implies that $(C_0^{(n)})_{n=N}^{\infty}$ converges to

$$P_0 + S_0 - e^{-rT} K = C_0.$$

We conclude this example by illustrating this convergence for the parameters of Example 6.31, namely $S_0 = 100$, $r = 0.1$, $\sigma = 0.15$ and $T = 1$. We consider put and call options with strike $K = 100$. Since

$$d_1 = \tfrac{1}{0.15 \times \sqrt{1}} \left(\ln \tfrac{100}{100} + \left(0.1 + \tfrac{1}{2}(0.15)^2\right) \times 1 \right) \approx 0.741667,$$

$$d_2 = \tfrac{1}{0.15 \times \sqrt{1}} \left(\ln \tfrac{100}{100} + \left(0.1 - \tfrac{1}{2}(0.15)^2\right) \times 1 \right) \approx 0.591667,$$

it follows that

$$C_0 = 100\Phi(0.741667) - 100e^{-0.1 \times 1}\Phi(0.591667)$$

$$\approx 100 \times 0.770855 - 100e^{-0.1} \times 0.722963 \approx 11.669128.$$

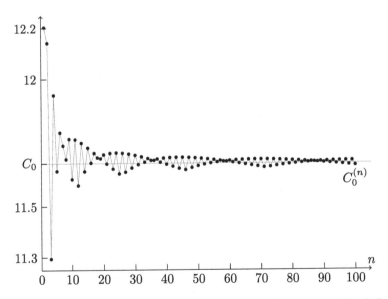

Figure 6.7 Convergence of Cox-Ross-Rubinstein call prices to Black-Scholes price in Example 6.33

Figure 6.7 shows how $(C_0^{(n)})_{n\in\mathbb{N}}$ converges to C_0. We also have

$$P_0 = 100e^{-0.1\times 1}\Phi(-0.591667) - 100\Phi(-0.741667)$$
$$\approx 100e^{-0.1} \times 0.277037 - 100 \times 0.229144 \approx 2.152870.$$

The convergence of $(P_0^{(n)})_{n\in\mathbb{N}}$ to P_0 is demonstrated in Figure 6.8. In both cases the convergence is fast. The oscillatory nature of the convergence of Cox-Ross-Rubinstein option prices is well known. Notice that put-call parity means that the graph in Figure 6.8 is a vertical shift of Figure 6.7 by $e^{-rT}K - S_0$, which is independent of n.

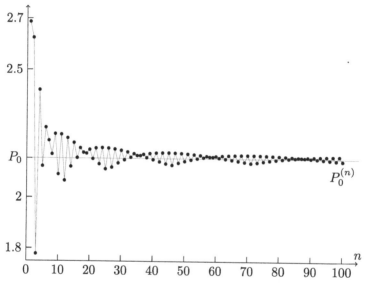

Figure 6.8 Convergence of Cox-Ross-Rubinstein put prices to Black-Scholes price in Example 6.33

This section concludes with an outline proof of the weak convergence of the sequence $(W_n^{(n)})_{n\geq N}$ of final values of the n-step random walks to the final value W_T of the Brownian motion. It uses the following important result from probability theory, which we quote without proof.

Theorem 6.34 (Central Limit Theorem)

Suppose that for each $n \geq N$ the random variables $X_1^{(n)}, \ldots, X_n^{(n)}$ are independent under a probability $\mathbb{Q}_n$, and that there exists a constant $c_n > 0$ such that

$$\left| X_t^{(n)} \right| < c_n \quad \text{for all } t. \tag{6.35}$$

Define

$$H^{(n)} := \sum_{t=1}^{n} X_t^{(n)}. \tag{6.36}$$

If

$$\lim_{n \to \infty} c_n = 0, \tag{6.37}$$

$$\lim_{n \to \infty} \mathbb{E}_{\mathbb{Q}_n} \left(H^{(n)} \right) = \mu, \tag{6.38}$$

$$\lim_{n \to \infty} \text{var}_{\mathbb{Q}_n} \left(H^{(n)} \right) = \nu, \tag{6.39}$$

then $(H^{(n)})_{n=N}^{\infty}$ converges weakly to a Gaussian random variable with mean μ and variance ν.

We will apply this to the random variables $\{C_t^{(n)} : n \geq N, 1 \leq t \leq n\}$ defined earlier to see that $(W_n^{(n)})_{n=N}^{N}$ converges weakly to W_T. For $n \geq N$ write $X_t^{(n)} := C_t^{(n)}$ for $1 \leq t \leq n$ and define $H^{(n)}$ by (6.36); then it is clear from (6.27) that $H^{(n)} = W_n^{(n)}$. Note moreover that from (6.23)

$$-\sqrt{\Delta_n} \leq C_t^{(n)} \leq \sqrt{\Delta_n}$$

for all n and t, so that (6.35) are (6.37) are satisfied if we put $c_n := \sqrt{\Delta_n} = \sqrt{\frac{T}{n}}$. To establish the weak convergence of $(H^{(n)})_{n=N}^{\infty}$ to W_T, which is a Gaussian random variable with mean 0 and variance T, it is therefore sufficient to show that (6.38)–(6.39) hold true with $\mu = 0$ and $\nu = T$.

For $n \geq N$, Exercise 6.30 assists the calculation of the mean and variance of $H^{(n)}$. Equation (6.25) gives

$$\mathbb{E}_{\mathbb{Q}_n} \left(H^{(n)} \right) = \sum_{t=1}^{n} \mathbb{E}_{\mathbb{Q}_n} \left(C_t^{(n)} \right) = 2n \left(q_n - \tfrac{1}{2} \right) \sqrt{\Delta_n} = \sqrt{2nT} \left(q_n - \tfrac{1}{2} \right).$$

Combining (6.16) and (6.26) leads to

$$\text{var}_{\mathbb{Q}_n} \left(H^{(n)} \right) = \sum_{t=1}^{n} \text{var}_{\mathbb{Q}_n} \left(C_t^{(n)} \right) = 4n q_n (1 - q_n) \Delta_n = 4T q_n (1 - q_n).$$

Exercise 6.35 below shows that

$$\lim_{n \to \infty} \mathbb{E}_{\mathbb{Q}_n} \left(H^{(n)} \right) = 0, \tag{6.40}$$

$$\lim_{n \to \infty} \text{var}_{\mathbb{Q}_n} \left(H^{(n)} \right) = T. \tag{6.41}$$

This completes the proof of the fact that $(W_n^{(n)})_{n=N}^{\infty} = (H^{(n)})_{n=N}^{\infty}$ converges weakly to the Gaussian random variable W_T with mean 0 and variance T.

Exercise 6.35‡

The purpose of this exercise is to establish (6.40) and (6.41). Fill in the gaps in the following argument.

(a) By putting $x = \sigma \sqrt{\Delta_n}$ show that

$$\lim_{n \to \infty} q_n = \lim_{x \to 0} \frac{e^{\frac{1}{2}x^2} - e^{-x}}{2 \sinh x}, \tag{6.42}$$

$$\lim_{n \to \infty} 2\sqrt{nT} \left(q_n - \tfrac{1}{2} \right) = \lim_{x \to 0} \frac{\sigma T (e^{\frac{1}{2}x^2} - \cosh x)}{x \sinh x}. \tag{6.43}$$

(b) Apply L'Hôpital's rule to (6.42) and obtain

$$\lim_{n \to \infty} q_n = \tfrac{1}{2}.$$

Deduce from this that

$$\lim_{n \to \infty} 4T q_n (1 - q_n) = T.$$

(c) Use the Taylor expansions of $e^{\frac{1}{2}x^2}$ and $\cosh x$ around $x = 0$ to show that

$$e^{\frac{1}{2}x^2} - \cosh x = \sum_{k=2}^{\infty} a_k x^{2k} \tag{6.44}$$

where $a_k = \frac{(2k)! - 2^k k!}{2^k k! (2k)!}$ for all k. Substitute this into (6.43) and use L'Hôpital's rule to show that

$$\lim_{n \to \infty} 2\sqrt{nT} \left(q_n - \tfrac{1}{2} \right) = 0.$$

American Options

All derivatives in this book so far have been *European* derivatives, which are simply random variables D that denote a payoff $D(\omega)$ for the owner at the fixed expiry time T. By contrast, *American options*, studied in this chapter, are derivatives that allow the owner to take the payoff (that is, to *exercise* the option) at any time up to and including the fixed expiry time T. The payoff may depend not only on the scenario ω but also on the *exercise time* $t \leq T$, so it can be represented as a stochastic process $X = (X_t)_{t=0}^{T}$. The process X is assumed to be adapted because when deciding whether or not to exercise an American option X in scenario ω at a given time t the payoff $X_t(\omega)$ must be known to the owner, so $X_t(\omega)$ must depend only on the history $\omega \uparrow t$.

The idea of American options is well illustrated by the particular case of American call and put options, the so-called *vanilla American options*, which we discuss in the next section before developing the general theory. These, and more general American options described below, are widely traded, raising important questions concerning the price to attach to such derivatives. The key questions are the same as for European derivatives: is there a fair price? If so, is it unique and how can it be found? A mathematical answer requires a precise definition of "fair price" for an American option, which is not as easy as for European options. Moreover, since an American option offers the choice of when to exercise, there is for the owner the additional question of when is the most profitable time to exercise.

To define the notion of a fair price for an American option we would expect to be guided by an appropriate extension of the principle of no-arbitrage.

N.J. Cutland, A. Roux, *Derivative Pricing in Discrete Time*,
Springer Undergraduate Mathematics Series,
DOI 10.1007/978-1-4471-4408-3_7, © Springer-Verlag London 2012

Initially we take an *informal* approach that is applicable in a complete model, without making a formal definition of extended arbitrage involving American options. The theory developed later in the chapter (see Section 7.3) will provide a precise mathematical definition of a fair (or arbitrage-free) price for a general American option. We will see that for a complete model the situation is similar to that for a European option: there is always a unique fair price, which has a characterization using equivalent martingale measures, and this agrees with the informal theory. The theory will also explain when it is sensible to exercise an American option, and when to wait. The last part of the chapter discusses the pricing and optimal exercise of American options in an incomplete model, where the formal pricing theory again has similarities to the corresponding theory for European derivatives.

For the whole chapter the underlying model is a general *viable* discrete time model with finitely many scenarios as in Chapters 4 and 5.

7.1 American Call and Put Options

For American call or put options there is, as with European calls and puts, a fixed *strike price* K; this is the price at which the owner of an option may buy (if it is a call) or sell (if it is a put) a single share of the underlying stock. The difference is that the owner may *exercise* the option at *any time* up to the expiry time T. Here is the formal definition.

Definition 7.1

(a) An *American call option on a stock S, with expiry time T and strike price K*, is a contract that gives the owner the right (but not the obligation) to buy one share for K at any time $t \leq T$.

(b) An *American put option on a stock S, with expiry time T and strike price K*, is a contract that gives the owner the right (but not the obligation) to sell one share for K at any time $t \leq T$.

When the owner of an American call or put option makes the decision to buy or sell the share as the case may be, this is known as *exercising* the option.

Consider the owner of an American call option, who can, at *any* time $t \leq T$ of his choosing, buy a share for K. Of course he may only do this once. The seller or *writer* of the option is obliged by the contract to sell a share for K at the time that the owner chooses to exercise.

If the current price S_t of the share is less than K the owner will naturally not exercise the option at time t, since if needed it would be cheaper to buy a share on the open market. But if the current price S_t is *greater* than K the owner might decide to exercise the option by buying a share for K; he would then own an asset worth S_t, which could be sold, giving an immediate profit of $S_t - K$. So the *exercise value*, or *cash-in value*, or *payoff* of an American call option at time t is $[S_t - K]_+$. However, the owner might believe that a *greater* profit could be made by waiting and exercising the option at a later time: perhaps he has a hunch that the market is going to rise even further. So even if $S_t > K$ it might be worth waiting to exercise at a later time.

At the expiry time T, if the call has not been exercised earlier then it must be exercised at that time. If $S_T > K$ the owner makes a profit of $S_T - K$. If on the other hand $S_T \leq K$, then the option is valueless and is discarded. Thus it is useful still to think of the owner as receiving $[S_T - K]_+ = 0$, which is the payoff of the option at that time.

The situation for an American put is similar: at *any* time $t \leq T$ the payoff of the option is $[K - S_t]_+$, because if $S_t < K$ then exercising it brings a profit of $K - S_t$, whereas if $S_t \geq K$ there is no benefit from exercising. However, even if $S_t < K$ the owner might believe that the stock price will fall in the future so it could be worth waiting in order to obtain a bigger profit.

Throughout this section we work with an informal commonsense notion of extended arbitrage involving American options and a corresponding informal notion of fair price. A formal definition will be given in Section 7.3.

7.1.1   Pricing American Calls and Puts: Informal Theory

We begin with an example to illustrate the informal approach to pricing an American option. For this we *assume* that there is a fair price for an American call or put, and develop an argument to find it.

Example 7.2

Consider Model 7.1 with a single stock S and interest rate $r = 10\%$. The model has a unique equivalent martingale measure $\mathbb{Q}$ with associated conditional probabilities as indicated. Consider an American put option on S with strike $K = 105$ and expiry date $T = 2$.

Assuming that there is a fair price P_t^A for this option at any time $t \leq 2$, we would like to find it. Clearly $P_t^A \geq [K - S_t]_+$ because otherwise a risk-free profit can be made by buying the option at time t and exercising it immediately. However, it could be that $P_t^A > [K - S_t]_+$ because of the possibility of a greater gain by exercising later.

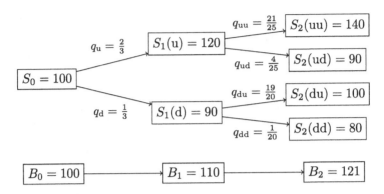

Model 7.1 Two-period single-stock binary model

At the expiry time 2 the choice is between exercising the option and receiving the payoff $K - S_2$, or discarding it (and receiving the payoff 0). A rational trader would only exercise if $K \geq S_2$, so the fair price of the option at that time is $[K - S_2]_+$.

At time 1, at the node u, the stock price is 120 so the payoff of the option is $[105 - 120]_+ = 0$. Since there is no benefit in exercising the option at this node, the only sensible choice for the owner is to keep it until time 2. Thus there is no difference *at this node* between the American option and a derivative with payoff P_2^A at time 2. This suggests that we should apply risk neutral pricing theory (Theorem 5.47) to give the unique value

$$P_1^A(\mathrm{u}) = \tfrac{1}{1+r}\mathbb{E}_{\mathbb{Q}}\big(P_2^A|\mathcal{F}_1\big)(\mathrm{u}) = \tfrac{1}{1.1}\big(\tfrac{21}{25} \times 0 + \tfrac{4}{25} \times 15\big) = 2\tfrac{2}{11}.$$

This is *greater than* $[K - S_1(\mathrm{u})]_+ = 0$, which is the payoff of the option here.

Now consider the node d. If the owner decides not to exercise, but to keep the option until time 2, then the discounted expected payoff at time 2 of the option is

$$\tfrac{1}{1+r}\mathbb{E}_{\mathbb{Q}}\big(P_2^A|\mathcal{F}_1\big)(\mathrm{d}) = \tfrac{1}{1.1}\big(\tfrac{19}{20} \times 5 + \tfrac{1}{20} \times 25\big) = 5\tfrac{5}{11}.$$

This is the value (to the owner) of the decision to keep it until time 2. On the other hand, exercising the option at this node yields an immediate payoff of $[105 - 90]_+ = 15$. Thus at the node d the owner is faced with the choice between

(a) Exercising and receiving a payoff of 15.

(b) Not exercising, and retaining an asset that is currently worth $5\tfrac{5}{11}$.

A rational person would prefer (a), and would therefore exercise the option at time 1 if he is at node d.

This suggests that the fair price of the option at the node d is 15. In fact, any other price at this node leads to arbitrage. If the price is less than 15 one could simply buy the option and exercise it immediately giving an instant profit. On the other hand, if the option is sold at a price π with $\pi > 15$ then a writer could make a guaranteed profit as follows.

(i) If the owner exercises the option immediately, the writer pays out 15 and so makes an immediate profit of $\pi - 15 > 0$.

(ii) If the owner chooses to delay exercising the option until time $t = 2$, the writer should form the portfolio $\varphi = ((90 + \pi)/B_1, -1)$. He then has an extended portfolio comprising $90 + \pi$ in bonds, 1 short share and 1 short option, with total outlay

$$(90 + \pi) - 90 - \pi = 0.$$

At time $t = 2$ in scenario du, the owner must exercise, so the writer must pay out $P_2^A(\text{du}) = [105 - 100]_+ = 5$. This means that the value of the writer's total portfolio at time 2 is

$$(1 + r)(90 + \pi) - S_2(\text{du}) - P_2^A(\text{du}) = 1.1(90 + \pi) - 100 - 5$$
$$> 1.1 \times 105 - 105 > 0.$$

At dd, the writer's liability is $P_2^A(\text{dd})$ so the value of his portfolio is

$$(1 + r)(90 + \pi) - S_2(\text{dd}) - P_2^A(\text{dd}) = 1.1(90 + \pi) - 80 - 25$$
$$> 1.1 \times 105 - 105 > 0.$$

Thus in every case the writer can make a risk free profit if the price of the option at the node d is different from 15, so there is extended arbitrage. Thus the fair price of the American call option at this node is

$$P_1^A(\text{d}) = 15.$$

Finally consider P_0^A. If the option is exercised at time 0 it yields the payoff $[105 - 100]_+ = 5$. If the owner decides not to exercise, then he retains an asset that will be worth its fair price at the next time, namely P_1^A as already calculated. The risk-neutral theory of pricing shows that the value of this asset at time 0 is

$$\tfrac{1}{1+r}\mathbb{E}_{\mathbb{Q}}(P_1^A) = \tfrac{1}{1.1}(\tfrac{2}{3} \times 2\tfrac{2}{11} + \tfrac{1}{3} \times 15) = 5\tfrac{105}{121} > 5.$$

That is, the value of *not* exercising at time $t = 0$ is greater than the value if exercised at that time. So a rational owner will prefer not to exercise immediately, but to keep it until a later time. So the fair price at time 0 of an American option with strike 105 and expiry 2 in this model is

$$P_0^A = \max\{5, 5\tfrac{105}{121}\} = 5\tfrac{105}{121} \tag{7.1}$$

which is again greater than the payoff.

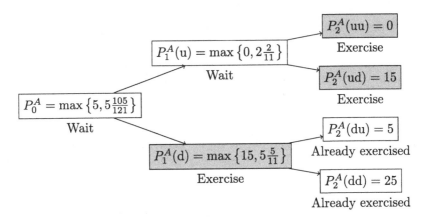

Figure 7.2 Price process and exercise decisions for the owner of an American put option

Figure 7.2 gives the fair price process of this American put option given by the above discussion, and indicates the rational exercise decision that would be made at each node by an owner of the option.

It is clear from this example that, informally at least, the fair price for an American put option at any time t will not necessarily be the same as the payoff at that time. This is the case at the root node and at the node u.

7.1.2 Comparison of American and European Calls and Puts

Continuing the informal approach to pricing, it is instructive to compare an American call or put with its European counterpart. Clearly the owner of an American call or put could simply opt to delay exercising until the expiry time T as if it were a European option. Thus an American call or put is *at least* as valuable as the European counterpart with the same strike price K and expiry time T. To clarify this, still assuming as above that the model is complete and that there is a well-defined unique fair price for American options, write P_t^A as before for the fair price of an American put and similarly write C_t^A, P_t^E and C_t^E

for the fair prices of an American call, and a European put and call respectively each with strike price K. Then it is easy to establish the following.

Theorem 7.3

$C_t^A \geq C_t^E$ and $P_t^A \geq P_t^E$, otherwise there is extended arbitrage.

Proof

If $C_t^A(\omega) < C_t^E(\omega)$ for some t and ω, we can obtain an extended arbitrage opportunity in the submodel at $\omega \uparrow t$ as follows. At time t in scenario ω buy an American call, sell a European call and invest the difference $C_t^E(\omega) - C_t^A(\omega)$ in bonds, so that there is no net outlay. Do not exercise the American option until time T, when the values of the two calls are equal and cancel each other. This guarantees a profit of $(C_t^E(\omega) - C_t^A(\omega))(1+r)^{T-t}$ from the bonds for any scenario ω' with $\omega' \uparrow t = \omega \uparrow t$. So to avoid extended arbitrage it is necessary that $C_t^A \geq C_t^E$.

The proof for put options is similar. □

Remark 7.4

We can avoid making any assumption about the existence of an as yet undefined fair price for American calls and puts by rephrasing the above theorem as follows: any price for an American call or put option that is *less* than the fair price for the corresponding European option is *unfair* in the sense that it presents an arbitrage opportunity involving the options, shares and bonds.

Example 7.2 above shows that there can be a genuine advantage in having the extra choice of exercise time for an American put. That is, the *current* payoff $[K - S_t]_+$ at time t can be greater than the value of choosing to exercise later rather than now. However, it is interesting to see that for an American call option the possible advantage of being able to exercise early is illusory; there is in fact no advantage in exercising early provided the interest rate is non-negative; that is, if $r \geq 0$ then $C_t^A = C_t^E$ for all t.

To see this informally, consider the choice at a time t between exercising an American call option or waiting until at least time $t+1$. The value of exercising is $[S_t - K]_+$ whereas the decision to wait means retaining an asset that is worth *at least* $[S_{t+1} - K]_+$ at time $t+1$. The risk neutral theory of pricing means that the value of waiting at time t is *at least*

$$\tfrac{1}{1+r}\mathbb{E}_{\mathbb{Q}}\big([S_{t+1}-K]_+|\mathcal{F}_t\big) \geq \tfrac{1}{1+r}\mathbb{E}_{\mathbb{Q}}\big((S_{t+1}-K)|\mathcal{F}_t\big)$$

$$= S_t - \tfrac{K}{1+r}$$

$$\geq S_t - K$$

since $r \geq 0$. The first term on the left is never negative, so in fact

$$\tfrac{1}{1+r}\mathbb{E}_{\mathbb{Q}}\big([S_{t+1}-K]_+|\mathcal{F}_t\big) \geq \max\{S_t - K, 0\} = [S_t - K]_+.$$

Hence the value at time t of the asset held, if the decision at that time is to wait, is at least as valuable as the payoff at time t. So there is no advantage gained by exercising the American call option at any time $t < T$; in fact, there may be a *dis*advantage (see Exercise 7.7). More formally we have the following remarkable result (still assuming of course that an American call has a fair price in an informal sense).

Theorem 7.5

If C^E and C^A are the fair prices of European and American call options with strike K and expiry T in a model with interest rate $r \geq 0$, then for all t

$$C_t^E = C_t^A. \tag{7.2}$$

Proof

We will prove this for $t = 0$. For $t > 0$ simply work with the full submodel based at any time-t node.

We know from Theorem 7.3 that $C_0^E \leq C_0^A$ so suppose that $C_0^E < C_0^A$. Then there is an informal extended arbitrage opportunity as follows.

Write and sell an American call for C_0^A, buy a European call for C_0^E, and invest the difference $C_0^A - C_0^E > 0$ in bonds; denote this extended portfolio by ψ. There are now two possibilities.

(i) The American option is not exercised until time T, when $C_T^E = C_T^A$. Then the final value of the portfolio ψ is

$$C_T^E - C_T^A + (1+r)^T\big(C_0^A - C_0^E\big) = (1+r)^T\big(C_0^A - C_0^E\big) > 0.$$

(ii) In some scenario ω the American option is exercised at some time $t < T$. To fulfil his obligation at that time, the writer could borrow one share, sell it for

K to the owner of the option, and invest the proceeds K in the bond. This amounts to modifying the portfolio ψ to ψ', say, consisting of 1 European option, -1 share, and bonds worth K in addition to the original investment in bonds, now worth $(1+r)^t(C_0^A - C_0^E)$. Make no further changes, so at expiry time T, the value of the extended portfolio ψ' for any scenario ω' with $\omega' \uparrow t = \omega \uparrow t$ is

$$[S_T - K]_+ - S_T + K(1+r)^{T-t} + \left(C_0^A - C_0^E\right)(1+r)^T$$

$$\geq [S_T - K]_+ - (S_T - K) + \left(C_0^A - C_0^E\right)$$

$$\geq C_0^A - C_0^E > 0$$

since $r \geq 0$ and because $[S_T - K]_+ \geq S_T - K$. This gives a risk-free profit for any scenario ω' passing through the node $\omega \uparrow t$.

In either case there is an extended arbitrage opportunity if $C_0^E < C_0^A$, and to preclude this we must have $C_0^E = C_0^A$.   □

Remark 7.6

The above result can be rephrased without mention of a fair price for an American call by saying that any price for an American call that is greater than the fair price for the corresponding European option presents an arbitrage opportunity involving the options, shares and bonds.

Exercise 7.7

Show that it can be disadvantageous to exercise an American call early, by considering the single period binary model with $r = 0.2$, $S_0 = 100$, $S_1(u) = 130$, $S_1(d) = 90$ and strike $K < 130$.

There is no counterpart to Theorem 7.5 for put options if $r > 0$ (but see Exercise 7.10 below for the situation when $r = 0$). In general an American put option is strictly more valuable than its European counterpart, as we can see in the following example.

Example 7.8

In Example 7.2 we calculated that $P_0^A = 5\frac{105}{121} = \frac{710}{121}$ for an American put with strike 105. The price P_0^E of the corresponding European put is

$$P_0^E = (1+r)^{-2}\mathbb{E}_{\mathbb{Q}}\left(P_2^A\right) = \frac{1}{(1.1)^2}\left(\frac{14}{25} \times 0 + \frac{8}{75} \times 15 + \frac{19}{60} \times 5 + \frac{1}{60} \times 25\right) = \frac{360}{121},$$

which is strictly less than P_0^A.

Here is another example.

Example 7.9

Consider an American put option with strike 170 and expiry time T on a stock S, in a model with interest rate $r = 0.15$. Suppose that the option has not been exercised before time $T - 1$, and that $S_{T-1} = 20$. If the owner exercises immediately, he receives 150, which may be invested in bonds to yield 172.5 at time T. If, on the other hand, he chooses to keep the option until time T, he will receive at most

$$[170 - S_T]_+ < 170.$$

In this case, it is clearly beneficial to the owner to exercise the option early.

Exercise 7.10‡

Show that if $r = 0$ then the fair prices P_0^A and P_0^E for European and American put options with the same strike and expiry date are equal; that is,

$$P_0^E = P_0^A.$$

Hint. Adapt the proof of Theorem 7.5.

7.2 General American Options in a Complete Model: Informal Theory

Before embarking on the formal theory of pricing for American options, to illustrate the ideas it is instructive to adopt an informal approach assuming that the model is complete. First, here is the precise mathematical definition of a general American option in *any* model.

Definition 7.11

An *American option* is a non-negative adapted stochastic process $X = (X_t)_{t=0}^T$. The value $X_t(\omega)$ is called the *payoff* of the option at time t in scenario ω.

The owner of an American option X has the right (but not the obligation) to receive the payoff X_t at any time $t \le T$. The payoff can only be received once; after that the option is discarded. If the owner opts to receive the payoff at time t he is said to *exercise* the option at that time.

Remarks 7.12

(1) The reason for restricting to non-negative X is that if $X_t < 0$ then no rational owner would consider exercising at that time.

(2) The requirement that X should be adapted means that it depends only on the history of prices to date; thus at each time t the decision whether to exercise the option or to wait is made in full knowledge of the payoff available at that time.

(3) The decision whether or not to exercise the option at time t must depend only on the partial history $\omega \uparrow t$ and not the whole of the scenario ω; this means that exercise decisions are made at *nodes*. This will be made mathematically precise in the formal theory developed in Section 7.3.

Assume for the rest of this section that we are in a complete viable model, so there is a unique equivalent martingale measure $\mathbb{Q}$ and every *European* derivative has a unique fair price. We continue to assume (without any rigorous justification as yet) that an American option X has a unique fair price, understood informally as a price that prevents arbitrage opportunities involving the trading of bonds, stock and the option. (In Section 7.3.2 the precise definition of fair price will enable us to prove a Law of One Price for American Options.)

Denote by $Z_t^X(\omega)$ the assumed unique fair price of the American option X at time t in scenario ω. Naturally this is at least as much as the payoff $X_t(\omega)$; that is

$$Z_t^X \geq X_t$$

although it may be worth more because of the possibility of a greater payoff in the future. At the final time T there is no question of waiting so we must have

$$Z_T^X = X_T.$$

The fair price for each earlier time can now be identified by a backwards recursion using the idea employed in the examples of the previous section.

Fix any $t < T$ and suppose that the fair price Z_{t+1}^X has been found for the option X. Consider the choices available to the holder of the option at time t at a node λ.

Choice (a). The option could be exercised to yield the payoff $X_t(\lambda)$.

Choice (b). The owner may decide not to exercise the option at time t. At time $t+1$ the market will move to one of the nodes $\mu \in \operatorname{succ} \lambda$ where the owner would then hold an asset worth the current fair price of the option, namely $Z_{t+1}^X(\mu)$. Thus at time t he is in the situation of a single-period model, with root node λ holding a derivative with payoff $D(\mu) = Z_{t+1}^X(\mu)$ at each of the successors of λ.

The risk neutral theory of pricing gives the fair price at time t of this derivative D as

$$\mathbb{E}_{\mathbb{Q}}\left(\tfrac{1}{1+r}D|\mathcal{F}_t\right)(\lambda) = \mathbb{E}_{\mathbb{Q}}\left(\tfrac{1}{1+r}Z_{t+1}^X|\mathcal{F}_t\right)(\lambda).$$

A rational owner at the node λ will decide between Choice (a): exercise and receive $X_t(\lambda)$, or Choice (b): wait and retain an asset worth $\mathbb{E}_{\mathbb{Q}}(\tfrac{1}{1+r}Z_{t+1}(\cdot)|\mathcal{F}_t)(\lambda)$, according to whichever has the greater value *now*. This suggests the following pricing theory for an American option in a complete model, where for convenience we write

$$\hat{Z}_t^X = \tfrac{1}{1+r}Z_t^X.$$

Theorem 7.13 (Informal pricing of an American option)

In a complete model the fair price for an American option X at time t is Z_t^X, where

$$Z_T^X := X_T,$$
$$Z_t^X := \max\{X_t, \mathbb{E}_{\mathbb{Q}}(\hat{Z}_{t+1}^X|\mathcal{F}_t)\} \quad \text{if } t < T. \tag{7.3}$$

Remark 7.14

For any $t < T$ and ω, if the option has not yet been exercised and

$$X_t(\omega) > \mathbb{E}_{\mathbb{Q}}(\hat{Z}_{t+1}^X|\mathcal{F}_t)(\omega)$$

then it is advantageous to exercise the option early at the node $\omega \uparrow t$ since the payoff is greater than the fair value of waiting. If, on the other hand, the option has not yet been exercised and

$$X_t(\omega) < \mathbb{E}_{\mathbb{Q}}(\hat{Z}_{t+1}^X|\mathcal{F}_t)(\omega),$$

then it is advantageous to wait at the node $\omega \uparrow t$ because the current fair value of the option *unexercised* is more than the payoff.

If these values are equal, then the payoff is identical to the current fair price of the option, so exercising and waiting have the same value.

Proof (of Theorem 7.13)

In the above discussion the derivation of the formula (7.3) for Z_t^X did not show how any other price would give extended arbitrage, so we show in detail that this is the case, proceeding by reverse induction. It is clear that $Z_T^X = X_T$ has to be the fair price for the option at the expiry time T.

Suppose now that for all $t' > t$ the fair price of an American option is given by $Z_{t'}^X$ as defined above, and that the option is priced at $\pi \neq Z_t^X$ at some time-t node λ. Here is how to obtain extended arbitrage in the submodel based at λ.

Since the model is complete the single step submodel at λ is complete so there is a portfolio $\varphi = (x, y)$ such that $xB_{t+1} + yS_{t+1}(\mu) = Z_{t+1}^X(\mu)$ for each $\mu \in \operatorname{succ} \lambda$. So

$$xB_t + yS_t(\lambda) = \mathbb{E}_{\mathbb{Q}}\big(\tfrac{1}{1+r}(xB_{t+1} + yS_{t+1})|\mathcal{F}_t\big)(\lambda)$$

$$= \mathbb{E}_{\mathbb{Q}}\big(\hat{Z}_{t+1}^X|\mathcal{F}_t\big)(\lambda)$$

$$\leq Z_t^X(\lambda)$$

from the definition (7.3). Now there are two cases to consider.

(i) $\pi < Z_t^X(\lambda)$. In this case buy one option for π and if $Z_t^X(\lambda) = X_t(\lambda)$, exercise it immediately and make a profit of $X_t(\lambda) - \pi$. Otherwise, $Z_t^X(\lambda) = xB_t + yS_t(\lambda)$, so short the portfolio φ and with the proceeds $Z_t^X(\lambda)$ pay for the option and invest the balance $Z_t^X(\lambda) - \pi$ in the bonds.

At time $t + 1$ sell the option for its fair price $Z_{t+1}^X(\mu) = xB_{t+1} + yS_{t+1}(\mu)$ and with the proceeds close the position in the portfolio φ. This gives a risk free profit of $(Z_t^X(\lambda) - \pi)(1 + r)$ from the bonds at time $t + 1$.

(ii) $\pi > Z_t^X(\lambda) \geq X_t(\lambda)$. In this case write an option and sell it for π. If the owner chooses to exercise immediately there is an immediate profit of $\pi - X_t(\lambda)$. Otherwise, note that $\pi > xB_t + yS_t(\lambda)$ so use the proceeds π of the sale to buy the portfolio $\varphi = (x, y)$ and invest $\pi - (xB_t + yS_t(\lambda)) > 0$ in bonds.

At time $t + 1$ the portfolio φ is worth $xB_{t+1} + yS_{t+1}(\mu) = Z_{t+1}^X(\mu)$ so liquidate it and with the proceeds buy one American option. Whenever the owner of the original option (sold for π at time t) chooses to exercise, we can meet our obligations by doing the same with our option. This gives a risk free profit from the funds $\pi - (xB_t + yS_t(\lambda))$ that were invested in bonds at time t.

In either case there is an extended arbitrage opportunity for one of the traders, so any price other than Z_t^X at time t is unfair; hence the unique fair price at this time is Z_t^X.   □

Remark 7.15

For any ω and t such that $X_t(\omega) < \mathbb{E}_{\mathbb{Q}}(\hat{Z}_{t+1}^X|\mathcal{F}_t)(\omega)$ there is of course no *guarantee* that by waiting there will be a greater payoff in the future. The point is that in this case the asset that consists of the American option with exercise restricted to times after t has fair price greater than the current payoff $X_t(\omega)$. So if certainty (for example in the form of cash in hand) is required it is better to sell the option for its fair price $\mathbb{E}_{\mathbb{Q}}(\hat{Z}_{t+1}^X|\mathcal{F}_t)$ than to exercise and receive only X_t.

Later we will see that in a complete model the price Z_t^X derived above agrees with the unique fair price given by the formal theory in Section 7.3.

First it is instructive to illustrate the above informal pricing theory with an example. We calculate $\mathbb{E}_{\mathbb{Q}}(\hat{Z}_{t+1}^X|\mathcal{F}_t)$ at each non-terminal node, working backwards through the price tree, and compare it with the payoff X_t.

Example 7.16

Consider the Cox-Ross-Rubinstein model with $T = 2$, $S_0 = 100$, $r = 0.25$, $u = 1.3$, $d = 1.1$ and an American option with payoff given by Figure 7.3. The risk neutral probabilities at each node are $(\frac{3}{4}, \frac{1}{4})$. The values of the process Z^X are found by backwards recursion using the definition (7.3). For example,

$$Z_1^X(\mathrm{u}) = \max\left\{90, \tfrac{1}{1.25}\left(\tfrac{3}{4} \times 100 + \tfrac{1}{4} \times 20\right)\right\} = \max\{90, 64\} = 90.$$

This gives the prices in Figure 7.4, and the fair price of this option at time $t = 0$ is $Z_0^X = 62.4$. It is advantageous to exercise early at the node u (shaded in the diagram) since there the payoff is 90, whereas the value of the choice to wait is only 64. At both of the other non-terminal nodes it is advantageous to wait.

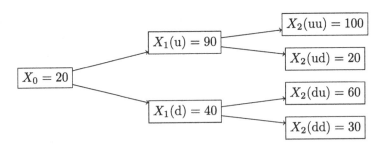

Figure 7.3 Payoff for an American option in Example 7.16

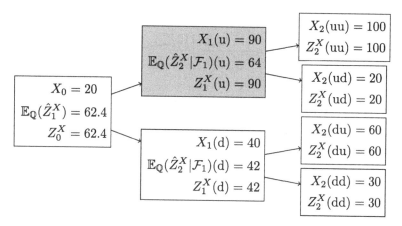

Figure 7.4 Pricing the American option in Example 7.16

Exercise 7.17[†]

(a) Let X be an American put option with strike price $K = 102$ in the Cox-Ross-Rubinstein model with parameters $T = 3$, $r = 0.1$, $S_0 = 100$, $u = 1.25$, $d = 0.5$. Find the fair prices $P_t^A = Z_t^X$ by backwards recursion and illustrate them together with the payoffs X_t on a tree diagram. Indicate the nodes where it is advantageous to exercise the option early if it is has not yet been exercised. Find the fair prices for the corresponding European put option with strike K and compare them with your answers.

(b) Illustrate the result of Exercise 7.10 by repeating (a) with the interest rate changed to zero.

Exercise 7.18[†]

Consider Model 7.5 with one stock. The initial bond price is $B_0 = 1$, and the interest rate is $r = 5\%$.

(a) Show that this model is viable and complete by constructing an equivalent martingale measure for it.

(b) Calculate the fair price at time 0 of an American call option with strike 95 in this model, by using the pricing procedure for a general American option. Verify that this price is the same as the price of a European call option with the same strike.

(c) Find the fair price at time 0 of an American and European put option, both with strike 110, in this model. Indicate the situations in which it is optimal (that is, not disadvantageous) to exercise the American put option early if it has not already been exercised.

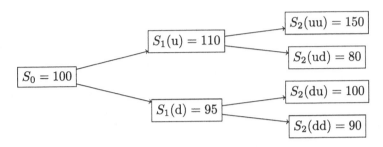

Model 7.5 Binary single-stock model in Exercise 7.18

7.3 Pricing American Options: Formal Theory

A rigorous discussion of fair pricing for American options in a general model (not necessarily complete as in the above informal theory) requires a precise definition of a fair price for these derivatives. For this it is necessary to think a little about what an American option X actually is for both owner and writer. Recall first the formal Definition 7.11 of a general American option in the previous section.

Definition 7.19

An *American option* is a non-negative adapted stochastic process $X = (X_t)_{t=0}^T$. The value $X_t(\omega)$ is called the *payoff* of the option at time t in scenario ω.

   The owner of an American option X has the right (but not the obligation) to receive the payoff X_t at any time $t \leq T$. The payoff can only be received once; after that the option is discarded. If the owner opts to receive the payoff at time t he is said to *exercise* the option at that time.

7.3.1 Stopping Times

The owner of an American option has purchased the opportunity to exercise the option at any time $\tau \leq T$ of his choosing. This *exercise time* may depend on the scenario ω; thus, either consciously or unconsciously an owner will operate an exercise strategy that is a *random time* τ. The idea is that $\tau(\omega)$ denotes the time the owner will exercise the option if in scenario ω.

By contrast, the liability of the writer of X to pay out when the owner exercises means that he must be prepared for any conceivable exercise strategy that the buyer may adopt.

Hence an important notion for both owner and writer is that of an *exercise time*, which is known as a *stopping time* in more general contexts. This is a random time τ that encapsulates the following important property: the choice whether or not to stop (that is, to exercise) at a given time t must depend only on the partial history $\omega \uparrow t$ and not the whole of the scenario ω. That is because it is only the partial history $\omega \uparrow t$ that is known to the owner at time t. This means that for any exercise time τ, if $\tau(\omega) = t$ and $\omega \uparrow t = \omega' \uparrow t$ then $\tau(\omega') = t$ also (because at time t it is impossible to distinguish between ω and ω').

The formal definition of a stopping time (which in our context we also call an *exercise time* or *exercise strategy* or *stopping rule*) for any discrete time setting is as follows.

Definition 7.20 (Stopping time)

A *stopping time* is a random variable τ that takes values in the set $\{0, 1, \ldots, T\}$, and is such that for every t

$$\{\omega \in \Omega : \tau(\omega) = t\} \in \mathcal{F}_t. \tag{7.4}$$

Remark 7.21

For all t, since $\mathcal{F}_t$ is an algebra, and $\mathcal{F}_s \subseteq \mathcal{F}_t$ if $s < t$, it follows that if τ is a stopping time then the sets $\{\tau \leq t\}$, and $\{\tau > t\}$ are also in $\mathcal{F}_t$ (where we use the probabilists' convention that $\{\tau \leq t\} = \{\omega : \tau(\omega) \leq t\}$ and similarly for $\{\tau > t\}$).

It is sufficient for τ to be a stopping time that either of these sets is in $\mathcal{F}_t$ for all t. To see the first of these for example, observe that

$$\{\tau \leq t\} = \bigcup_{s=0}^{s=t} \{\tau = s\}.$$

Conversely, if $\{\tau \leq t\} \in \mathcal{F}_t$ for all t then since $\{\tau = t\} = \{\tau \leq t\} \setminus \{\tau \leq t-1\}$ we have $\{\tau = t\} \in \mathcal{F}_t$.

Exercise 7.22[‡]

Prove that τ is a stopping time if and only if $\{\tau \geq t\} \in \mathcal{F}_{t-1}$ for each $t > 0$.

The simplest stopping time is a constant time $\tau(\omega) = t$ for all ω. Here is an example of a non-constant stopping time.

Example 7.23

Consider Model 7.1. Writing $\Omega := \{uu, ud, du, dd\}$, the filtration connected with this model is $\mathcal{F}_0, \mathcal{F}_1, \mathcal{F}_2$ where $\mathcal{F}_0 = \{\emptyset, \Omega\}$,

$$\mathcal{F}_1 = \{\emptyset, \{uu, ud\}, \{du, dd\}, \Omega\}$$

and $\mathcal{F}_2$ is the collection of all subsets of Ω.

Consider the random variable τ defined by

$$\tau(\omega) := \min\{t : S_t(\omega) \leq 95 \text{ or } t = 2\}.$$

It is easy to verify that $\tau(uu) = \tau(ud) = 2$ and $\tau(du) = \tau(dd) = 1$. Consequently,

$$\{\omega \in \Omega : \tau(\omega) = 0\} = \emptyset \in \mathcal{F}_0,$$

$$\{\omega \in \Omega : \tau(\omega) = 1\} = \{du, dd\} \in \mathcal{F}_1,$$

$$\{\omega \in \Omega : \tau(\omega) = 2\} = \{ud, uu\} \in \mathcal{F}_2.$$

So the random variable τ is a stopping time. On the other hand, defining

$$\tau'(\omega) := \max\{t \in \{0, 1, 2\} : S_t(\omega) \leq 105\}$$

for $\omega \in \Omega$ we see that $\tau'(uu) = 0$, $\tau'(ud) = \tau'(du) = 2$, and $\tau'(dd) = 2$. So τ' is *not* a stopping time because

$$\{\omega \in \Omega : \tau'(\omega) = 0\} = \{uu\} \notin \mathcal{F}_0.$$

Exercise 7.24[†]

Consider a Cox-Ross-Rubinstein model with $T = 3$. Determine whether each of the random times τ_1, τ_2 given by the following table is a stopping time.

ω	uuu	uud	udu	udd	duu	dud	ddu	ddd
$\tau_1(\omega)$	2	2	3	3	1	1	1	1
$\tau_2(\omega)$	2	2	3	1	1	1	1	1

The stopping time τ_* given by the next result was shown in Theorem 7.13 to be the first time where it is not disadvantageous to exercise an American option in a complete model. It plays an important role in subsequent development when the model is complete.

Theorem 7.25

Let X be an American option in a complete model, with corresponding process Z^X as defined in Theorem 7.13. Then the random time

$$\tau_*(\omega) = \min\{t : X_t(\omega) = Z^X_t(\omega)\}$$

is a stopping time.

Proof

Suppose that $\tau_*(\omega) = t$ for some t and ω. Then $X_s(\omega) < Z^X_s(\omega)$ for all $s < t$. If $\omega' \uparrow t = \omega \uparrow t$ then since Z^X and X are adapted, $X_s(\omega') = X_s(\omega)$ and $Z^X_s(\omega') = Z^X_s(\omega)$ for $s \le t$. Thus $X_s(\omega') < Z^X_s(\omega')$ if $s < t$ and $X_t(\omega') = Z^X_t(\omega')$. This means that $\tau_*(\omega') = t$. So $\{\omega \in \Omega | \tau_*(\omega) = t\} \in \mathcal{F}_t$ and τ_* is a stopping time. $\square$

We will need the following technical result concerning stopping times and martingales, which is a special case of a more general result. It tells us that for a $\mathbb{Q}$-martingale $M = (M_t)_{t=0}^T$ the equality $\mathbb{E}_\mathbb{Q}(M_t) = M_0$ extends to stopping times τ. We need the following notation: if Y is a process then Y_τ is the random variable defined for all ω by

$$Y_\tau(\omega) = Y_{\tau(\omega)}(\omega).$$

Theorem 7.26 (Doob's Optional Sampling Theorem: simple version)

If M_t is a $\mathbb{Q}$-martingale and τ is a stopping time, then $\mathbb{E}_\mathbb{Q}(M_\tau) = M_0$.

This result is a particular case of Theorem 7.70, which is proved in Section 7.6. The following exercise gives an easy corollary which will be needed on several occasions below.

Exercise 7.27[‡]

Suppose that Φ is a trading strategy in a viable model and τ is a stopping time such that $V^\Phi_\tau \ge 0$.

(a) Show that $V^\Phi_0 \ge 0$.

(b) Show that if in addition $V^\Phi_\tau(\omega) > 0$ for some ω then $V^\Phi_0 > 0$.

Hint. Apply Theorem 7.26 to the discounted value process $\bar{V}^\Phi$.

7.3.2 Definition of a Fair Price for an American Option

It is enough to define the fair price of an American option only at the initial time $t = 0$. The price at later nodes is then defined by considering the appropriate submodel. For the sake of simplicity, the definition will involve strategies that involve the purchase or sale of American options only at time $t = 0$, in contrast to the notion of a fair price for a European derivative, which involved extended strategies that allowed trading derivatives at all times. Note however that Exercise 5.71 shows that for European derivatives a similar restriction would have made no difference, giving further justification for the current restriction.

The idea underlying the definition is that a fair price for an American option is one that prevents extended arbitrage for either the owner or writer. For the *owner* extended arbitrage means a self-financing trading strategy in bond and stock together with an exercise strategy for the option that will give a risk free profit with positive probability. For the *writer* it means a self-financing strategy in bond and stock that will give a risk free profit with positive probability no matter what exercise strategy is adopted by the buyer. With this in mind we can move directly to the definition of a fair price for an American option without formally defining the notions of extended trading strategy or extended arbitrage opportunity (which is possible, but somewhat convoluted and not really helpful).

Definition 7.28 (Fair price of an American option)

Suppose that X is an American option priced at π.

(a) The price π is *unfair to the advantage of the writer* of X if there exists a trading strategy $\Phi = (\varphi_t)_{t=1}^{T}$ such that $V_0^{\Phi} = \pi$ and if τ is any stopping time, then $V_\tau^{\Phi} \geq X_\tau$ and $V_\tau^{\Phi}(\omega) > X_\tau(\omega)$ for at least one ω.

(b) The price π is *unfair to the advantage of the owner* of X if there exists a trading strategy $\Phi = (\varphi_t)_{t=1}^{T}$ and a stopping time τ such that $V_0^{\Phi} = \pi$ and $V_\tau^{\Phi} \leq X_\tau$, and $V_\tau^{\Phi}(\omega) < X_\tau(\omega)$ at least one ω.

(c) The price π is a *(simple) fair price* for X if it is not unfair to the advantage of either buyer or seller.

For the writer the idea underlying the above definition is this. Suppose that π is unfair to the advantage of the writer using the strategy $\Phi = (\varphi_t)_{t=0}^{T}$ with $V_0^{\Phi} = \pi$. If the option is priced at π, then a writer may sell it for π at time 0 and with the proceeds initiate the strategy Φ. At any stopping time τ chosen by the owner, the writer can liquidate the current portfolio φ_τ, with value $V_\tau^\Phi \geq X_\tau$, to meet his obligation without any loss. Since $V_\tau^\Phi(\omega) > X_\tau(\omega)$ for some ω this gives a profit of $V_\tau^\Phi(\omega) - X_\tau(\omega)$ in that scenario.

For the owner the idea is similar. Suppose that π is unfair to the advantage of the owner with trading strategy $\Phi = (\varphi_t)_{t=0}^T$ and stopping time τ as in the definition. If the option is bought at the price π, the owner can finance it by shorting the strategy Φ, with total outlay $\pi - V_0^\Phi = 0$. The owner should then exercise the option at time τ, and use the payoff X_τ to close the current portfolio $-\varphi_\tau$ without loss since $X_\tau - V_\tau^\Phi \geq 0$. Since $X_\tau(\omega) > V_\tau^\Phi(\omega)$ for some ω there is actually a profit of $X_\tau(\omega) - V_\tau^\Phi(\omega)$ in that scenario.

7.3.3 Replication and Hedging

Recall that for pricing a European derivative a key notion is replication (if it is attainable), and, in general, super- or sub-replication. For an American option X (even in a complete model) there is no question of replication, since the discounted value process $\bar{V}^\Phi$ of a self-financing trading strategy Φ is a martingale whereas in general $\bar{X}$ is not. We will see that the above definition of a fair price for an American option is, however, closely related to super- and sub-replication.

A writer wishing to protect or *hedge* his position will seek to construct a strategy Φ that will guard against the liability inherent in the option, namely his obligation to payout X_t whenever the owner (to whom he has sold the option) chooses to exercise. A good hedge would thus be a self-financing strategy Φ with minimum outlay such that $V_t^\Phi \geq X_t$ for all times t (equivalently, $V_\tau^\Phi \geq X_\tau$ for every stopping time τ); that is, a trading strategy that *super-replicates* X.

Hedging for an owner will be different. He will, implicitly or explicitly, adopt an exercise strategy τ, and this effectively makes the option into a derivative X_τ that pays out $X_\tau(\omega)$ in scenario ω; this is similar to a European derivative but with a random expiry date. A good hedge for any chosen τ will involve a self-financing strategy Ψ with minimum outlay such that $V_\tau^\Psi + X_\tau \geq 0$ for this stopping time τ, since by adopting such a strategy the owner will, at the exercise time, own assets with value $V_\tau^\Psi + X_\tau \geq 0$ and be protected; this motivates the idea of a trading strategy that *sub-replicates* X.

These considerations are made precise as follows.

Definition 7.29

Let X be an American option.

(a) A self-financing strategy Φ *super-replicates* X if $V_\tau^\Phi \geq X_\tau$ for every stopping time τ. It *strictly super-replicates* X if, in addition, for every stopping time τ, $V_\tau^\Phi(\omega) > X_\tau(\omega)$ for some ω.

(b) A self-financing strategy Φ together with a stopping time τ *sub-replicates* X if $V_\tau^\Phi \leq X_\tau$. We say that the pair (Φ, τ) is a *sub-replicating strategy* or a *sub-replicating pair* for X. The pair (Φ, τ) *strictly sub-replicates* X if $V_\tau^\Phi \leq X_\tau$ and $V_\tau^\Phi(\omega) < X_\tau(\omega)$ for some ω.

(c) A self-financing strategy Φ *sub-replicates* X if there is a stopping time τ such that (Φ, τ) *sub-replicates* X; Φ *strictly sub-replicates* X if there is a stopping time τ such that (Φ, τ) *strictly sub-replicates* X.

These notions provide an alternative way to define the fair price of an American option using the following observation, whose proof is left as an exercise.

Proposition 7.30

The following are equivalent.

(1) The price π for an American option X is unfair (that is, it is not fair).

(2) $\pi = V_0^\Phi$ for a strategy Φ that either strictly super-replicates X or strictly sub-replicates X.

Exercise 7.31[‡]

Prove Proposition 7.30.

In the case of an American option the counterpart of a replicating strategy is a *hedging strategy*; that is, one that is good for both writer and owner.

Definition 7.32

Let X be an American option.

(a) A pair (Φ, τ) is a *hedging pair* for X if Φ super-replicates X and (Φ, τ) sub-replicates X; that is

  (i) $V_\sigma^\Phi \geq X_\sigma$ for all stopping times σ.

  (ii) $V_\tau^\Phi = X_\tau$.

(b) A strategy Φ is a *hedging strategy* for X if there is a stopping time τ such that (Φ, τ) is a hedging pair for X.

(c) X is said to be *attainable* if it has a hedging strategy.

Remarks 7.33

(1) We do not use the term *replicating strategy*, because that suggests a strategy Φ such that $V_t^\Phi = X_t$ for all t, which is both unlikely to exist and (as we shall see) is not the appropriate notion needed.

(2) If (Φ, τ) is a hedging pair for an attainable American option X, then the writer would use Φ to protect his position. The owner would actually use $-\Phi$ (that is, he would short the strategy Φ) and exercise at the associated stopping time τ.

(3) We will see below that in a well defined sense, if (Φ, τ) is a hedging pair for X then Φ is a *minimal super-replicating strategy* for X, and τ is an *optimal stopping time* for X.

A simple example of an attainable American option is $X = V^\Phi$ where Φ is any trading strategy. A hedging pair is given by Φ along with *any* stopping time. Here is another example.

Example 7.34

Consider the model in Example 7.16 and the American option X discussed there. Then X is attainable using the strategy $\Phi = (\varphi_1, \varphi_2)$ and stopping time τ defined as follows.

$$\varphi_1 = \left(-177\tfrac{3}{5}, 2\tfrac{2}{5}\right), \qquad \varphi_2(u) = \left(-248, 3\tfrac{1}{13}\right), \qquad \varphi_2(d) = \left(-86\tfrac{2}{5}, 1\tfrac{4}{11}\right),$$
$$\tau(uu) = \tau(ud) = 1, \qquad \tau(du) = \tau(dd) = 2.$$

Then we have, for example

$$V_0^\Phi = -177\tfrac{3}{5} + 2\tfrac{2}{5} \times 100 = 62.4,$$
$$V_2^\Phi(ud) = -248 \times (1.25)^2 + 3\tfrac{1}{13} \times 143 = 52.5.$$

Calculating the other values of the process V^Φ similarly we can represent them in the tree given by Figure 7.6.

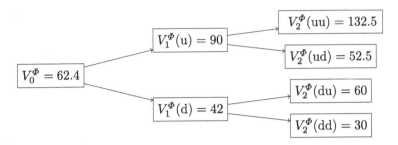

Figure 7.6 Value of the trading strategy Φ in Example 7.34

It is routine to check that Φ is self-financing and that $V^\Phi \geq X$ so Φ super-replicates X. To see that Φ also sub-replicates X by means of τ, we have

$$\tau(\omega) = 1 \quad \text{and} \quad V_1^\Phi(\omega) = 90 = X_1(\omega) \quad \text{if } \omega \in \{\text{uu}, \text{ud}\};$$

$$\tau(\text{du}) = 2 \quad \text{and} \quad V_2^\Phi(\text{du}) = 60 = X_2(\text{du});$$

$$\tau(\text{dd}) = 2 \quad \text{and} \quad V_2^\Phi(\text{dd}) = 30 = X_2(\text{dd}).$$

Thus $V_\tau^\Phi = X_\tau$ and so (Φ, τ) is a hedging pair for X.

Pricing theory for attainable American options follows a pattern similar to that for pricing attainable European derivatives; and we will see later that in a complete model *every* American option is attainable. The theory will also show how a hedging pair for an attainable American option (such as that in the above example) can be found.

7.4 Attainable American Options

7.4.1 Pricing Attainable American Options

In preparation for the main pricing result for attainable American options we have the following.

Theorem 7.35

Suppose that X is an attainable American option. If (Φ, τ) and (Φ', τ') are hedging pairs for X and $\mathbb{Q}, \mathbb{Q}'$ are equivalent martingale measures then

$$V_0^{\Phi'} = \mathbb{E}_{\mathbb{Q}'}(\bar{X}_{\tau'}) = V_0^\Phi = \mathbb{E}_{\mathbb{Q}}(\bar{X}_\tau).$$

Proof

Since $V_\tau^\Phi = X_\tau$, Theorem 7.26 shows that $V_0^\Phi = \mathbb{E}_\mathbb{Q}(\bar{V}_\tau^\Phi) = \mathbb{E}_\mathbb{Q}(\bar{X}_\tau)$ and $V_0^{\Phi'} = \mathbb{E}_{\mathbb{Q}'}(\bar{V}_{\tau'}^{\Phi'}) = \mathbb{E}_{\mathbb{Q}'}(\bar{X}_{\tau'})$

To see that $V_0^\Phi = V_0^{\Phi'}$ we have, using the fact that Φ' super-replicates X,

$$V_0^\Phi = \mathbb{E}_\mathbb{Q}(\bar{X}_\tau) \le \mathbb{E}_\mathbb{Q}(\bar{V}_\tau^{\Phi'}) = V_0^{\Phi'}$$

and similarly $V_0^{\Phi'} \le V_0^\Phi$ which gives $V_0^{\Phi'} = V_0^\Phi$.   □

The Law of One Price extends to attainable American options as follows.

Theorem 7.36 (Law of One Price for American options)

An attainable American X option has a unique fair price π_X given by

$$\pi_X := V_0^\Phi = \mathbb{E}_\mathbb{Q}(\bar{X}_\tau) \tag{7.5}$$

for any hedging pair (Φ, τ) and any equivalent martingale measure $\mathbb{Q}$.

Proof

Note that Theorem 7.35 shows that π_X is well defined by (7.5).

Let (Φ, τ) be a hedging pair for X. By definition $V_\sigma^\Phi \ge X_\sigma$ for every stopping time σ and $V_\tau^\Phi = X_\tau$; and $\pi_X = V_0^\Phi = \mathbb{E}_\mathbb{Q}(\bar{X}_\tau)$.

First we see that any price other than π_X is unfair. If $\pi' > \pi_X$ then increase the bond holding in Φ to give a strictly super-replicating strategy Φ' for X with $V_0^{\Phi'} = \pi'$, so π' is unfair. In detail, if $\Phi = (x_t, y_t)_{t=0}^T$ then set $\Phi' = (x_t', y_t)_{t=0}^T$ with $x_t' = x_t + (\pi' - \pi_X)/B_0$.

If $\pi'' < \pi_X$ then *decreasing* the bond holding in Φ by $(\pi_X - \pi'')/B_0$ gives a strictly sub-replicating strategy Φ'' with $V_0^{\Phi''} = \pi''$, so π'' is unfair.

It remains to show that the price π_X is fair. First, if it is unfair to the advantage of the owner, then there is a strictly sub-replicating pair (Ψ, σ) for X such that $V_0^\Psi = \pi_X$ and $V_\sigma^\Psi \le X_\sigma \le V_\sigma^\Phi$ and $V_\sigma^\Psi(\omega) < X_\sigma(\omega) \le V_\sigma^\Phi(\omega)$ for some ω. Thus

$$V_\sigma^{\Phi-\Psi} = V_\sigma^\Phi - V_\sigma^\Psi \ge 0$$

and

$$V_\sigma^{\Phi-\Psi}(\omega) = V_\sigma^\Phi(\omega) - V_\sigma^\Psi(\omega) > 0$$

whereas $V_0^{\Phi-\Psi} = V_0^\Phi - V_0^\Psi = 0$. From Exercise 7.27 this is impossible.

By a similar argument, if the price π_X is unfair to the advantage of the writer, take a strictly super-replicating strategy Ψ with $\pi_X = V_0^\Psi$. Then $V_\tau^\Psi \geq X_\tau = V_\tau^\Phi$ and $V_\tau^\Psi(\omega) > X_\tau(\omega)$ for some ω. So

$$V_\tau^{\Psi-\Phi} = V_\tau^\Psi - V_\tau^\Phi \geq 0$$

and

$$V_\tau^{\Psi-\Phi}(\omega) = V_\tau^\Psi(\omega) - V_\tau^\Phi(\omega) > 0$$

which is impossible since $V_0^{\Psi-\Phi} = V_0^\Psi - V_0^\Phi = 0$. Thus the price π_X is fair. $\square$

Example 7.37

Consider the attainable American option in Example 7.34. According to the above result its unique fair price is $V_0^\Phi = 62.4$, which agrees with the fair price for this option found informally in Example 7.16.

7.4.2 Optimal Stopping and Minimal Super-replication for Attainable American Options

The results in this section will show that a hedging pair (Φ, τ) for an attainable American option combines the ideas of a *minimal super-replicating strategy* and an *optimal stopping time*.

Theorem 7.38

If Ψ super-replicates an attainable American option X then $V_0^\Psi \geq \pi_X$.

Proof

Take any hedging pair (Φ, τ) and equivalent martingale measure $\mathbb{Q}$. Then

$$V_0^\Psi = \mathbb{E}_\mathbb{Q}\big(\bar{V}_\tau^\Psi\big) \geq \mathbb{E}_\mathbb{Q}(\bar{X}_\tau) = \pi_X.$$

$\square$

This motivates the following definition.

Definition 7.39 (Minimal super-replicating strategy)

Let X be an attainable American option. A trading strategy Ψ is a *minimal super-replicating strategy for* X if it super-replicates X and $V_0^\Psi = \pi_X$.

The notion of an optimal stopping time for an attainable American option (defined below) derives from the following easy result.

Theorem 7.40

Let X be an attainable American option and σ any stopping time. Then for any equivalent martingale measure $\mathbb{Q}$

$$\mathbb{E}_\mathbb{Q}(\bar{X}_\sigma) \leq \pi_X.$$

Proof

Fix a hedging pair (Φ, τ) for X. Then

$$\mathbb{E}_\mathbb{Q}(\bar{X}_\sigma) \leq \mathbb{E}_\mathbb{Q}(\bar{V}_\sigma^\Phi) = V_0^\Phi = \pi_X.$$

$\square$

Theorem 7.40 motivates the following definition.

Definition 7.41 (Optimal stopping time)

Let X be an attainable American option. A stopping time τ is said to be *optimal* for X if $\mathbb{E}_\mathbb{Q}(\bar{X}_\tau) = \pi_X$ for some equivalent martingale measure $\mathbb{Q}$.

The next result shows that a hedging pair results from any choice of minimal super-replicating Φ and any choice of optimal stopping time.

Theorem 7.42

If X is an attainable American option then the following are equivalent for any strategy Φ and stopping time τ:

(1) (Φ, τ) is a hedging pair for X.

(2) Φ is a minimal super-replicating strategy for X and τ is an optimal stopping time for X.

Proof

We know from Theorem 7.35 and the definition of π_X that (1) implies (2).

Conversely, suppose that Φ is a minimal super-replicating strategy and τ is an optimal stopping time for X. Take an equivalent martingale measure $\mathbb{Q}$ such that $\pi_X = \mathbb{E}_{\mathbb{Q}}(\bar{X}_\tau)$. Using $X_\tau \le V_\tau^\Phi$ gives

$$\pi_X = \mathbb{E}_{\mathbb{Q}}(\bar{X}_\tau) \le \mathbb{E}_{\mathbb{Q}}(\bar{V}_\tau^\Phi) = \pi_X$$

and so $\mathbb{E}_{\mathbb{Q}}(\bar{X}_\tau) = \mathbb{E}_{\mathbb{Q}}(\bar{V}_\tau^\Phi)$. Since $X_\tau \le V_\tau^\Phi$ we must have $X_\tau = V_\tau^\Phi$, so (Φ, τ) is a hedging pair for X. $\qquad\qquad\square$

The following corollary is immediate.

Corollary 7.43

If X is an attainable American option and Φ is a strategy, then the following are equivalent:

(1) Φ is a minimal super-replicating strategy for X.

(2) Φ is a hedging strategy for X.

We can now see that there are several alternative ways to characterize optimal stopping times for an attainable American option X.

Theorem 7.44

Let X be an attainable American option and τ a stopping time. The following are equivalent.

(1) τ is optimal for X (that is, $\mathbb{E}_{\mathbb{Q}}(\bar{X}_\tau) = \pi_X$ for *some* equivalent martingale measure $\mathbb{Q}$).

(2) $\mathbb{E}_{\mathbb{Q}}(\bar{X}_\tau) = \pi_X$ for *every* equivalent martingale measure $\mathbb{Q}$.

(3) $X_\tau = V_\tau^\Phi$ for every hedging strategy Φ for X.

(4) $X_\tau = V_\tau^\Phi$ for some hedging strategy Φ for X.

Proof

Suppose that τ is optimal. Take any hedging strategy Φ. Then $\mathbb{E}_{\mathbb{Q}}(\bar{X}_\tau) = \pi_X$ for *every* equivalent martingale measure $\mathbb{Q}$ (from Theorem 7.35 and the definition of π_X), so (1) implies (2).

Theorem 7.42 showed that (Φ, τ) is a hedging pair for X, so $X_\tau = V_\tau^\Phi$, which shows that (1) implies (3).

Clearly (2) implies (1) and (3) implies (4). Finally, if $X_\tau = V_\tau^\Phi$ for some Φ that super-replicates X then (Φ, τ) is a hedging pair, so τ is optimal. This shows that (4) implies (1). □

To conclude the discussion of attainable American options we show that there is always a least and (in Exercise 7.48) a greatest optimal stopping time. First note that all hedging strategies have the same value process up to any optimal stopping time.

Theorem 7.45

Suppose that τ is an optimal stopping time and Φ and Φ' are hedging strategies for an attainable American option X. Then for all ω, if $t \leq \tau(\omega)$ then

$$V_t^\Phi(\omega) = V_t^{\Phi'}(\omega).$$

Proof

From Theorem 7.44 we know that $V_\tau^\Phi = X_\tau = V_\tau^{\Phi'}$. Fix ω and $t \leq \tau(\omega)$. Then if $\omega' \uparrow t = \omega \uparrow t$ we must have $t \leq \tau(\omega')$ also. Now we can apply Theorem 7.26 to the submodel Ω_λ based at the node $\lambda = \omega \uparrow t$, which has initial time t and induced probability $\mathbb{Q}_\lambda$ given by $\mathbb{Q}_\lambda(\omega') = \mathbb{Q}(\omega')/\mathbb{Q}(\Omega_\lambda)$ for $\omega' \in \Omega_\lambda$. Since $t \leq \tau(\omega')$ for all $\omega' \in \Omega_\lambda$ the restriction of τ to Ω_λ gives a stopping time on Ω_λ. Then, considering the restriction of Φ to Ω_λ, applying Theorem 7.26 to this submodel gives

$$\bar{V}_t^\Phi(\omega) = \mathbb{E}_{\mathbb{Q}_\lambda}\left(\bar{V}_\tau^\Phi\right) = \mathbb{E}_{\mathbb{Q}_\lambda}\left(\bar{V}_\tau^{\Phi'}\right) = \bar{V}_t^{\Phi'}(\omega).$$

□

Theorem 7.46

Let X be an attainable American option. Then

$$\tau_{\min} := \min\{t : X_t = V_t^\Phi \text{ for every hedging strategy } \Phi\} \tag{7.6}$$

is the smallest optimal stopping time. We also have

$$\tau_{\min} = \min\{t : X_t = V_t^\Phi \text{ for some hedging strategy } \Phi\} \tag{7.7}$$

and, for every hedging strategy Ψ

$$\tau_{\min} = \min\{t : X_t = V_t^\Psi\}. \tag{7.8}$$

Each of the minima on the right is of course a *random* time. In the second definition, the minimum may be attained by different Φ for different ω.

Proof

Showing that $\tau_{\min}$ is a stopping time is left as an exercise. Theorem 7.44 shows that it is optimal. Let τ be any other optimal stopping time. Then by Theorem 7.44, $V_\tau^\Phi = X_\tau$ for every hedging strategy Φ, so $\tau_{\min} \leq \tau$. Hence $\tau_{\min}$ is the smallest optimal stopping time.

If we write $\tau'_{\min}$ for the alternative definition (7.7), then clearly $\tau'_{\min} \leq \tau_{\min}$. Now consider any time $t < \tau_{\min}$. Then there is some hedging strategy Φ such that $X_t \neq V_t^\Phi$. Since $\tau_{\min}$ is optimal, Theorem 7.45 shows that $X_t \neq V_t^\Psi$ for *every* hedging strategy Ψ and so $t < \tau'_{\min}$. Hence $\tau_{\min} \leq \tau'_{\min}$ and so $\tau'_{\min} = \tau_{\min}$.

If we write $\tau^\Psi_{\min}$ for the alternative definition (7.8) then we have $\tau'_{\min} \leq \tau^\Psi_{\min} \leq \tau_{\min}$, so $\tau^\Psi_{\min} = \tau_{\min}$ also. $\square$

Exercise 7.47‡

Show that if X is an attainable American option the random time

$$\tau := \min\{t : V_t^\Phi = X_t \text{ for every hedging strategy } \Phi\}$$

is a stopping time.

Exercise 7.48‡

Show that if τ_1 and τ_2 are optimal stopping times for an attainable American option X then the maximum $\sigma = \max\{\tau_1, \tau_2\}$ is also a stopping time that is optimal. Deduce that there is a greatest optimal stopping time for X.

Exercise 7.49‡

Let D be a nonnegative European derivative and define a corresponding American option X^D by $X_T^D = D$ and $X_t^D = 0$ if $t < T$.

(a) Show that X_D is attainable if and only if D is attainable.

(b) Show that if X_D is attainable then its unique fair price π_X is the unique fair price D_0 for D.

(c) Investigate the optimal stopping times for X_D if it is attainable and compute the least and greatest stopping times.

Fair pricing for non-attainable American options will be discussed in the final section of this chapter, after investigating the special case of complete models, where it turns out that all American options are attainable.

7.5 Pricing American Options in a Complete Model

Here we show that in a *complete* model every American X option is attainable, and its unique fair price π_X is the price Z_0^X derived informally in Section 7.2. The completeness of the model allows the definition of the process Z^X that was introduced in the informal discussion in Section 7.2, and this gives extra information, including information about optimal exercise times.

Throughout this section we assume that the model is complete, with unique equivalent martingale measure $\mathbb{Q}$, and that an American option X is given. Recall the process $Z^X = (Z_t^X)_{t=0}^T$ defined in Section 7.2, which we denote by Z^X here to emphasize that it depends on X:

$$Z_T^X := X_T \tag{7.9}$$

$$Z_t^X := \max\{X_t, \mathbb{E}_{\mathbb{Q}}(\hat{Z}_{t+1}^X | \mathcal{F}_t)\} \quad \text{if } t < T. \tag{7.10}$$

The discounted version of this process has an important property that is a generalization of the notion of a martingale: it is a *supermartingale*.

Definition 7.50

A *supermartingale* is an adapted stochastic process $Y = (Y_t)_{t=0}^T$ such that whenever $s \le t$

$$\mathbb{E}_{\mathbb{Q}}(Y_t | \mathcal{F}_s) \le Y_s.$$

Strictly we should call this a $\mathbb{Q}$-supermartingale, but since there is only one relevant probability at the moment it is unambiguous. It is helpful to think of a supermartingale as a process that is "stochastically decreasing". In a discrete time setting, for an adapted process to be a supermartingale it is sufficient that $\mathbb{E}_{\mathbb{Q}}(Y_{t+1} | \mathcal{F}_t) \le Y_t$ for all $t < T$. Of course a martingale is also a supermartingale.

Here is why supermartingales are relevant for our discussion.

Theorem 7.51

The process $\bar{Z}^X$ is a supermartingale and is the smallest supermartingale that dominates $\bar{X}$; that is, such that $\bar{Z}^X \ge \bar{X}$.

Proof

Clearly $\bar{Z}_T^X = \bar{X}_T$. For $t < T$ the definition of Z_t^X is equivalent to

$$\bar{Z}_t^X = \max\{\bar{X}_t, \mathbb{E}_{\mathbb{Q}}(\bar{Z}_{t+1}^X | \mathcal{F}_t)\}. \tag{7.11}$$

Thus $\bar{Z}_t^X \geq \mathbb{E}_{\mathbb{Q}}(\bar{Z}_{t+1}^X | \mathcal{F}_t)$ so that $\bar{Z}$ is a supermartingale, and $\bar{Z}^X \geq \bar{X}$.

We show now by backwards induction that if Y is any other supermartingale that dominates $\bar{X}$ then $Y_t \geq \bar{Z}_t^X$ for all t. At time T

$$Y_T \geq \bar{X}_T = \bar{Z}_T^X.$$

For any $t < T$, the inequality $Y_{t+1} \geq \bar{Z}_{t+1}^X$ means that

$$Y_t \geq \mathbb{E}_{\mathbb{Q}}(Y_{t+1} | \mathcal{F}_t) \geq \mathbb{E}_{\mathbb{Q}}(\bar{Z}_{t+1}^X | \mathcal{F}_t).$$

Since $Y_t \geq \bar{X}_t$ it follows that

$$Y_t \geq \max\{\bar{X}_t, \mathbb{E}_{\mathbb{Q}}(\bar{Z}_{t+1}^X | \mathcal{F}_t)\} = \bar{Z}_t^X$$

which concludes the inductive step.   □

Remark 7.52

In the general theory of supermartingales the process $\bar{Z}^X$ is called the *Snell envelope of* $\bar{X}$ (named after the mathematician who first used it).

It is worth noting the following corollary, which tells us in particular that the initial value of a super-replicating strategy is at least as great as Z_0^X.

Corollary 7.53

A trading strategy Φ super-replicates X if and only if $V^\Phi \geq Z^X$.

Exercise 7.54[‡]

Prove Corollary 7.53.

The following technical result applied to the supermartingale $\bar{Z}^X$ will enable us to show that any American option in a complete model X is attainable. It is a special case of a more general result, and easy to prove for any discrete time situation.

Theorem 7.55 (Doob Decomposition)

If Y is a supermartingale there is a unique martingale $M = (M_t)_{t=0}^T$ and a predictable non-decreasing process $A = (A_t)_{t=0}^T$ such that $A_0 = 0$ and

$$Y = M - A.$$

In particular $Y_0 = M_0$.

Proof

Set $A_0 := 0$ and for $t > 0$ define

$$A_t := A_{t-1} + Y_{t-1} - \mathbb{E}_{\mathbb{Q}}(Y_t | \mathcal{F}_{t-1}). \qquad (7.12)$$

Then $A_t \geq A_{t-1}$ since Y is a supermartingale, so A is increasing. Clearly A_t is $\mathcal{F}_{t-1}$-measurable so A is predictable. Now define

$$M := Y + A.$$

To see that M is a martingale, take any $t > 0$. Using (7.12) we have

$$\begin{aligned}
\mathbb{E}_{\mathbb{Q}}(M_t | \mathcal{F}_{t-1}) &= \mathbb{E}_{\mathbb{Q}}(Y_t + A_t | \mathcal{F}_{t-1}) \\
&= \mathbb{E}_{\mathbb{Q}}(Y_t | \mathcal{F}_{t-1}) + \mathbb{E}_{\mathbb{Q}}(A_t | \mathcal{F}_{t-1}) \\
&= \mathbb{E}_{\mathbb{Q}}(Y_t | \mathcal{F}_{t-1}) + A_t \\
&= (A_{t-1} + Y_{t-1} - A_t) + A_t \\
&= M_{t-1}.
\end{aligned}$$

□

Exercise 7.56[‡]

Show that the above decomposition, known as the *Doob decomposition* of Y, is unique.

The Doob decomposition of $\bar{Z}^X$ is the key to the following theorem. First recall the stopping time

$$\tau_*(\omega) = \min\{t : Z_t^X(\omega) = X_t(\omega)\} \qquad (7.13)$$

defined in Theorem 7.25.

Theorem 7.57

Let X be an American option in a complete model. Then X is attainable. Its unique fair price is $\pi_X = Z_0^X$ and $\tau_* = \tau_{\min}$.

Proof

Take the martingale M and predictable increasing process A given by the Doob decomposition of $\bar{Z}^X$ (Theorem 7.55). Then

$$M = \bar{Z}^X + A.$$

The model is complete so there is a strategy Φ with $V_T^\Phi = (1 + r)^T M_T$. Then, for any t

$$\bar{V}_t^\Phi = \mathbb{E}_\mathbb{Q}\big(\bar{V}_T^\Phi | \mathcal{F}_t\big) = \mathbb{E}_\mathbb{Q}(M_T | \mathcal{F}_t) = M_t. \tag{7.14}$$

This gives

$$V_t^\Phi = (1 + r)^t M_t = Z_t^X + (1 + r)^t A_t \geq Z_t^X \geq X_t \tag{7.15}$$

so Φ super-replicates X. Note that $V_0^\Phi = Z_0^X$ since $A_0 = 0$.

For any ω, if $t < \tau_*$ the definition of Z^X gives

$$Z_t^X(\omega) > X_t(\omega)$$

and so

$$\bar{Z}_t^X(\omega) = \mathbb{E}_\mathbb{Q}\big(\bar{Z}_{t+1}^X | \mathcal{F}_t\big)(\omega).$$

Consequently, whenever $t < \tau_*$, using the fact that $A_{t+1} = M_{t+1} - \bar{Z}_{t+1}^X$ and A_{t+1} is $\mathcal{F}_t$-measurable we have

$$A_{t+1} = \mathbb{E}_\mathbb{Q}(A_{t+1} | \mathcal{F}_t) = \mathbb{E}_\mathbb{Q}\big(M_{t+1} - \bar{Z}_{t+1}^X | \mathcal{F}_t\big) = M_t - \bar{Z}_t^X = A_t.$$

Hence $0 = A_0 = A_1 = \cdots = A_{\tau_*}$ and (7.15) gives $V_t^\Phi = Z_t^X$ whenever $t \leq \tau_*$. In particular

$$V_{\tau_*}^\Phi = Z_{\tau_*}^X = X_{\tau_*},$$

so (Φ, τ_*) is a hedging pair for X. Thus X is attainable, and its fair price is $\pi_X = V_0^\Phi = Z_0^X$.

From Theorem 7.42, τ_* is optimal. Any other optimal stopping time has $X_\tau = V_\tau^\Phi \geq Z_\tau^X \geq X_\tau$ and so $X_\tau = Z_\tau^X$, which means that $\tau \geq \tau_*$ from the definition of τ_*. Hence $\tau_* = \tau_{\min}$. $\qquad\square$

Example 7.58

Consider the American option X discussed in Example 7.16, which was shown to be attainable in Example 7.34. Here is how we can use Theorem 7.57 to find a hedging pair (Φ, τ) for X. Note that the risk neutral probabilities at each node are $\left(\frac{3}{4}, \frac{1}{4}\right)$.

First we must obtain the Doob decomposition

$$\bar{Z}^X = M - A$$

of the discounted process $\bar{Z}^X$, using the formula (7.12). We found Z^X in Example 7.16. Then we obtain $A_0 = 0$ and

$$A_1 = A_0 + \bar{Z}_0^X - \mathbb{E}_\mathbb{Q}\big(\bar{Z}_1^X | \mathcal{F}_0\big) = 0 + 62.4 - \left(\tfrac{3}{4} \times 72 + \tfrac{1}{4} \times 33.6\right) = 0,$$
$$A_2(\mathrm{u}) = A_1 + \bar{Z}_1^X(\mathrm{u}) - \mathbb{E}_\mathbb{Q}\big(\bar{Z}_2^X | \mathcal{F}_1\big)(\mathrm{u}) = 0 + 72 - \left(\tfrac{3}{4} \times 64 + \tfrac{1}{4} \times 12.8\right) = 20.8,$$
$$A_2(\mathrm{d}) = A_1 + \bar{Z}_1^X(\mathrm{d}) - \mathbb{E}_\mathbb{Q}\big(\bar{Z}_2^X | \mathcal{F}_1\big)(\mathrm{d}) = 0 + 33.6 - \left(\tfrac{3}{4} \times 38.4 + \tfrac{1}{4} \times 19.2\right) = 0.$$

From this we obtain $M = \bar{Z}^X + A$. Figure 7.7 gives the values of $\bar{Z}^X$, A and M.

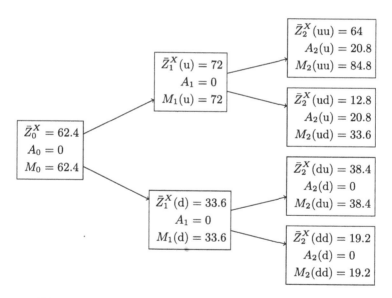

Figure 7.7 The Doob decomposition in Example 7.58

We can now use the techniques of Chapter 4 to obtain the unique self-financing strategy that replicates $(1 + r)^2 M_2$, and this is the strategy Φ that was given in Example 7.34. The above theorem shows that

$$V_t^\Phi = (1 + r)^t M_t \geq Z_t^X \geq X_t$$

for all t, which can be checked with the values in Example 7.34.

We have $Z_0^X \neq X_0$, $Z_1^X(\mathrm{u}) = X(\mathrm{u})$ and $Z_1^X(\mathrm{d}) \neq X(\mathrm{d})$ so the stopping time $\tau_{\min} = \tau_*$ defined by (7.13) is $\tau(\mathrm{uu}) = \tau(\mathrm{ud}) = 1$; $\tau(\mathrm{du}) = \tau(\mathrm{dd}) = 2$ as given in Example 7.34.

Exercise 7.59[†]

For the model of Exercise 7.17 and a put option X with strike $K = 102$ as in that exercise, construct the Doob decomposition of the process $\bar{Z}^X$. Use it to obtain the hedging strategy $\varPhi$ and stopping time τ_* that is given by the proof of Theorem 7.57.

Hint. Although the stock price process, the option and the process Z^X may be represented by a recombinant tree, the processes A and M will not necessarily be recombinant, so you should work with the full tree with eight branches.

We conclude this section with some further results that can be established with the extra machinery available in a complete model.

The proof of Theorem 7.57 gave a trading strategy $\varPhi$ that replicates the derivative X_{τ_*}; that is, $V_{\tau_*}^\varPhi = X_{\tau_*}$. The completeness of the model means that for any stopping time the derivative X_τ can be replicated in the following sense.

Theorem 7.60

If X is an American option in a complete model, for any stopping time τ there exists a self-financing trading strategy $\varPhi$ such that

$$V_\tau^\varPhi = X_\tau. \tag{7.16}$$

The initial value of this strategy $V_0^\varPhi$ satisfies

$$V_0^\varPhi = \mathbb{E}_\mathbb{Q}(\bar{X}_\tau) \leq Z_0^X.$$

Proof

Consider the derivative with payoff D at time T given by

$$D := (1+r)^{T-\tau} X_\tau.$$

Since the model is complete, there is a replicating strategy $\varPhi$ for D. This means that

$$V_T^\varPhi = (1+r)^{T-\tau} X_\tau.$$

To verify (7.16), fix any ω, and suppose that $\tau(\omega) = t$ and $\lambda = \omega \uparrow t$. Then for all $\omega' \in \Omega_\lambda$ we have $\tau(\omega') = \tau(\omega) = t$ since τ is a stopping time, and $X_t(\omega') = X_t(\omega)$ since X is adapted. Thus, for all $\omega' \in \Omega_\lambda$

$$D(\omega') = (1+r)^{T-t} X_t(\omega) = D(\omega) \qquad (7.17)$$

meaning that D is constant with value $(1+r)^{T-t} X_t(\omega)$ on Ω_λ. Hence

$$\begin{aligned}
V_\tau^\Phi(\omega) = V_t^\Phi(\lambda) &= \mathbb{E}_{\mathbb{Q}}\big((1+r)^{t-T} V_T^\Phi | \mathcal{F}_t\big)(\lambda) \\
&= \mathbb{E}_{\mathbb{Q}}\big((1+r)^{t-T} D | \mathcal{F}_t\big)(\lambda) \\
&= X_t(\omega) = X_\tau(\omega)
\end{aligned}$$

using (7.17). Theorem 7.40 then gives

$$V_0^\Phi = \mathbb{E}_{\mathbb{Q}}\big(\bar{V}_\tau^\Phi\big) = \mathbb{E}_{\mathbb{Q}}(\bar{X}_\tau) \le \pi_X = Z_0^X.$$

$\square$

In a complete model the process Z^X allows us to identify the greatest optimal stopping time that was shown to exist in Exercise 7.48 for an attainable American option.

Theorem 7.61

Let X be an American option in a complete model.

(1) The random time

$$\tau_{\max} := \begin{cases} T & \text{if } \mathbb{E}_{\mathbb{Q}}(\bar{Z}_{t+1}^X | \mathcal{F}_t) = \bar{Z}_t^X \text{ for all } t < T, \\ \min\{t : \mathbb{E}_{\mathbb{Q}}(\bar{Z}_{t+1}^X | \mathcal{F}_t) < \bar{Z}_t^X\} & \text{otherwise,} \end{cases}$$

(7.18)

is the greatest optimal stopping time for X.

(2) A stopping time τ is optimal if and only if $\tau \le \tau_{\max}$ and $Z_\tau^X = X_\tau$.

Proof

Exercise 7.62 below shows that $\tau_{\max}$ is a stopping time. We show next that $\tau_{\max}$ is optimal.

Note that from the definition we must have $\bar{Z}_{\tau_{\max}}^X = \bar{X}_{\tau_{\max}}$ and, for $t < \tau_{\max}$

$$\bar{Z}_t^X = \mathbb{E}_{\mathbb{Q}}\big(\bar{Z}_{t+1}^X | \mathcal{F}_t\big).$$

Take the Doob decomposition $\bar{Z}^X = M - A$ and let Φ be the hedging strategy for X given in the proof of Theorem 7.57. Arguing as in the proof of Theorem 7.57, we deduce that $A_t = 0$ for $t \leq \tau_{\max}$ and so

$$V_{\tau_{\max}}^\Phi = Z_{\tau_{\max}}^X = X_{\tau_{\max}}$$

so that $(\Phi, \tau_{\max})$ is a hedging pair and, by Theorem 7.42, $\tau_{\max}$ is optimal.

The proof of (1) is completed by showing that $\tau_{\max}$ is the greatest optimal stopping time. If τ is any optimal stopping time then $X_\tau = V_\tau^\Phi$ and so

$$X_\tau = V_\tau^\Phi \geq Z_\tau^X \geq X_\tau.$$

Hence $V_\tau^\Phi = Z_\tau^X$ and $A_\tau = 0$, so $A_t = 0$ and $\bar{Z}_t^X = M_t$ for all $t \leq \tau$. Consequently, if $t < \tau$ then $\bar{Z}_t^X = M_t = \mathbb{E}_\mathbb{Q}(M_{t+1}|\mathcal{F}_t) = \mathbb{E}_\mathbb{Q}(\bar{Z}_{t+1}^X|\mathcal{F}_t)$ and so $t < \tau_{\max}$. Hence $\tau \leq \tau_{\max}$.

For (2), suppose that $\tau \leq \tau_{\max}$ and $Z_\tau^X = X_\tau$. We saw above that for $t \leq \tau_{\max}$ we have $V_t^\Phi = Z_t^X$. The extra condition $Z_\tau^X = X_\tau$ means that $V_\tau^\Phi = X_\tau$ and so τ is optimal.   □

Exercise 7.62[‡]

Show that the random time defined by equation (7.18) is a stopping time.

Remark 7.63

The definition of $\tau_{\max}$ means that $\bar{Z}^X$ is actually a *martingale* "up to the stopping time $\tau_{\max}$". By this we mean that for any ω and $0 < t \leq \tau_{\max}(\omega)$

$$\mathbb{E}_\mathbb{Q}(\bar{Z}_t^X|\mathcal{F}_{t-1})(\omega) = \bar{Z}_t^X(\omega).$$

A more sophisticated way to say this is that the stopped process $(\bar{Z}^X)^{\tau_{\max}}$ is a martingale where, for any process Y and stopping time τ, the *stopped process* $Y^\tau = (Y_t^\tau)_{t=0}^T$ is defined by $Y_t^\tau(\omega) = Y_{\min\{t, \tau(\omega)\}}(\omega)$ for all t. That is, Y^τ is the same as Y for $t \leq \tau$ and is constant for $t \geq \tau$.

Exercise 7.64[†]

Consider Model 7.8 with two periods, one stock and a risk-free asset with initial price $B_0 = 1$ and interest rate $r = 5\%$.

In this model, find the fair price of an American option X with payoff function

$$X_t(\omega) := \left[S_t(\omega) - \tfrac{1}{t+1} \sum_{s=0}^{t} S_s(\omega)(1+r)^{t-s} \right]_+$$

for $\omega \in \Omega$ and $t \leq T$. Give the optimal exercise strategy $\tau_{\max}$ for the holder of this option, and construct a self-financing strategy Φ with $V^{\Phi}_{\tau_{\max}} = X_{\tau_{\max}}$.

Hint. The payoff is not path-independent.

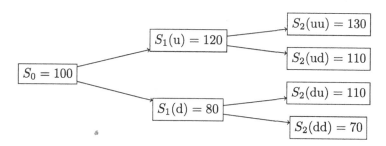

Model 7.8   Binary single-stock model in Exercise 7.64

Exercise 7.65[†]

Find Z_0^X, a hedging strategy and the stopping times $\tau_{\min}$ and $\tau_{\max}$ for the American option given in Figure 7.9 in the Cox-Ross-Rubinstein model with parameters $T = 3$, $r = 0$, $S_0 = 100$, $u = 1.2$, $d = 0.6$.

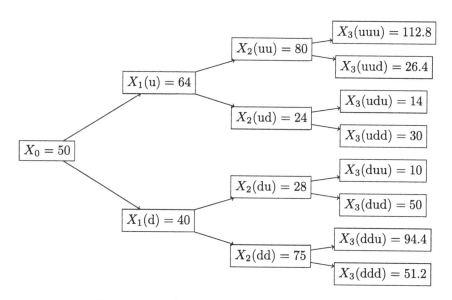

Figure 7.9   American option in Exercise 7.65

7.6 Pricing American Options in an Incomplete Model

In an incomplete model there will be American options that are not attainable; for these the pricing theory for European derivatives leads us to expect that there is not a unique fair price. In fact, there is the question of whether there is a fair price at all. We will see in this section that as with European derivatives the set of fair prices F_X for a non-attainable American option X is a non-empty interval; but we can only show that $F_X = (\pi_X^b, \pi_X^a)$ or $F_X = [\pi_X^b, \pi_X^a)$ and both of these possibilities can occur. Here of course the numbers π_X^a and π_X^b have to be defined appropriately.

For this section the model is a general viable discrete time model as in Chapters 4 and 5, so it may be incomplete and there will be a collection Q of equivalent martingale measures. Fundamental to the theory of pricing an American option X in any such model is the following process that is a natural generalization of the process Z_t^X in the previous section.

Definition 7.66

Let X be an American option. Define the adapted process $Z^X = (Z_t^X)_{t=0}^T$ by

$$Z_T^X = X_T$$

and, for $t < T$

$$Z_t^X = \max\left\{ X_t, \sup_{\mathbb{Q} \in \mathcal{Q}} \mathbb{E}_{\mathbb{Q}}(\hat{Z}_{t+1}^X | \mathcal{F}_t) \right\}.$$

Let us call a process Y a $\mathcal{Q}$-*supermartingale* if it is a $\mathbb{Q}$-supermartingale for every $\mathbb{Q} \in \mathcal{Q}$. Theorem 7.51 and its proof can be generalized to give the following result.

Theorem 7.67

The process $\bar{Z}^X$ is a $\mathcal{Q}$-supermartingale and is the smallest $\mathcal{Q}$-supermartingale that dominates $\bar{X}$.

Exercise 7.68[‡]

Prove Theorem 7.67.

The process $\bar{Z}^X$ is the *Snell envelope* of $\bar{X}$ with respect to the family $\mathcal{Q}$. An elementary but important corollary to Theorem 7.67 is the following.

Corollary 7.69

If Φ super-replicates an American option X then $V^\Phi \geq Z^X$ and in particular

$$V_0^\Phi \geq Z_0^X.$$

Proof

The process $\bar{V}^\Phi$ is a $\mathbb{Q}$-martingale for every $\mathbb{Q} \in \mathcal{Q}$ so it is a $\mathbb{Q}$-supermartingale that dominates $\bar{X}$. $\qquad\square$

For an *attainable* X the process Z^X plays the same role as the corresponding process in a complete model (where there is just one equivalent martingale measure and *every* American option is attainable). We will need the following natural generalization of Theorem 7.26 to supermartingales.

Theorem 7.70 (Doob's Optional Stopping Theorem: Supermartingales)

Let $\mathbb{Q}$ be any probability measure. If Y is a $\mathbb{Q}$-supermartingale and τ is a stopping time, then $\mathbb{E}_\mathbb{Q}(Y_\tau) \leq Y_0$.

Proof

Define a new process Y^τ (called *the process Y stopped at τ*) by

$$Y_t^\tau(\omega) = \begin{cases} Y_t(\omega) & \text{if } t \leq \tau(\omega), \\ Y_{\tau(\omega)}(\omega) & \text{if } t \geq \tau(\omega) \end{cases}$$

$$= Y_{\min\{t,\tau(\omega)\}}(\omega).$$

Then Y^τ is a $\mathbb{Q}$-supermartingale. To see this check that

$$\mathbb{E}_\mathbb{Q}\left(Y_{t+1}^\tau | \mathcal{F}_t\right) \leq Y_t^\tau$$

by taking the cases $t < \tau(\omega)$ and $t \geq \tau(\omega)$ separately as follows.

Suppose first that $t < \tau(\omega)$. Then $Y_t^\tau = Y_t$ and, if $\omega' \uparrow t = \omega \uparrow t$ then $t < \tau(\omega')$ also since τ is a stopping time. Thus $Y_{t+1}^\tau(\omega') = Y_{t+1}(\omega')$ if $\omega' \uparrow t = \omega \uparrow t$, and the definition of conditional expectation gives

$$\mathbb{E}_\mathbb{Q}\left(Y_{t+1}^\tau | \mathcal{F}_t\right)(\omega) = \mathbb{E}_\mathbb{Q}(Y_{t+1} | \mathcal{F}_t)(\omega) \leq Y_t(\omega) = Y_t^\tau(\omega).$$

The other case is when $t \geq \tau(\omega)$. In this case, if $\omega' \uparrow t = \omega \uparrow t$ then $Y_{t+1}^\tau(\omega') = Y_{\tau(\omega)}(\omega) = Y_{t+1}^\tau(\omega)$ so

$$\mathbb{E}_\mathbb{Q}\left(Y_{t+1}^\tau | \mathcal{F}_t\right)(\omega) = \mathbb{E}_\mathbb{Q}\left(Y_{\tau(\omega)}(\omega) | \mathcal{F}_t\right)(\omega) = Y_{\tau(\omega)}(\omega) = Y_t^\tau(\omega).$$

In either case $\mathbb{E}_{\mathbb{Q}}(Y_{t+1}^\tau | \mathcal{F}_t) \leq Y_t^\tau$, so Y^τ is a supermartingale and hence

$$\mathbb{E}_{\mathbb{Q}}\left(Y_T^\tau\right) \leq Y_0^\tau = Y_0. \tag{7.19}$$

The result follows from (7.19) by noting that $Y_T^\tau = Y_\tau$ and so $\mathbb{E}_{\mathbb{Q}}(Y_T^\tau) = \mathbb{E}_{\mathbb{Q}}(Y_\tau)$.
$\qquad\qquad\qquad\qquad\qquad\qquad\qquad\qquad\qquad\qquad\qquad\qquad\qquad\qquad$ □

Theorem 7.26 follows immediately since if M is a martingale then both M and $-M$ are supermartingales, as we now note.

Corollary 7.71 (= Theorem 7.26)

If M is a $\mathbb{Q}$-martingale and τ is a stopping time, then $\mathbb{E}_{\mathbb{Q}}(M_\tau) = M_0$.

Here is how Z^X works for an attainable American option X.

Theorem 7.72

Let X be an attainable American option with unique fair price π_X, and let (Φ, τ) be a hedging pair for X. The following hold true.

(1) $Z_t^X = V_t^\Phi$ for all $t \leq \tau$, and in particular $V_0^\Phi = Z_0^X = \pi_X$.

(2) For any ω and $0 < t \leq \tau(\omega)$

$$\mathbb{E}_{\mathbb{Q}}\left(\bar{Z}_t^X | \mathcal{F}_{t-1}\right)(\omega) = \bar{Z}_t^X(\omega)$$

for every $\mathbb{Q} \in \mathcal{Q}$ (so $\bar{Z}^X$ is a $\mathcal{Q}$-martingale "up to the random time τ").

Proof

Corollary 7.69 tells us that $V_0^\Phi \geq Z_0^X$. On the other hand if τ is any optimal stopping time then $V_\tau^\Phi = X_\tau$ so for any $\mathbb{Q} \in \mathcal{Q}$

$$V_0^\Phi = \mathbb{E}_{\mathbb{Q}}\left(\bar{V}_\tau^\Phi\right) = \mathbb{E}_{\mathbb{Q}}(\bar{X}_\tau) \leq \mathbb{E}_{\mathbb{Q}}\left(\bar{Z}_\tau^X\right) \leq Z_0^X$$

where we have applied Theorem 7.70 to the $\mathbb{Q}$-supermartingale $\bar{Z}^X$. So we have $V_0^\Phi = Z_0^X$.

For (2), if τ is optimal then $V_\tau^\Phi = X_\tau \leq Z_\tau^X \leq V_\tau^\Phi$ so $V_\tau^\Phi = Z_\tau^X$. Fix ω and $t \leq \tau(\omega)$ and argue as in the proof of Theorem 7.45 in the submodel Ω_λ rooted at $\lambda = \omega \uparrow t$. Since $\tau \geq t$ on Ω_λ we may apply Theorem 7.70 to the submodel Ω_λ to give

$$\bar{V}_t^\Phi(\omega) = \mathbb{E}_{\mathbb{Q}_\lambda}\left(\bar{V}_\tau^\Phi\right) = \mathbb{E}_{\mathbb{Q}_\lambda}\left(\bar{Z}_\tau^X\right) \leq Z_t^X(\omega) \leq \bar{V}_t^\Phi(\omega)$$

where $\mathbb{Q} \in \mathcal{Q}$ and $\mathbb{Q}_\lambda$ is the probability induced on Ω_λ by $\mathbb{Q}$. So $Z_t^X(\omega) = V_t^\Phi(\omega)$ for all $t \leq \tau(\omega)$. Now $\bar{V}_t^\Phi$ is a martingale up to τ (in the sense defined earlier), hence so is $\bar{Z}^X$.   □

The identification of the optimal time $\tau_{\min}$ given by Theorem 7.57 in a complete model also generalizes to an attainable American option in any model.

Theorem 7.73

Let X be an attainable American option. The smallest optimal stopping time $\tau_{\min}$ for X is given by

$$\tau_{\min} = \min\{t : X_t = Z_t^X\}.$$

Proof

Theorem 7.46 tells us that

$$\tau_{\min} = \min\{t : X_t = V_t^\Phi\}.$$

Thus $X_{\tau_{\min}} = V_{\tau_{\min}}^\Phi$ and so $X_{\tau_{\min}} = Z_{\tau_{\min}}^X$ (because $X \leq Z \leq V^\Phi$). If $t < \tau_{\min}$ then $X_t < V_t^\Phi = Z_t^X$ (using Theorem 7.72) and so

$$\tau_{\min} = \min\{t : X_t = Z_t^X\}.$$

□

The ask and bid prices for an American option are appropriate generalizations of those for European options discussed in Chapter 5.

Definition 7.74

Let X be an American option. Its *ask price* π_X^a and its *bid price* π_X^b are defined by

$$\pi_X^a := \inf\{V_0^\Phi : \Phi \text{ super-replicates } X\},$$

$$\pi_X^b := \sup\{V_0^\Phi : \Phi \text{ sub-replicates } X\}.$$

Recall that Φ *sub-replicates* X means that there is a stopping time τ such that $V_\tau^\Phi \leq X_\tau$.

We have seen (Corollary 7.69) that if Φ super-replicates an American option X then $V_t^\Phi \geq Z_t^X$, and in particular $V_0^\Phi \geq Z_0^X$. Thus the ask price π_X^a for X is at least as big as Z_0^X. We will see that in fact $\pi_X^a = Z_0^X$.

Exercise 7.75‡

Show that if Φ super-replicates X and Φ' sub-replicates X then there is a stopping time τ such that $V_\tau^\Phi \geq V_\tau^{\Phi'}$. Deduce that $V_0^\Phi \geq V_0^{\Phi'}$ and hence $\pi_X^a \geq \pi_X^b$.

For attainable American options the ask and bid prices are the same, as might be expected.

Theorem 7.76

If X is an attainable American option then its unique fair price is

$$\pi_X = \pi_X^a = \pi_X^b = Z_0^X = V_0^\Phi$$

for any minimal super-replicating strategy Φ for X.

Proof

If X is attainable with hedging strategy Φ we know (Theorem 7.36) that its unique fair price is $\pi_X = V_0^\Phi$. Since Φ both super- and sub-replicates X this means that $\pi_X^a = \pi_X^b = V_0^\Phi = \pi_X = Z_0^X$ (using Theorem 7.72). $\square$

We will see later that $\pi_X^a = \pi_X^b$ is characteristic of attainable American options.

Exercise 7.77‡

Let D be a non-negative European derivative and let X^D be the American option defined in Exercise 7.49. Show that $\pi_{X^D}^a = \pi_D^a$ and $\pi_{X^D}^b = \pi_D^b$.

For any European derivative an important fact (essentially proved in Proposition 5.67) is that there is a super-replicating strategy with initial value equal to the ask price. This is generalized to American options in the next theorem. Of course we already know this for attainable American options; see Theorem 7.76.

Theorem 7.78

Let X be an American option. The following are equivalent:

(1) $\pi \geq Z_0^X$.

(2) $\pi = V_0^\Phi$ for some super-replicating strategy Φ for X.

Hence the ask price for X coincides with the initial value of the Snell envelope; that is, $\pi_X^a = Z_0^X$.

Proof

We already know (Corollary 7.69) that $V_0^\Phi \geq Z_0^X$ for any super-replicating strategy giving (2) $\Rightarrow$ (1).

The proof of the converse is almost identical to that of Proposition 5.67, and proceeds by induction on the number of steps in the model.

To begin suppose then that $\pi \geq Z_0^X$ in a single-period model. Consider the derivative $D = X_1$. Applying the theory of Section 3.6 we have $Z_0^X \geq \pi_D^+ \geq \pi_D^a$ so there is a portfolio φ with $V_0^\varphi = \pi \geq X_0$ and $V_1^\varphi \geq D = X_1$ as required.

Now suppose that $\pi \geq Z_0^X$ in a model with $T > 1$ time steps. Again applying the single step theory as above, this time to the single period submodel at the root node and the derivative $D = Z_1^X$ there is a portfolio φ with $V_0^\varphi = \pi \geq Z_0^X \geq X_0$ and $V_1^\varphi \geq Z_1^X \geq X_1$.

Now let $\lambda_1, \dots, \lambda_k$ be the nodes at time 1. Each node λ_i is the root node of a submodel with $T - 1$ steps, we call the ith submodel, whose price histories may be identified with the set $\Omega_{\lambda_i} = \{\omega : \omega \uparrow 1 = \lambda_i\}$. Assuming that (1) $\Rightarrow$ (2) holds in each model with $T - 1$ steps, and taking $\pi_i = V_1^\varphi(\lambda_i) \geq Z_1^X(\lambda_i)$ there is a strategy Φ_i in the ith submodel that super-replicates X in this model, with initial value $V_1^\varphi(\lambda_i)$. Combine the strategies Φ_i with the portfolio φ over the first time interval to give a self-financing trading strategy Φ with $V_0^\Phi = V_0^\varphi = \pi$, which is as required.

The conclusion $\pi_X^a = Z_0^X$ is immediate from the definition of π_X^a. $\qquad\square$

In view of this result it is appropriate to make the following definition, which, by Theorem 7.76, is consistent with the definition given earlier for an *attainable* American option.

Definition 7.79

A super-replicating strategy Φ for an American option X is *minimal* if $V_0^\Phi = \pi_X^a$, the ask price of X.

The previous theorem showed that any American option has a minimal super-replicating strategy.

The ask and bid prices can alternatively be characterized using the notion of an unfair price as given in Definition 7.28.

Proposition 7.80

Let X be an American option.

(1) The ask price for X is given by

$$\pi_X^a = \inf\{\pi : \pi \text{ is an unfair price to the advantage of the writer of } X\}$$
$$= \inf\{V_0^\Phi : \Phi \text{ strictly super-replicates } X\}.$$

(2) The bid price for X is given by

$$\pi_X^b = \sup\{\pi : \pi \text{ is an unfair price to the advantage of the owner of } X\}$$
$$= \sup\{V_0^\Phi : \Phi \text{ strictly sub-replicates } X\}.$$

(3) $(\pi_X^b, \pi_X^a) \subseteq F_X \subseteq [\pi_X^b, \pi_X^a]$.

Proof

Define $\hat{\pi} := \inf\{V_0^\Phi : \Phi \text{ strictly super-replicates } X\}$. It is clear that $\hat{\pi} \geq \pi_X^a$. If $\hat{\pi} > \pi_X^a$ take π with $\hat{\pi} > \pi > \pi_X^a$ and Φ that super-replicates X with $V_0^\Phi < \pi$. Now increase the bond holding in Φ by $(\hat{\pi} - \pi)/B_0$ to give Φ' with $V_t^{\Phi'} > V_t^\Phi$ for all t. This means that Φ' strictly super-replicates X. On the other hand, $V_0^{\Phi'} = V_0^\Phi + \hat{\pi} - \pi < \hat{\pi}$, which is a contradiction. Thus $\hat{\pi}_X^a \leq \hat{\pi}$ and so $\hat{\pi}_X^a = \hat{\pi}$, giving (1), after recalling Proposition 7.30.

The proof of (2) is similar. Exercise 7.75 shows that $\pi_X^b \leq \pi_X^a$ and the assertion (3) follows. □

We do not yet know whether $F_X \neq \emptyset$ if X is non-attainable, but this will follow from the previous result if we can show that $\pi_X^b < \pi_X^a$. To this end, we first pin down the set F_X a little more by showing that, as with European derivatives, if the ask price is fair then X is attainable.

Proposition 7.81

If π_X^a is a fair price for an American option X then it is attainable.

Proof

Let Φ be a minimal super-replicating strategy Φ for X, so that $V_0^\Phi = Z_0^X = \pi_X^a$. If π_X^a is not unfair to the advantage of the writer then Φ is not strictly super-replicating, so there is a stopping time τ with $V_\tau^\Phi = X_\tau$ and X is attainable. □

Corollary 7.82

If an American option X is not attainable then π_X^a is not a fair price and $F_X = (\pi_X^b, \pi_X^a)$ or $F_X = [\pi_X^b, \pi_X^a)$.

The goal now is to show that if X is not attainable then $\pi_X^b < \pi_X^a$ and both of the possibilities in Corollary 7.82 can occur. For this we turn to the equivalent martingale measure approach to pricing, beginning with the following considerations.

If the owner of an American option adopts the exercise strategy τ, his asset then becomes a derivative with payoff X_τ, which is just like a European derivative but with the payoff being available at the random time τ rather than T. Let us define the *European* derivative D_X^τ by

$$D_X^\tau = (1+r)^{T-\tau} X_\tau \qquad (7.20)$$

which is the result of taking the payoff of the derivative X_τ at time τ and investing it in bonds until the time T. Since X_τ is a new kind of derivative, we may *define* a price to be fair for X_τ if and only if it is fair for D_X^τ, as follows.

Definition 7.83

Let X be an American derivative and τ a stopping time, with D_X^τ the European derivative defined by (7.20).

(a) A *fair price* for the derivative X_τ is any price that is fair for D_X^τ.

(b) Write F_{X_τ} for the set of fair prices for X_τ; that is $F_{X_\tau} := F_{D_X^\tau}$.

(c) Write $\pi_{X_\tau}^+ := \sup F_{X_\tau} = \sup F_{D_X^\tau}$ and $\pi_{X_\tau}^- := \inf F_{X_\tau} = \inf F_{D_X^\tau}$.

Then we have the following consequence of pricing theory for European derivatives, which further justifies the above definition of a fair price for X_τ.

Proposition 7.84

(1) The set of fair prices F_{X_τ} for the derivative X_τ is given by

$$F_{X_\tau} = \left\{ \mathbb{E}_{\mathbb{Q}}(\bar{X}_\tau) : \mathbb{Q} \in \mathcal{Q} \right\}.$$

(2) If $\varPhi$ super-replicates D_X^τ then $V_\tau^\varPhi \geq X_\tau$ and if $\varPhi$ sub-replicates D_X^τ then $V_\tau^\varPhi \leq X_\tau$.

Proof

(1) Note that $\bar{D}_X^\tau := (1+r)^{-T} D_X^\tau = (1+r)^{-\tau} X_\tau = \bar{X}_\tau$. This together with Theorem 5.64 gives

$$F_{X_\tau} = F_{D_X^\tau} = \{\mathbb{E}_\mathbb{Q}(\bar{D}_X^\tau) : \mathbb{Q} \in \mathcal{Q}\} = \{\mathbb{E}_\mathbb{Q}(\bar{X}_\tau) : \mathbb{Q} \in \mathcal{Q}\}.$$

(2) Suppose that Φ super-replicates D_X^τ; that is, $V_T^\Phi \geq D_X^\tau$. Then for any $\mathbb{Q} \in \mathcal{Q}$ and all t

$$\bar{V}_t^\Phi = \mathbb{E}_\mathbb{Q}(\bar{V}_T^\Phi | \mathcal{F}_t) \geq \mathbb{E}_\mathbb{Q}(\bar{D}_X^\tau | \mathcal{F}_t).$$

Now fix ω and let $t = \tau(\omega)$. Then $D_X^\tau = (1+r)^{T-t} X_t$ on the set $\Omega_{\omega \uparrow t}$ so for this t

$$\mathbb{E}_\mathbb{Q}(\bar{D}_X^\tau | \mathcal{F}_t)(\omega) = \bar{X}_t(\omega) = \bar{X}_\tau(\omega).$$

Hence $\bar{V}_\tau^\Phi \geq \bar{X}_\tau$ and so $V_\tau^\Phi \geq X_\tau$.

The proof for sub-replication is similar. □

Now consider any price π for X. We might guess that if $\pi > F_{X_\tau}$ (meaning that $\pi > \pi'$ for every $\pi' \in F_{X_\tau}$) for every stopping time τ then the writer can hedge against any exercise strategy of the owner, so that π is unfair. Similarly, if $\pi < F_\tau$ for some τ then we may suspect that π is also unfair, this time to the advantage of the owner. Below we will see that this is correct, and this discussion gives the following natural definitions that arise from the equivalent martingale measure approach to fair pricing, remembering that we have defined $\pi_{X_\tau}^+ := \sup F_{X_\tau}$ and $\pi_{X_\tau}^- := \inf F_{X_\tau}$.

Definition 7.85

For any American option the prices π_X^+ and π_X^- are defined as

$$\pi_X^+ := \inf\{\pi : \pi > F_{X_\tau} \text{ for all stopping times } \tau\}$$

$$= \sup\{\pi_{X_\tau}^+ : \tau \text{ a stopping time}\} = \sup_\tau \sup_{\mathbb{Q} \in \mathcal{Q}} \mathbb{E}_\mathbb{Q}(\bar{X}_\tau),$$

$$\pi_X^- := \sup\{\pi : \pi < F_{X_\tau} \text{ for some stopping time}\}$$

$$= \sup\{\pi_{X_\tau}^- : \tau \text{ a stopping time}\} = \sup_\tau \inf_{\mathbb{Q} \in \mathcal{Q}} \mathbb{E}_\mathbb{Q}(\bar{X}_\tau).$$

Exercise 7.86[‡]

Let D be a nonnegative European derivative and let X^D be the American option as defined in Exercise 7.49. Show directly from the definitions that $\pi_{X^D}^+ = \pi_D^+$ and $\pi_{X^D}^- = \pi_D^-$.

The aim now is to show that $\pi_X^a = \pi_X^+$ and $\pi_X^b = \pi_X^-$, so that as with European derivatives in an incomplete model the super- and sub-replication approach to fair pricing coincides with the equivalent martingale approach. The following helps to establish this.

Proposition 7.87

Let X be an American option.

(1) If a price π is unfair to the advantage of the writer then $\pi > F_{X_\tau}$ for every stopping time τ.

(2) A price π is unfair to the advantage of the owner if and only if $\pi < F_{X_\tau}$ for some τ.

Proof

If π is unfair to the advantage of the writer then there is a trading strategy Φ with $V_0^\Phi = \pi$ that strictly super-replicates X. Then, by Theorem 7.26, for every τ and equivalent martingale measure $\mathbb{Q}$ we have $\pi = \mathbb{E}_\mathbb{Q}(\bar{V}_\tau^\Phi) > \mathbb{E}_\mathbb{Q}(\bar{X}_\tau)$ and so $\pi > F_{X_\tau}$. This give (1).

For (2) take (Φ, τ) with $V_0^\Phi = \pi$ that strictly sub-replicates X. Then for every equivalent martingale measure $\mathbb{Q}$ we have $\pi = \mathbb{E}_\mathbb{Q}(\bar{V}_\tau^\Phi) < \mathbb{E}_\mathbb{Q}(\bar{X}_\tau)$ and so $\pi < F_{X_\tau}$.

Conversely, suppose that $\pi < F_{X_\tau}$ for some τ; then $\pi < \mathbb{E}_\mathbb{Q}(\bar{X}_\tau)$ for every $\mathbb{Q}$.

Since $F_{X_\tau} = F_{D_X^\tau}$, where D_X^τ is the derivative defined by (7.20) above, the theory of pricing for European derivatives shows (see Proposition 5.67) that there is a strategy Φ that sub-replicates D_X^τ, with $V_0^\Phi = \pi$. Proposition 7.84 shows that $V_\tau^\Phi \leq X_\tau$.

But for any $\mathbb{Q}$ we have $\mathbb{E}_\mathbb{Q}(\bar{V}_\tau^\Phi) = V_0^\Phi = \pi < \mathbb{E}_\mathbb{Q}(\bar{X}_\tau)$ and so $V_\tau^\Phi(\omega) < X_\tau(\omega)$ for some ω. Thus π is unfair, to the advantage of the owner. $\square$

The following corollary to Proposition 7.87 is straightforward.

Corollary 7.88

If X is an American option then $\pi_X^a \geq \pi_X^+$ and $\pi_X^b = \pi_X^-$.

Proof

If $\pi > \pi_X^a$ then π is unfair to the advantage of the writer, so by Proposition 7.87(1) we have $\pi > F_{X_\tau}$ for every stopping time τ. Hence $\pi \geq \pi_X^+$ and so $\pi_X^a \geq \pi_X^+$. The proof of $\pi_X^b = \pi_X^-$ is similar using Proposition 7.87(2). $\square$

From the second assertion of this corollary we can see as follows that there is a sub-replicating strategy with initial value equal to π_X^b, which may be called a *maximal sub-replicating strategy*.

Corollary 7.89

For any American option X there is a sub-replicating strategy Φ such that $V_0^\Phi = \pi_X^b$.

Proof

Since Ω and T are finite there are only finitely many stopping times. Hence $\pi_X^b = \pi_X^- = \pi_{X_\tau}^-$ for some stopping time τ. Pricing for European derivatives shows (see Proposition 5.67) that there is a trading strategy Φ that sub-replicates the derivative D_X^τ, with $V_0^\Phi = \pi_{D_X^\tau}^- = \pi_{X_\tau}^- = \pi_X^b$. Proposition 7.84 shows that $V_\tau^\Phi \leq X_\tau$. $\qquad\qquad\square$

The following result almost completes the pricing story for American options in an incomplete model.

Theorem 7.90

For an American option X

$$\pi_X^+ = Z_0^X = \pi_X^a.$$

Proof

We already know (Theorem 7.78) that $Z_0^X = \pi_X^a$. The proof that $\pi_X^+ = Z_0^X$ is by induction on the number of periods in the model. For a single period model there are only the two stopping times $\tau_0 = 0$ and $\tau_1 = 1$. Let $\mathcal{Q}_0$ denote the set of risk neutral probabilities $\mathbf{q} = (q_1, \ldots, q_k)$ for such a model; then

$$\pi_X^+ = \sup_\tau \pi_{X_\tau}^+ = \max\{\pi_{X_{\tau_0}}^+, \pi_{X_{\tau_1}}^+\}$$

$$= \max\left\{ \sup_{\mathbf{q}\in\mathcal{Q}_0} \mathbb{E}_\mathbf{q}(X_0), \sup_{\mathbf{q}\in\mathcal{Q}_0} \mathbb{E}_\mathbf{q}\left(\frac{X_1}{1+r}\right) \right\}$$

$$= \max\left\{ X_0, \sup_{\mathbf{q}\in\mathcal{Q}_0} \mathbb{E}_\mathbf{q}\left(\frac{Z_1^X}{1+r}\right) \right\} = Z_0^X.$$

Now suppose that X is an American option in a model with $T > 1$ periods, and denote by $\lambda_1, \ldots, \lambda_k$ the nodes at time 1. Each of these nodes corresponds to a submodel Ω_{λ_i} with $T - 1$ steps and root node λ_i. Note that each $\mathbb{Q} \in \mathcal{Q}$ can be viewed as a risk-neutral probability $\mathbf{q} \in \mathcal{Q}_0$ on the single-step submodel at the root node, together with a collection $\mathbb{Q}_1, \ldots, \mathbb{Q}_k$ of equivalent martingale measures in the submodels $\Omega_{\lambda_1}, \ldots, \Omega_{\lambda_k}$. Moreover any stopping time τ other than τ_0 has $\tau \geq 1$ and may be viewed as a family $\tau_1, \tau_2, \ldots, \tau_k$ with τ_i a stopping time in Ω_{λ_i}. Suppose as induction hypothesis that in each submodel Ω_{λ_i} the theorem holds. That means that if we write $\pi_X^+(\lambda_i)$ to denote the price π_X^+ interpreted in the submodel Ω_{λ_i} then

$$Z_1^X(\lambda_i) = \pi_X^+(\lambda_i) = \sup_{\tau_i \geq 1} \sup_{\mathbb{Q}_i \in \mathcal{Q}_i} \mathbb{E}_{\mathbb{Q}_i}(\bar{X}_{\tau_i}), \tag{7.21}$$

where $\mathbb{E}_{\mathbb{Q}_i}(\bar{X}_{\tau_i})$ is to be interpreted in the model Ω_{λ_i}; that is, $\bar{X}_{\tau_i} = \frac{X_{\tau_i}}{(1+r)^{\tau_i - 1}}$ and $\mathbb{E}_{\mathbb{Q}_i}(\bar{X}_{\tau_i}) = \sum_{\omega \in \Omega_{\lambda_i}} \mathbb{Q}_i(\omega) \bar{X}_{\tau_i}(\omega)$.

In the following τ will always be a stopping time, and τ_i a stopping time in Ω_{λ_i} (so that $\tau_i \geq 1$). Taking $\tau = 0$ and $\tau \geq 1$ separately we have

$$\pi_X^+ = \sup_\tau \sup_{\mathbb{Q} \in \mathcal{Q}} \mathbb{E}_{\mathbb{Q}}(\bar{X}_\tau)$$

$$= \max\left\{\sup_{\mathbb{Q} \in \mathcal{Q}} \mathbb{E}_{\mathbb{Q}}(\bar{X}_0), \sup_{\tau \geq 1} \sup_{\mathbb{Q} \in \mathcal{Q}} \mathbb{E}_{\mathbb{Q}}(\bar{X}_\tau)\right\}. \tag{7.22}$$

Now $\mathbb{E}_{\mathbb{Q}}(\bar{X}_0) = X_0$ for all $\mathbb{Q}$. If we let $\mathbf{q} = (q_1, \ldots, q_k)$ range over risk neutral probabilities in $\mathcal{Q}_0$ then, using (7.21) gives

$$\sup_{\tau \geq 1} \sup_{\mathbb{Q} \in \mathcal{Q}} \mathbb{E}_{\mathbb{Q}}(\bar{X}_\tau) = \sup_{(\tau_1, \tau_2, \ldots, \tau_k)} \sup_{(\mathbf{q}, \mathbb{Q}_1, \ldots, \mathbb{Q}_k)} \sum_i q_i \mathbb{E}_{\mathbb{Q}_i}\left(\frac{X_{\tau_i}}{(1+r)^{\tau_i}}\right)$$

$$= \sup_{\mathbf{q}} \sum_i q_i \sup_{\tau_i} \sup_{\mathbb{Q}_i} \frac{1}{1+r} \mathbb{E}_{\mathbb{Q}_i}(\bar{X}_{\tau_i})$$

$$= \sup_{\mathbf{q}} \sum_i q_i \frac{1}{1+r} \sup_{\tau_i} \sup_{\mathbb{Q}_i} \mathbb{E}_{\mathbb{Q}_i}(\bar{X}_{\tau_i})$$

$$= \sup_{\mathbf{q}} \sum_i q_i \frac{1}{1+r} Z_1^X(\lambda_i)$$

$$= \sup_{\mathbf{q}} \mathbb{E}_{\mathbf{q}}\left(\frac{1}{1+r} Z_1^X\right).$$

Combining this with (7.22) gives

$$\pi_X^+ = \max\left\{\mathbb{E}_\mathbb{Q}(\bar{X}_0), \sup_\mathbf{q} \mathbb{E}_\mathbf{q}\left(\frac{1}{1+r}Z_1^X\right)\right\} = Z_0^X.$$

$\square$

Corollary 7.91

For an American option X, write $\pi_{X,t}^+(\lambda)$ to denote the price π_X^+ evaluated in the submodel at the time t node λ. Then $\pi_{X,t}^+ = Z_t^X$.

We can now establish that $F_X \neq \emptyset$ if X is non-attainable.

Theorem 7.92

Let X be an American option. If $\pi_X^- = \pi_X^+$ then X is attainable, so if X is non-attainable then $\pi_X^- < \pi_X^+$ and $F_X \neq \emptyset$.

Proof

Suppose that $\pi_X^- = \pi_X^+$. Since the model is finite there are only finitely many stopping times. Thus, taking a minimal super-replicating strategy Φ and an equivalent martingale measure $\mathbb{Q}$

$$\pi_X^- = \max\left\{\pi_{X_\sigma}^- : \sigma \text{ a stopping time}\right\} = \pi_{X_\tau}^-$$

for some τ. Now $\pi_{X_\tau}^- \leq \mathbb{E}_\mathbb{Q}(\bar{X}_\tau) \leq \mathbb{E}_\mathbb{Q}(\bar{V}_\tau^\Phi) = \pi_X^a = \pi_X^+ = \pi_X^-$ which means that $\mathbb{E}_\mathbb{Q}(\bar{X}_\tau) = \mathbb{E}_\mathbb{Q}(\bar{V}_\tau^\Phi)$. Together with $V_\tau^\Phi \geq X_\tau$ this means that $V_\tau^\Phi = X_\tau$, so (Φ, τ) is a hedging pair for X, so X is attainable. $\square$

We have seen (Proposition 7.81) that if X is not attainable then π_X^a is not a fair price, but we have not yet established whether or not π_X^b is fair. The following example shows that it can be fair or unfair, depending on the model and the option X.

Example 7.93

Consider American options X and X' in the single period model with three scenarios u, m and d and one stock in Figure 7.10, with interest rate $r = 0$. The equivalent martingale measures are simply the risk neutral probabilities of

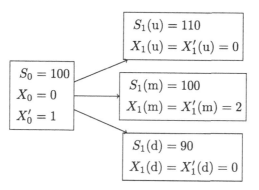

Figure 7.10 Stock and option price in Example 7.93

the form $(\frac{1}{2}(1-p), p, \frac{1}{2}(1-p)) = \mathbb{Q}_p$, say, with $0 < p < 1$. We write $\mathbb{E}_p$ for $\mathbb{E}_{\mathbb{Q}_p}$
There are just two possible stopping times, namely $\tau_0 \equiv 0$ and $\tau_1 \equiv 1$.

Let us find the set of fair prices for X and X'. For X we have $\mathbb{E}_p(X_{\tau_0}) = 0$
for all p so $F_{X_{\tau_0}} = \{0\}$. On the other hand $\mathbb{E}_p(X_{\tau_1}) = 2p$ and so $F_{X_{\tau_1}} = (0, 2)$.
Thus $\pi_X^- = 0 < F_{X_{\tau_1}} = (0, 2)$ and $\pi_X^+ = 2$. Proposition 7.87(2) shows that $\pi_X^- = \pi_X^b = 0$ is unfair for X, and hence $F_X = (0, 2)$.

For X' we have $\mathbb{E}_p(X'_{\tau_0}) = 1$ for all p so $F_{X'_{\tau_0}} = \{1\}$. On the other hand
$\mathbb{E}_p(X'_{\tau_1}) = 2p$ and so $F_{X'_{\tau_1}} = (0, 2)$. Thus $\pi_{X'}^- = 1$, $\pi_{X'}^+ = 2$, and Proposition 7.87(2) shows that $\pi_{X'}^- = \pi_{X'}^b = 1$ is fair for X'. So $F_{X'} = [1, 2)$.

These examples involving an American option in a model with only one time
step may seem unrealistic. However they could easily be incorporated within a
model having any number of time steps to give a similar pair of examples.

The following theorem gathers together what has been established concerning the pricing of American options in the models we are working with.

Theorem 7.94

Let X be an American option.

(1) If X is attainable then it has a unique fair price

$$\pi_X := \pi_X^a = \pi_X^b = \pi_X^- = \pi_X^+.$$

(2) If X is not attainable then $\pi_X^b = \pi_X^- < \pi_X^a = \pi_X^+$. The set of fair prices F_X
is given by

$$F_X = \left(\pi_X^b, \pi_X^a\right) \quad \text{or} \quad F_X = [\pi_X^b, \pi_X^a)$$

and both possibilities can occur.

There is one loose end to tie up before we conclude: we have the converse to Proposition 7.87(1).

Proposition 7.95

A price π is unfair for an American option X, to the advantage of the writer if and only if $\pi > F_{X_\tau}$ for every stopping time τ.

Proof

Proposition 7.87 showed one half of this. For the converse, suppose that $\pi > F_{X_\tau}$ for every stopping time τ. Then $\pi \geq \pi_X^+ = \pi_X^a$ and so there is a super-replicating strategy Φ with $V_0^\Phi = \pi$, and $V_\tau^\Phi \geq X_\tau$ for every stopping time τ. If there is τ with $V_\tau^\Phi = X_\tau$ then X is attainable, and Φ is a hedging strategy for X. In that case $\pi = \pi_X$ and τ is an optimal stopping time for X, so $F_{X_\tau} = \{\pi\}$ contrary to the hypothesis.

Hence, for every τ there is ω such that $V_\tau^\Phi(\omega) > X_\tau(\omega)$ for all τ and so Φ is strictly super-replicating, which means that π is unfair to the advantage of the writer. □

Remark 7.96

Föllmer and Schied (2011) define a fair price for an American option X by saying that a price π is unfair if either $\pi > F_{X_\tau}$ for every stopping time τ (π is "too big") or $\pi < F_{X_\tau}$ for some stopping time τ (π is "too small"). Propositions 7.87 and 7.95 show that their definition is equivalent to ours.

We conclude with an example and an exercise to illustrate the theory of pricing an American option an incomplete model.

Example 7.97

Consider the following 2-period model with one stock, which has binary branching followed by ternary branching at the second period. Thus the set of scenarios may be designated $\Omega = \{uu, um, ud, du, dm, dd\}$. The interest rate is zero, and the stock prices together with an American option X are specified in Figure 7.11.

Since the model is ternary at the second step, if it is viable it is incomplete. Routine calculations of risk neutral probabilities give $q_u = q_d = \frac{1}{2}$, and at the

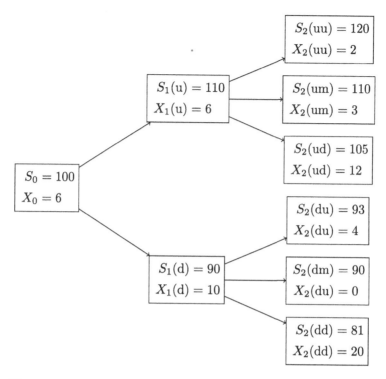

Figure 7.11 Stock price and American option in Example 7.97

other nodes they take the form $(q_{uu}, q_{um}, q_{ud}) = (\frac{1}{3}(1 - \lambda_u), \lambda_u, \frac{2}{3}(1 - \lambda_u))$ and $(q_{du}, q_{dm}, q_{dd}) = (\frac{3}{4}(1 - \lambda_d), \lambda_d, \frac{1}{4}(1 - \lambda_d))$ where $0 < \lambda_u, \lambda_d < 1$.

Let us find the interval of fair prices for X. The theory that has been developed shows that we need to find the set of fair prices for each derivative X_τ where τ is a stopping time. There are five stopping times in all, namely the constant stopping times $\tau_0 \equiv 0$, $\tau_1 \equiv 1$, $\tau_2 \equiv 2$ and two non-constant times $\tau_{1,2}$ and $\tau_{2,1}$ given by

$$\tau_{1,2}(\omega) = \begin{cases} 1 & \text{if } \omega = uu, um, ud, \\ 2 & \text{if } \omega = du, dm, dd, \end{cases} \qquad \tau_{2,1}(\omega) = \begin{cases} 2 & \text{if } \omega = uu, um, ud, \\ 1 & \text{if } \omega = du, dm, dd. \end{cases}$$

The calculations are now routine but a little tedious. First note that $X_{\tau_0} = X_0 = 6$ and so $F_{X_{\tau_0}} = \{6\}$. Next we have $X_{\tau_1}(u) = 6$ and $X_{\tau_1}(d) = 10$ and for any equivalent martingale measure $\mathbb{Q}$ we have $\mathbb{E}_\mathbb{Q}(X_{\tau_1}) = \frac{1}{2}X_{\tau_1}(u) + \frac{1}{2}X_{\tau_1}(d) = 8$, and so $F_{X_{\tau_0}} = \{8\}$. That leaves $F_{X_{\tau_2}}$, $F_{X_{\tau_{1,2}}}$ and $F_{X_{\tau_{2,1}}}$ to find. We begin with $F_{X_{\tau_2}}$. Now $X_{\tau_2} = X_2$ and so for any choice of λ_u, λ_d the corresponding equivalent martingale measure $\mathbb{Q}$ gives

$$\mathbb{E}_{\mathbb{Q}}(X_{\tau_2}) = \tfrac{1}{6}(1-\lambda_u)X_2(uu) + \tfrac{1}{2}\lambda_u X_2(um) + \tfrac{1}{3}(1-\lambda_u)X_2(ud)$$
$$+ \tfrac{3}{8}(1-\lambda_d)X_2(du) + \tfrac{1}{2}\lambda_d X_2(dm) + \tfrac{1}{8}(1-\lambda_d)X_2(dd)$$
$$= \tfrac{1}{6}(1-\lambda_u)\times 2 + \tfrac{1}{2}\lambda_u \times 3 + \tfrac{1}{3}(1-\lambda_u)\times 12$$
$$+ \tfrac{3}{8}(1-\lambda_d)\times 4 + \tfrac{1}{2}\lambda_d \times 0 + \tfrac{1}{8}(1-\lambda_d)\times 20$$
$$= \tfrac{25}{3} - \tfrac{17}{6}\lambda_u - 4\lambda_d.$$

Thus $\sup F_{X_{\tau_2}} = 8\tfrac{1}{3}$ (given by $\lambda_u = \lambda_d = 0$) and $\inf F_{X_{\tau_2}} = 1\tfrac{1}{2}$ (given by $\lambda_u = \lambda_d = 1$), so X_{τ_2} is not attainable and $F_{X_{\tau_2}} = (1\tfrac{1}{2}, 8\tfrac{1}{3})$.

Similarly we have

$$\mathbb{E}_{\mathbb{Q}}(X_{\tau_{1,2}}) = \tfrac{1}{6}(1-\lambda_u)X_1(u) + \tfrac{1}{2}\lambda_u X_1(u) + \tfrac{1}{3}(1-\lambda_u)X_1(u)$$
$$+ \tfrac{3}{8}(1-\lambda_d)X_2(du) + \tfrac{1}{2}\lambda_d X_2(dm) + \tfrac{1}{8}(1-\lambda_d)X_2(dd)$$
$$= \tfrac{1}{6}(1-\lambda_u)\times 6 + \tfrac{1}{2}\lambda_u \times 6 + \tfrac{1}{3}(1-\lambda_u)\times 6$$
$$+ \tfrac{3}{8}(1-\lambda_d)\times 4 + \tfrac{1}{2}\lambda_d \times 0 + \tfrac{1}{8}(1-\lambda_d)\times 20$$
$$= 7 - 4\lambda_d$$

giving $F_{X_{\tau_{1,2}}} = (3,7)$. Finally

$$\mathbb{E}_{\mathbb{Q}}(X_{\tau_{2,1}}) = \tfrac{1}{6}(1-\lambda_u)X_2(uu) + \tfrac{1}{2}\lambda_u X_2(um) + \tfrac{1}{3}(1-\lambda_u)X_2(ud)$$
$$+ \tfrac{3}{8}(1-\lambda_d)X_1(d) + \tfrac{1}{2}\lambda_d X_1(d) + \tfrac{1}{8}(1-\lambda_d)X_1(d)$$
$$= \tfrac{1}{6}(1-\lambda_u)\times 2 + \tfrac{1}{2}\lambda_u \times 3 + \tfrac{1}{3}(1-\lambda_u)\times 12$$
$$+ \tfrac{3}{8}(1-\lambda_d)\times 10 + \tfrac{1}{2}\lambda_d \times 10 + \tfrac{1}{8}(1-\lambda_d)\times 10$$
$$= \tfrac{28}{3} - \tfrac{17}{6}\lambda_u$$

giving $F_{X_{\tau_{2,1}}} = (6\tfrac{1}{2}, 9\tfrac{1}{3})$. From Propositions 7.87 we see that a price π is unfair to the advantage of the writer if $\pi > 6$, $\pi > 8$, $\pi \ge 8\tfrac{1}{3}$, $\pi \ge 7$ and $\pi \ge 9\tfrac{1}{3}$, which is equivalent to $\pi \ge 9\tfrac{1}{3}$. Similarly π is unfair to the advantage of the owner if either $\pi < 6$ or $\pi < 8$ or $\pi \le 1\tfrac{1}{2}$ or $\pi \le 3$ or $\pi \le 6\tfrac{1}{2}$, which is equivalent to $\pi < 8$. Hence $\pi_X^+ = 9\tfrac{1}{3}$, $\pi_X^- = 8$ and π_X^- is a fair price; thus $F_X = [8, 9\tfrac{1}{3})$.

Exercise 7.98[†]

Figure 7.12 gives the stock price and payoff of an American option X in a two-period model with zero interest rate. Find the interval of fair prices for X.

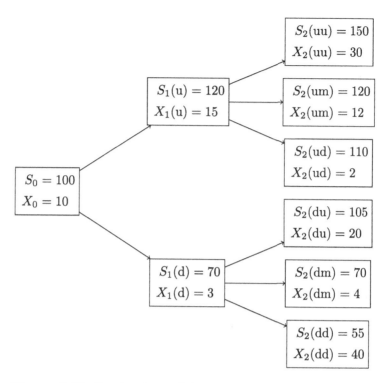

Figure 7.12 Stock price and American option in Exercise 7.98

Advanced Topics

This final chapter is devoted to some more advanced topics, beginning with futures, which are derivatives related to forwards but rather more sophisticated. After that we take a brief look at models where some of the assumptions we have made are modified, namely models with variable interest rates and models with transaction costs. The final section indicates directions for further study that follow on naturally from where we end.

8.1 Futures

In order to understand futures and futures contracts it is helpful to recall the earlier discussion of *forwards* and their pricing. We saw in Section 4.5.2 that a *forward contract* with delivery date T on a stock is an agreement at some time $t < T$ to purchase a share at time T at a *forward price* F_t that is written into the contract. No money changes hands at the time that the forward contract is made. The party who contracts to buy at the forward price is said to be *long in the forward*, while the party who agrees to sell is said to be *short in the forward*. In Theorem 4.55 it was shown that to avoid extended arbitrage the forward price must be

$$F_t = (1 + r)^{T-t} S_t.$$

N.J. Cutland, A. Roux, *Derivative Pricing in Discrete Time*,
Springer Undergraduate Mathematics Series,
DOI 10.1007/978-1-4471-4408-3_8, © Springer-Verlag London 2012

A forward contract written at time t with the forward price F_t written into it has zero value *at time* t, since this was the basis of the calculation of the price F_t. However, at any later time the value of this contract may well be different from zero. Intuitively, if the stock price goes up significantly then the forward contract will begin to look very attractive to the person who is long in the forward; however, it will be unwelcome news to the trader with the short position since, if the price rise is maintained, he would be able to sell at a higher price on the open market. If, on the other hand, the stock price falls significantly then the forward contract will become less desirable to the trader with the long position, since it would be better to buy a share at the lower market price than at the price agreed in the contract. For the opposite reason, a fall in the stock price will be welcome to the trader who is short in the forward.

To make all this precise mathematically we have the following result, proved by generalizing the solution to Exercise 4.56.

Theorem 8.1

Let $0 \le t \le s \le T$. The value (or fair price) to pay at time s for a long position in a forward contract written at time t (with forward price F_t written in the contract) is

$$(1+r)^{s-T}(F_s - F_t) = S_s - (1+r)^{s-t}S_t.$$

Taking $t = 0$ for example, this result shows that although a forward written at time 0 has zero value at its inception date, as time progresses the value of the contract will change. To be precise, it shows that by time s the long forward party has made a theoretical "profit" of $S_s - (1+r)^s S_0$, which is of course a loss if $S_s < (1+r)^s S_0$ (and the corresponding "profit" to the short forward party is $(1+r)^s S_0 - S_s$). This is only in theory because under the terms of the contract nothing actually happens until the delivery date, when the contract is fulfiled, giving a profit of $S_T - F_0$ to the long forward party and $F_0 - S_T$ to the owner of the short forward contract.

One practical problem with forwards is the possibility of default by the long forward party when $S_T < F_t$ or by the short forward party if $S_T > F_t$. A *futures contract* is a more complex instrument designed to eliminate this possibility, by a process that involves settling the contract in instalments rather than all at once at the delivery time. The instalments are designed to reflect the changes in the value of the contract, and bring its current value back to zero. Here is how it works.

Initially a futures contract (for delivery of one share at time T) taken out at time $t = 0$ costs nothing to enter into, as with a forward; associated with it is the *futures price* f_0 that is written into the contract. As the market evolves there will be a futures price f_t for each subsequent t, which will be the price named in any futures contract made at time t for delivery of one share at time T. It is clear that f_t must be $\mathcal{F}_t$-measurable because it can only depend on price histories up to the present time t.

Now, by contrast with a forward, something *does* happen at each time step $t > 0$ with a futures contract entered into at time 0. *The delivery price in the contract is changed to reflect the current market condition*; that is, the contract is changed to give the futures price f_t as the new delivery price. In effect, the contract from the previous date $t - 1$ is closed and a new contract issued. To compensate for this there is an exchange of cash equivalent to the change in value of the contract caused by the change of delivery price.

The process of exchanging cash at each time step of a futures contract is called *marking to market*. Once marking to market has taken place neither party has any *current* financial obligation so defaulting is not an issue. The value of the new or revised contract is zero as at the beginning. Marking to market involves sums of cash that are random, so in effect a futures contract is a derivative that is a *random cash flow* rather than a single random payoff. This cashflow depends on the futures prices f_t for $t \leq T$.

This is precisely how it works for the trader in the long position. At time 0, he agrees to buy a share at time T for the futures price f_0. At time 1 the new futures price f_1 becomes known. Since the new contract will have f_1 as the futures price, compared with the earlier price of f_0 there is a loss of $f_1 - f_0$ (which is actually a gain if $f_1 < f_0$) because under the old contract he would pay only f_0 for a share that is now priced at f_1 in the new contract. To compensate for this he receives a payment of $f_1 - f_0$ (or pays $f_0 - f_1$ if $f_1 < f_0$). Similarly at each subsequent time step $t < T$, marking to market involves the long futures party in being issued with a new contract specifying the futures price f_t and receiving compensation of $f_t - f_{t-1}$.

At the final time T the trader in the long position pays the futures price f_{T-1} in the contract issued at time $T - 1$ and receives one share valued at S_T. So in effect he receives $S_T - f_{T-1} = f_T - f_{T-1}$ if we put $f_T = S_T$. In summary, a trader who takes a long futures position has entered a contract that involves receiving a sequence of random payments $f_t - f_{t-1}$ (which may be negative) at each time $t > 0$, with $f_T = S_T$. The amount $\Delta f_t := f_t - f_{t-1}$ is called the *cash flow in* at time t. At any time t, once the cash-flow in has been received the value of the contract is zero, so it can be closed without any further obligation, and in practice this is what usually happens: it is rare for a futures contract to reach its expiry date.

Remarks 8.2

(1) In this discussion, for simplicity we have omitted an aspect of futures that is important in practice. In order to avoid default by either party to a futures contract, an investor is obliged to pay a deposit to the clearing house to provide a fund from which to pay any money that might be due when marking to market. This deposit is called the *initial margin*. If depleted below a certain level it must be replenished; on the other hand, any money received from marking to market is added to this account and the excess over the initial deposit may be withdrawn. At closure any money in this account is returned to the investor.

(2) From the discussion about marking to market, it is clear that the futures price f_t named at time t during this procedure reflects the current market view of the time T delivery prices discounted to the present time, because the sum $f_t - f_{t-1}$ is the compensation at today's prices for the supposed loss involved in the change of the time T delivery price. In the real world this is essential because interest rates are not constant and their future evolution is unknown. The discussion below shows that in the case of constant interest rates, the fair futures prices are the same whether or not they reflect discounting to the present.

If a futures position is held from time 0 to time T then the total cash flow is

$$\sum_{t=1}^{T}(f_t - f_{t-1}) = f_T - f_0 = S_T - f_0,$$

so (ignoring discounting) the total cash flow of a long futures position is the same as the payoff of a long forward position entered into at time 0 and with forward price f_0. The difference here is that in a forward contract the payoff $S_T - f_0$ is paid at time T, whereas in a futures contract it is paid in instalments, with $f_t - f_{t-1}$ being paid at each time $t > 0$.

A key question is now whether there is a sequence of futures prices $(f_t)_{t=0}^{T}$ that is fair to all traders—and if so how can it be determined? We will show below that in the multi-period models of Chapters 4 and 5 the fair futures price f_t at any time t is the same as the fair forward price F_t. Thus the total cash flow for a futures contract entered into at time 0 is $S_T - F_0$ which is the same as the final payout of a fairly priced forward. This is *not* entirely obvious because payments or receipts by instalment (that is, early) would not necessarily be expected to amount to the same as if they were paid or received all at time T.

For the rest of this section we assume we are in a viable multi-period model with a futures contract for delivery of a single share at time T.

Theorem 8.3

The fair futures price f_t is the same as the forward price F_t, namely

$$f_t = (1+r)^{T-t} S_t = F_t.$$

Proof

We proceed by backward induction in a manner similar to the informal derivation of the fair price of an American option.

We have $f_T = S_T$ by definition. Suppose now that $f_t = (1+r)^{T-t} S_t$ and consider f_{t-1}. At time $t-1$, once the future has been marked to market, the investor who is long in the future owns an asset with zero current value, that will provide the payoff $f_t - f_{t-1} = -f_{t-1} + (1+r)^{T-t} S_t$ at time t. This asset may be replicated at time t by the portfolio $(\frac{-f_{t-1}}{B_t}, (1+r)^{T-t})$ which has the value

$$-f_{t-1} \frac{B_{t-1}}{B_t} + (1+r)^{T-t} S_{t-1}$$

at time $t-1$. Since marking to market is designed to make this value zero we have

$$f_{t-1} = \frac{B_t}{B_{t-1}} (1+r)^{T-t} S_{t-1} = (1+r)^{T-t+1} S_{t-1}.$$

□

The following is immediate since $f_t = (1+r)^T \tilde{S}_t$.

Corollary 8.4

For any equivalent martingale measure $\mathbb{Q}$ the process of futures prices $(f_t)_{t=0}^T$ is a $\mathbb{Q}$-martingale.

A futures contract is a commitment to a (random) cash flow until the time of closure, so it is instructive to examine this cash flow in more detail. The *cumulative cash-flow in up to time t* of a futures contract entered into at time 0 is defined by $H_0 := 0$ and for $t > 0$

$$H_t := \sum_{s=1}^t \Delta f_s = f_t - f_0 = (1+r)^{T-t} S_t - (1+r)^T S_0 = (1+r)^T (\tilde{S}_t - S_0).$$

The next result follows immediately.

Corollary 8.5

For any equivalent martingale measure $\mathbb{Q}$ the cumulative cash flow $H = (H_t)_{t=0}^{T}$ is a $\mathbb{Q}$-martingale and $\mathbb{E}_{\mathbb{Q}}(H_t) = 0$ for all t.

This corollary shows that, whereas a forward involves no exchange of cash until T, while a futures contract involves exchange of cash at each intermediate step before expiry, the expected value of this exchange with respect to any equivalent martingale measure $\mathbb{Q}$ is zero. Thus, as with a fairly priced forward, with respect to $\mathbb{Q}$ the expected net flow of cash is zero.

We conclude the discussion of futures with two further observations concerning the situation in the case of the constant interest rate models we have discussed.

Remarks 8.6

(1) If we define the *discounted cash flow in* at time t to be $(1+r)^{-t}\Delta f_t$ and the corresponding *discounted cumulative cash flow in* by $\hat{H}_0 := 0$ and $\hat{H}_t := \sum_{s=1}^{t}(1+r)^{-t}\Delta f_s$ then $\hat{H} = (\hat{H}_t)_{t=0}^{T}$ is a $\mathbb{Q}$-martingale and $\mathbb{E}_{\mathbb{Q}}(\hat{H}_t) = 0$ for all t for any equivalent martingale measure.

(2) If a futures contract specified a "futures price" g_t that is designed to reflect the actual contractual price of delivery at time T in a market with known constant interest rate r, then when marking to market at time t the loss to be compensated for would be today's value of the change in delivery price from g_{t-1} to g_t, which is $(1+r)^{t-T}(g_t - g_{t-1})$. Clearly $g_T = S_T$. Since the contract has zero value after marking to market, the same inductive argument as before shows that $g_t = (1+r)^{T-t}S_t = f_t$ for all t.

8.2 Stochastic Interest Rates

One of the important simplifying assumptions made in earlier chapters was that interest rates were constant and known at time 0. This section gives a brief indication of how the theory can be adapted to allow the interest rate to vary over time and also according to the scenario. In other words, it evolves stochastically. In keeping with what happens in reality, the interest rate declared at any given time will remain in force for some time interval.

This can be made precise by modifying the multi-period models studied in Chapters 4 and 5, as follows. Fix any time $t > 0$, and let r_t be the interest rate over the time step $[t-1,t]$, which, from Remark 4.1(1) corresponds to a real time interval of length τ. From the remarks above it is natural to require that r_t should be known at time $t-1$ and held constant on the interval $[t-1,t]$. Thus, in any scenario ω the value $r_t(\omega)$ should depend only on $\omega \uparrow (t-1)$, or equivalently the random variable r_t must be $\mathcal{F}_{t-1}$-measurable. This leads to the following, which should be compared with Assumption 4.1(2).

Assumption 8.1 (Locally risk-free asset)

The model contains a *bond* with initial bond price $B_0 > 0$ at time 0. The value of the bond at any time $t > 0$ is

$$B_t = B_0 \prod_{s=1}^{t} (1 + r_s), \tag{8.1}$$

where the *interest rate process* $(r_t)_{t=1}^{T}$ is predictable and $r_t(\omega) > -1$ for all t and ω.

For convenience define $r_0 := 0$ and

$$\beta_t := \prod_{s=0}^{t} (1 + r_s).$$

Thus $B_t = B_0 \beta_t$ for all t. The predictability of $(r_t)_{t=0}^{T}$ means that $\beta \equiv (\beta_t)_{t=0}^{T}$ and $B \equiv (B_t)_{t=0}^{T}$ are both predictable. Moreover, $B_t > 0$ for all t.

Remark 8.7

The predictability of B means that a trader investing in bonds at time $t-1$ will know in advance what the value of that investment will be at time t; however bond prices beyond time t are not known at time $t-1$ because they depend on future interest rate movements. Thus the bond is risk-free only over short (single) periods; this is described by saying that the bond is *locally* (but not globally) *risk-free*.

In a single-period model, Assumption 8.1 means that the interest rate r_1 from time 0 to time 1 is $\mathcal{F}_0$-measurable, hence constant, and the theory of Chapters 2 and 3 remains unchanged after defining $r := r_1$.

Let us now consider the multi-period models of Chapters 4 and 5 with Assumption 4.1(2) replaced by Assumption 8.1. As might be expected, the most notable change is in the definition of discounted value.

Definition 8.8 (Discounted value)

Given a stochastic process $X = (X_t)_{t=0}^T$, the *discounted process* $\bar{X} = (\bar{X}_t)_{t=0}^T$ corresponding to X is defined for all t by

$$\bar{X}_t := \frac{X_t}{\beta_t}.$$

Remarks 8.9

(1) Let $\mathbb{Q}$ be any probability measure. Since β_t is a random variable, if $t > 1$ then in general

$$\mathbb{E}_{\mathbb{Q}}(\bar{X}_t) \neq \frac{1}{\beta_t} \mathbb{E}_{\mathbb{Q}}(X_t).$$

However, the predictability of β gives

$$\mathbb{E}_{\mathbb{Q}}(\bar{X}_1) = \frac{1}{\beta_1} \mathbb{E}_{\mathbb{Q}}(X_1) = \frac{1}{1 + r_1} \mathbb{E}_{\mathbb{Q}}(X_1).$$

(2) Note that $\bar{B}_t = B_0$ for all t, so $\Delta \bar{B}_t = \bar{B}_t - \bar{B}_{t-1} = 0$ as before.

All other definitions in Chapters 4 and 5 adapt very easily to accommodate the fact that the bond price B_t is random. For example, the value at time t of a trading strategy $\Phi = (x_t, y_t)_{t=0}^T$ is

$$V_t^{\Phi}(\omega) = x_t(\omega) B_t(\omega) + y_t(\omega) \cdot S_t(\omega)$$

for any t and ω. The only change from Definition 4.9 is that B_t now depends on ω; note that the $\mathcal{F}_{t-1}$-measurability of B_t means that the process $V^{\Phi} = (V_t^{\Phi})_{t=0}^T$ is adapted, just as in Chapter 4. It is also straightforward to check that the main results of Sections 4.2–4.4, 4.6–5.3 and 5.5 hold true as stated, with minor adjustments to their proofs to take into account the fact that the bond price process B is predictable rather than deterministic. This includes the First Fundamental Theorem of Asset Pricing (Theorem 5.44), the Second Fundamental Theorem of Asset Pricing (Theorem 5.57) and the Law Of One Price (Theorem 4.46).

Theorem 5.47 (the main pricing result of Section 5.4, which shows how to use equivalent martingale measures to find fair prices) extends easily as follows.

Theorem 8.10

Suppose that D is an attainable European derivative in a viable multi-period model with stochastic interest rates and equivalent martingale measure $\mathbb{Q}$. At any time t the unique fair price D_t of D is

$$D_t = \mathbb{E}_{\mathbb{Q}}\left(\frac{\beta_t}{\beta_T}D\Big|\mathcal{F}_t\right) = \beta_t\mathbb{E}_{\mathbb{Q}}(\bar{D}|\mathcal{F}_t),$$

where $\bar{D} := \dfrac{D}{\beta_T}$. In particular, the fair price at time 0 is

$$D_0 = \mathbb{E}_{\mathbb{Q}}(\bar{D}).$$

Proof

Since D is attainable, it has a replicating strategy Φ with $V_T^\Phi = D$, and by the Law Of One Price its unique fair price at time t is $D_t = V_t^\Phi$. Theorem 5.40 (which shows that $\bar{V}^\Phi$ is a martingale) extends easily to show that

$$\bar{V}_t^\Phi = \mathbb{E}_{\mathbb{Q}}\left(\bar{V}_T^\Phi|\mathcal{F}_t\right).$$

Thus, since β_t is $\mathcal{F}_{t-1}$-measurable,

$$D_t = V_t^\Phi = \beta_t\bar{V}_t^\Phi = \beta_t\mathbb{E}_{\mathbb{Q}}\left(\bar{V}_T^\Phi|\mathcal{F}_t\right) = \beta_t\mathbb{E}_{\mathbb{Q}}(\bar{D}|\mathcal{F}_t) = \mathbb{E}_{\mathbb{Q}}\left(\frac{\beta_t}{\beta_T}D\Big|\mathcal{F}_t\right).$$

□

The extension of Corollaries 5.48 and 5.49 of Theorem 5.47 to stochastic interest rate models is straightforward, and is omitted.

For models with stochastic interest rates, Theorem 8.10 shows that the pricing machinery is similar to that for models where the interest rate is constant; however the class of attainable derivatives is in general different. For example, a derivative with constant payoff is always attainable in the models studied in Chapters 4 and 5, but it is not always attainable if interest rates are stochastic, as seen in the following example.

Example 8.11

Consider the European derivative with constant payoff $D = 121$ in Model 5.4, where $B_0 = 1$ and the constant interest rate r is replaced by a predictable interest rate process $(r_t)_{t=1}^T$.

Suppose first that $r_1 = r_2 = r = 0.1$, so that $B_1 = 1.1$ and $B_2 = 1.21$. It was shown in Example 5.69 that the model is viable with this constant interest rate process. Since D is constant it is replicated by the constant strategy $\Phi = (\varphi_t)_{t=1}^2$

with $\varphi_t = (\frac{D}{B_2}, 0) = (100, 0)$, giving the unique fair price

$$D_0 = V_0^\Phi = 100$$

at time 0.

Let us now consider the case where the interest rate is stochastic. Define

$$r_1 := 0.1, \qquad r_2(\omega) = \begin{cases} 0.15 & \text{if } \omega \uparrow 1 = \text{u}, \\ 0.1 & \text{if } \omega \uparrow 1 = \text{m}, \\ 0.05 & \text{if } \omega \uparrow 1 = \text{d}, \end{cases}$$

so that

$$B_0 = \beta_0 = 1, \qquad B_1 = \beta_1 = 1.1, \qquad B_2(\omega) = \beta_2(\omega) = \begin{cases} 1.265 & \text{if } \omega \uparrow 1 = \text{u}, \\ 1.21 & \text{if } \omega \uparrow 1 = \text{m}, \\ 1.155 & \text{if } \omega \uparrow 1 = \text{d}. \end{cases}$$

Note that

$$S_1(\text{d}) = 90 < 110 = (1 + r_1)S_0 < 120 = S_1(\text{u}),$$

$$S_2(\text{ud}) = 108 < 138 = (1 + r_2(\text{u}))S_1(\text{u}) < 144 = S_2(\text{uu}),$$

$$S_2(\text{md}) = 99 < 121 = (1 + r_2(\text{m}))S_1(\text{m}) < 132 = S_2(\text{mu}),$$

$$S_2(\text{dd}) = 81 < 94.5 = (1 + r_2(\text{d}))S_1(\text{d}) < 108 = S_2(\text{du}),$$

so the viability of this model follows from an extended version of Theorem 4.29, which says that a multi-period model is viable if and only if all its single-step submodels are viable.

Let us now try to find a replicating strategy $\Phi = (x_t, y_t)_{t=1}^T$ for D in this model. In the single-step submodel at the node u the portfolio

$$(x_2(\text{u}), y_2(\text{u})) := \left(\frac{D}{B_2(\text{u})}, 0 \right) = \left(\frac{121}{1.265}, 0 \right)$$

replicates D so it has a unique fair price given by

$$D_1(\text{u}) = \frac{D}{B_2(\text{u})} \times B_1 = \frac{D}{1 + r_2(\text{u})} = \frac{121}{1.15}.$$

Similar arguments at the nodes m and d show that

$$(x_2(\omega), y_2(\omega)) = \left(\frac{121}{B_2(\omega)}, 0 \right)$$

for all ω, and the unique fair price at time 1 of D is

$$D_1(\omega) = \begin{cases} \frac{121}{1.15} & \text{if } \omega \uparrow 1 = \mathrm{u}, \\ \frac{121}{1.1} & \text{if } \omega \uparrow 1 = \mathrm{m}, \\ \frac{121}{1.05} & \text{if } \omega \uparrow 1 = \mathrm{d}. \end{cases}$$

The next step would be to try and find a portfolio (x_1, y_1) at time 0 such that $x_1 B_1 + y_1 S_1 = D_1$; that is,

$$1.1x_1 + 120y_1 = \tfrac{121}{1.15}, \qquad 1.1x_1 + 110y_1 = \tfrac{121}{1.1}, \qquad 1.1x_1 + 90y_1 = \tfrac{121}{1.05}.$$

It is straightforward to check that this system does not have a solution. In detail, the first two equations can be solved to yield $(x_1, y_1) \approx (147.826, -0.478)$ but

$$1.1x_1 + 90y_1 \approx 1.1 \times 147.826 + 90 \times -0.478 \approx 119.565 \neq 115.238 \approx \tfrac{121}{1.05},$$

which means that the third equation is not satisfied. This means that there is no replicating strategy for the constant derivative D, and therefore it is not attainable in the model with stochastic interest rates.

The results of Section 4.5 (concerning put-call parity and forward prices) depended on the fact that a European derivative with payoff $S_T - c$ (where c is a constant) is attainable. Since we cannot take it for granted that a derivative with constant payoff c is attainable if interest rates are random, it is also not necessarily true that $S_T - c$ is attainable (though of course S_T is attainable). For this reason, the proofs of the results below depend on the fact that the set of fair prices at time 0 of any European derivative D is still given by

$$F_D = \{ \mathbb{E}_{\mathbb{Q}}(\bar{D}) : \mathbb{Q} \text{ an equivalent martingale measure} \}.$$

The proof is a straightforward extension of the proof of Proposition 5.64. Extending the rest of Section 5.6 to models with stochastic interest rates is left as an exercise for the reader.

The following result should be compared with put-call parity in constant interest models (Theorem 4.49).

Theorem 8.12

Consider a viable model with a single stock S. To prevent extended arbitrage, the prices C_0 of a European call and P_0 of a European put at time 0, both with strike K and expiry date T, should satisfy

$$C_0 - P_0 = S_0 - K \mathbb{E}_{\mathbb{Q}} \left(\frac{1}{\beta_T} \right)$$

for some equivalent martingale measure $\mathbb{Q}$.

Proof

Consider the European derivative with payoff $D := C_T - P_T = S_T - K$. If $\mathbb{Q}$ is any equivalent martingale measure, then

$$\mathbb{E}_{\mathbb{Q}}(\bar{D}) = \mathbb{E}_{\mathbb{Q}}\left(\bar{S}_T - \frac{K}{\beta_T}\right) = S_0 - K\mathbb{E}_{\mathbb{Q}}\left(\frac{1}{\beta_T}\right).$$

Thus

$$F_D = \left\{S_0 - K\mathbb{E}_{\mathbb{Q}}\left(\frac{1}{\beta_T}\right) : \mathbb{Q} \text{ an equivalent martingale measure}\right\},$$

and any prices C_0, P_0 for the call and put at time 0 that do not satisfy $C_0 - P_0 \in F_D$ allow extended arbitrage (involving trading in the stock, bond and the options). $\qquad\square$

To conclude this brief discussion of models with stochastic interest rates we take a look at fair forward prices in such models. The following result should be compared with Theorem 4.55.

Theorem 8.13

Consider a viable model with a stock S. To prevent extended arbitrage, the forward price F_0 for a forward contract made at time 0 on S with delivery time T must satisfy

$$F_0 = \frac{S_0}{\mathbb{E}_{\mathbb{Q}}\left(\frac{1}{\beta_T}\right)}$$

for some equivalent martingale measure $\mathbb{Q}$.

Proof

A forward with forward price F_0 is a European derivative with payoff $D = S_T - F_0$ at time T for the trader in the long forward position. The set of fair prices at time 0 for this derivative is

$$F_D = \left\{S_0 - F_0\mathbb{E}_{\mathbb{Q}}\left(\frac{1}{\beta_T}\right) : \mathbb{Q} \text{ an equivalent martingale measure}\right\}.$$

The price at time 0 for a forward contract is 0, and it is fair if and only if $0 \in F_D$; that is, if

$$S_0 - F_0\mathbb{E}_{\mathbb{Q}}\left(\frac{1}{\beta_T}\right) = 0$$

for some equivalent martingale measure $\mathbb{Q}$. $\qquad\square$

8.3 Transaction Costs

Another simplifying assumption made in earlier chapters is that there are no transaction costs involved with trading, so that the buying and selling price of a stock is the same. Models with no transaction costs are called *friction-free*. In real-world markets there are different types of transaction costs, the simplest being fixed costs (where a fixed fee is payable at each trade, irrespective of the size of the transaction) and proportional costs (where the transaction fee is directly proportional to the size of the transaction).

In this section we investigate briefly how the theory in earlier chapters may be adapted to include transaction costs involved in trading in the risky asset. For simplicity the discussion is restricted to proportional transaction costs. Thus, in the multi-period single stock models considered in Chapters 4 and 5, Assumption 4.1(3) is replaced by the following.

Assumption 8.2 (Proportional transaction costs)

The model contains a single stock S whose price is non-negative and evolves over time. Trading in the stock is subject to a proportional transaction cost $\gamma \in [0,1)$ which is fixed for the model. At any time t a share can be *sold* for its *bid price* $S_t^b = S_t(1 - \gamma)$ and *bought* for its *ask price* $S_t^a = S_t(1 + \gamma)$.

Thus the transaction cost paid in any transaction is proportional to both the stock price and the quantity of shares traded; that is, a trader wishing to buy y shares at time t would pay yS_t^a, consisting of yS_t for his stock holding, plus $y\gamma S_t$ in transaction costs. Similarly, a trader selling y shares would receive yS_t^b, which is yS_t minus $y\gamma S_t$ transaction costs.

The most important difference between the friction-free models studied in Chapters 4 and 5 and models with proportional transaction costs is that the bond and stock are no longer freely exchangeable. To see this, suppose that $v \geq 0$ and define two portfolios at time t by $\varphi := (\frac{v}{B_t}, 0)$ and $\varphi' := (0, \frac{v}{S_t})$. Since φ does not contain any stock, the value of φ is equal to v irrespective of whether or not the model includes transaction costs. If there are no transaction costs on the stock (that is, $\gamma = 0$), then the value of φ' is the same as φ, namely v. On the other hand, if $\gamma > 0$, then a trader wishing to *create* the portfolio φ' would expect to pay $\frac{v}{S_t}S_t^a = v(1 + \gamma) > v$ in cash, and a trader wishing to *liquidate* φ' would expect to receive $\frac{v}{S_t}S_t^b = v(1 - \gamma) < v$. Thus, there are two possible ways to understand the value of the portfolio φ', and neither of them is equal to the value v of φ.

These two notions of the value of a portfolio are important, so they are defined as follows.

Definition 8.14

Let $\varphi = (x, y)$ be any portfolio at time t.

(a) The *liquidation value* of φ is defined as

$$\vartheta_t^\varphi := xB_t + y_+ S_t^b - y_- S_t^a = xB_t + yS_t - \gamma|y|S_t;$$

this is the amount of cash that a trader would expect to receive when converting φ into cash at time t.

(b) The *cost of creating* φ at time t is

$$xB_t + y_+ S_t^a - y_- S_t^b = xB_t + yS_t + \gamma|y|S_t = -\vartheta_t^{-\varphi};$$

this is the amount in cash that a trader would have to invest to create φ at time t.

Remarks 8.15

(1) If $\gamma = 0$, then $\vartheta_t^\varphi = -\vartheta_t^{-\varphi} = xB_t + yS_t$, but if $\gamma > 0$ then $\vartheta_t^\varphi \leq -\vartheta_t^{-\varphi}$ and $\vartheta_t^\varphi(\omega) < -\vartheta_t^{-\varphi}(\omega)$ whenever $y(\omega) \neq 0$. Hence, if there are transaction costs, care is needed when defining the value of a trading strategy, on which a number of important theoretical notions (such as self-financing trading, arbitrage and replication) studied in Chapters 4 and 5 depend.

(2) An important property of ϑ_t to note for future use is that

$$\vartheta_t^{(x,y)} \leq xB_t + ys \leq -\vartheta_t^{-(x,y)}$$

whenever $S_t^b \leq s \leq S_t^a$.

Suppose that $\Phi = (\varphi_t)_{t=0}^T$ is a trading strategy as in Definition 4.8; that is, a predictable process with $\varphi_0 = \varphi_1$. Consider the notion of self-financing trading when proportional transaction costs are involved. At any time t, when changing from the portfolio $\varphi_t = (x_t, y_t)$ to $\varphi_{t+1} = (x_{t+1}, y_{t+1})$, if $\Delta y_{t+1} := y_{t+1} - y_t > 0$ then the increase in the stock holding requires a cash payment of $\Delta y_{t+1} S_t^a$, which must be financed by reducing the bond position if the adjustment to the portfolio is to be self-financing. That is, writing $\Delta x_{t+1} := x_{t+1} - x_t$, we require $\Delta x_{t+1} < 0$ and

$$-\Delta x_{t+1} B_t = \Delta y_{t+1} S_t^a.$$

Writing $\Delta \varphi_{t+1} := (\Delta x_{t+1}, \Delta y_{t+1})$, this is equivalent to

$$0 = \vartheta_t^{-\Delta \varphi_{t+1}}.$$

If on the other hand $\Delta y_{t+1} < 0$, then the trader will receive $-\Delta y_{t+1} S_t^b$ from the reduction in the stock holding. For the adjustment to be self-financing, the bond holding must be increased, so $\Delta x_{t+1} > 0$ and

$$\Delta x_{t+1} B_t = -\Delta y_{t+1} S_t^b,$$

which is also equivalent to $0 = \vartheta_t^{-\Delta \varphi_{t+1}}$. This motivates the following definition.

Definition 8.16 (Self-financing trading strategy)

A trading strategy $\Phi = (\varphi_t)_{t=0}^T$ is *self-financing* if $\vartheta_t^{-\Delta \varphi_{t+1}} = 0$ for all $t < T$.

Remark 8.17

It is easy to check that, under proportional transaction costs, a trading strategy Φ being self-financing does not mean that $-\Phi$ is self-financing. This should be contrasted with Exercise 4.24(a).

It turns out that the appropriate definition of an equivalent martingale measure under proportional transaction costs is as follows.

Definition 8.18 (Equivalent martingale measure)

An equivalent probability $\mathbb{Q}$ is an *equivalent martingale measure* for a model with proportional transaction costs if there exists an adapted process $S' = (S_t')_{t=0}^T$ such that $S_t^b \le S_t' \le S_t^a$ for all t and the discounted value $\bar{S}'$ of S' is a martingale under $\mathbb{Q}$.

Thus $\mathbb{Q}$ is an equivalent martingale measure in the model with proportional transaction costs if and only if it is an equivalent martingale measure for a friction-free model whose stock price process S' takes its values within the bid-ask spreads. Any such viable friction-free model is said to be *embedded* in the model with transaction costs. It is important to note that a model with proportional transaction costs can have an equivalent martingale measure even if the friction-free model with the same stock price process S is not viable. This is illustrated in the following example.

Example 8.19

Consider a Cox-Ross-Rubinstein model with parameters $T = 1$, $r = 0.1$, $S_0 = 100$, $u = 1.05$, $d = 1$. Since $1 + r > u$, by Theorem 6.5 the friction-free Cox-Ross-Rubinstein model is not viable.

Suppose that trading in the stock is subject to proportional transaction costs at the rate $\gamma = 0.03$. The evolution of the bid and ask prices of the stock appear in Model 8.1. There are many equivalent martingales in this model. One possibility is to choose S_1' and $\mathbb{Q}$ such that

$$S_1'(u) := S_1^a(u) = 108.15, \qquad S_1'(d) := S_1^a(d) = 103, \qquad \mathbb{Q}(u) \equiv q := \tfrac{9}{10},$$

and define

$$S_0' := \mathbb{E}_{\mathbb{Q}}\big(\bar{S}_1'\big) = q\frac{S_1'(u)}{1+r} + (1-q)\frac{S_1'(d)}{1+r} = 97.85 \in \big[S_0^b, S_0^a\big].$$

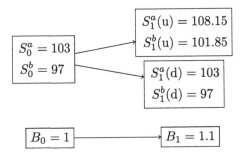

Model 8.1 One-period model with transaction costs in Example 8.19

The following result should be compared with Theorem 5.40.

Proposition 8.20

If $\mathbb{Q}$ is an equivalent measure with associated process S', then the process $(x_t B_0 + y_t \bar{S}_t')_{t=0}^T$ is a $\mathbb{Q}$-supermartingale for any self-financing trading strategy $\Phi = (x_t, y_t)_{t=0}^T$.

Proof

The self-financing property together with Remark 8.15(2) gives

$$\Delta x_{t+1} B_t + \Delta y_{t+1} S_t' \leq -\vartheta_t^{-(\Delta x_{t+1}, \Delta y_{t+1})} = 0.$$

After rearranging and discounting this leads to

$$x_{t+1} B_0 + y_{t+1} \bar{S}_t' \leq x_t B_0 + y_t \bar{S}_t'.$$

The martingale property of $\bar{S}'$ and the predictability of Φ means that

$$\mathbb{E}_\mathbb{Q}\big(x_{t+1}B_0 + y_{t+1}\bar{S}'_{t+1}\big|\mathcal{F}_t\big) = x_{t+1}B_0 + y_{t+1}\bar{S}'_t \le x_t B_0 + y_t \bar{S}'_t,$$

so $(x_t B_0 + y_t \bar{S}'_t)_{t=0}^T$ is a supermartingale under $\mathbb{Q}$. $\qquad\square$

The following definitions are important in the study of arbitrage and replication.

Definition 8.21

Suppose that $\Phi = (\varphi_t)_{t=0}^T$ is a trading strategy.

(1) The *cost of creating* Φ at time 0 is the initial investment required to create the initial portfolio φ_0, namely

$$C_0^\Phi := -\vartheta_0^{-\varphi_0}.$$

(2) The *liquidation value* of Φ at time T is the liquidation value of the final portfolio φ_T, namely

$$L_T^\Phi := \vartheta_T^{\varphi_T}.$$

An arbitrage opportunity is a self-financing trading strategy requiring zero investment at time 0, that yields nonnegative wealth when liquidated at time T. It is formally defined as follows.

Definition 8.22 (Arbitrage opportunity)

A self-financing trading strategy $\Phi_{t=0}^T$ is an *arbitrage opportunity* in bond and stock if $C_0^\Phi = 0$, $L_T^\Phi \ge 0$ and $L_T^\Phi(\omega) > 0$ for some ω.

Suppose that $\mathbb{Q}$ is an equivalent martingale measure with associated process S', and that $\Phi = (x_t, y_t)_{t=0}^T$ is a self-financing trading strategy with $L_T^\Phi \ge 0$ and $L_T^\Phi(\omega) > 0$ for some ω. It is straightforward to deduce from Remark 8.15(2) and Proposition 8.20 that

$$C_0^\Phi \ge x_0 B_0 + y_0 \bar{S}'_0$$
$$\ge \mathbb{E}_\mathbb{Q}\big(x_T B_0 + y_T \bar{S}'_T\big)$$
$$\ge (1+r)^{-T}\mathbb{E}_\mathbb{Q}\big(L_T^\Phi\big) > 0,$$

so Φ cannot be an arbitrage opportunity. This is one half of the generalization of the First Fundamental Theorem of Asset Pricing, where, as usual, a model with proportional transaction costs is said to be *viable* if there are no arbitrage opportunities.

Theorem 8.23 (First Fundamental Theorem of Asset Pricing)

A multi-period model with proportional transaction costs is viable if and only if it admits an equivalent martingale measure.

Theorem 8.23 says that a model with proportional transaction costs is viable if and only if one can embed in it a viable friction-free model with stock price process S'. The construction of $\mathbb{Q}$ and S' in a viable model depends on the separation of convex polyhedral cones and is omitted.

Let us now consider the pricing and hedging of European derivatives. Under proportional transaction costs the payoff of a derivative is not always uniquely defined; it could be either cash or a physical portfolio (and since the bond and stock cannot be exchanged freely these payoffs are different). For example the payoff of a call option with strike K and exercise date T and cash delivery can be thought of as either a cash payment of $S_T^a - K$, which reflects the benefit of buying at the strike K instead of the ask price S_T^a, or as physical delivery of a portfolio $(-\frac{K}{B_T}, 1)$, which reflects the action of purchasing one share for the strike K. In each case a rational owner would only exercise the option at time T if it is beneficial—for example a call with cash delivery of $S_T^a - K$ would only be exercised if $S_T^a \geq K$. The payoff of any particular derivative is always specified in the contract. For simplicity we consider only derivatives with cash delivery in this discussion; the notions below extend easily to the pricing of derivatives with physical delivery. Formally, we continue to think of a European derivative D as a cash payoff at time T.

Definition 8.24 (Replication)

Let D be a derivative in a multi-period model with proportional transaction costs. A self-financing trading strategy Φ *replicates* D if $L_T^\Phi = D$.

Thus a strategy Φ replicates a derivative D if its cash liquidation value at time T is equal to D.

Definition 8.25 (Super-replication)

Let D be a European derivative in a viable model with proportional transaction costs.

(a) A self-financing trading strategy Φ *super-replicates D for its writer* if
$$L_T^\Phi \geq D.$$

(b) A self-financing trading strategy Φ *super-replicates D for its owner* if
$$L_T^\Phi \geq -D.$$

Remarks 8.26

(1) It is clear from Definition 5.62 that the notions of super-replication and *super-replication for the writer* coincide in a friction-free model.

(2) The notion of *super-replication for the owner* is related but not equivalent to sub-replication. In a friction-free model $L_T^\Phi = x_T B_T + y_T S_T = V_T^\Phi$ for any trading strategy $\Phi = (x_t, y_t)_{t=0}^T$, so Φ super-replicates D for its owner if and only if $-\Phi$ sub-replicates D.

Definition 8.25 is the right one to use in models with proportional transaction costs, because a trader who owns a derivative D (and who expects to *receive* D, equivalently "pay" $-D$, at time T) would use a super-replicating strategy Φ to hedge his *liability* $-D$ at that time. The notion of sub-replication is convenient to use in friction-free models because Φ is self-financing if and only if $-\Phi$ is, and $L_T^\Phi = -L_T^{-\Phi}$; but if $\gamma > 0$ then these properties hold true only when Φ has no stock holding.

(3) Super-replication for the writer of D is the same as super-replication for the owner of $-D$.

The notions in the following definition should be compared with those in Exercise 5.71.

Definition 8.27

Let D be a European derivative with price π at time 0 in a model with transaction costs.

(a) A *simple extended trading strategy* in D is a pair (Φ, z) where $z \in \mathbb{R}$ is the holding in D from time 0 to time T, and Φ is a self-financing trading strategy.

(b) A *simple extended arbitrage opportunity* in D at π is a simple extended trading strategy (Φ, z) with $C_0^\Phi + z\pi = 0$, $L_T^\Phi + zD \geq 0$ and $L_T^\Phi(\omega) + zD(\omega) > 0$ for some ω.

(c) The price π is (simply) *fair* for D if there exists no simple extended arbitrage opportunity in D.

Remark 8.28

An extended arbitrage opportunity in D priced at π is a pair (Φ, z) where the trading strategy Φ super-replicates zD for the owner, but does not replicate it (equivalently where Φ super-replicates $-zD$ for the writer without replicating it), and $C_0^{\Phi} + z\pi = 0$.

We now see in the following example a significant difference between models with and without transaction costs when it comes to the question of replication.

Example 8.29

Consider a call option with strike K offering a cash payoff of $D = [S_1^a - K]_+$ at time 1 in Model 8.1, which is viable by Theorem 8.23. A trading strategy in this model is just a portfolio φ, so define the cost to create φ as $C_0^{\varphi} := -\vartheta_0^{-\varphi}$ and its liquidation value as $L_1^{\varphi} := \vartheta_1^{\varphi}$.

To find a replicating portfolio (x, y) for the call, it is necessary to solve the system

$$L_1^{(x,y)}(\mathrm{u}) = D(\mathrm{u}), \qquad L_1^{(x,y)}(\mathrm{d}) = D(\mathrm{d}),$$

which is

$$1.1x + 101.85y_+ - 108.15y_- = 3.15, \qquad 1.1x + 97y_+ - 103y_- = 0.$$

It can be shown that this system has a unique solution so that the call option has a unique replicating portfolio $(x, y) = (-\frac{63}{1.1}, \frac{63}{97}) \approx (-57.273, 0.649)$. The cost of creating (x, y) at time 0 is

$$C_0^{(x,y)} = xB_0 + y_+ S_0^a - y_- S_0^b = -\frac{63}{1.1} \times 1 + \frac{63}{97} \times 103 \approx 9.624.$$

The cost $C_0^{(x,y)}$ of creating (x, y) is *not* a fair price for D. This is because the portfolio

$$\varphi' = (x', y') := \left(\frac{C_0^{(x,y)}}{B_0}, 0 \right) \approx (9.624, 0)$$

has the same initial cost $C_0^{(x,y)}$ as (x, y) but

$$L_1^{(x',y')} - D = x'B_1 - D \approx 10.587 - D > 0$$

because $D(\mathrm{u}) = 3.15$ and $D(\mathrm{d}) = 0$. Thus $(x', y', -1)$ is an extended arbitrage opportunity in D at the price $C_0^{(x,y)}$. Observe that the portfolio φ' super-replicates D for its writer but does not replicate it.

Moreover, it is even possible to super-replicate D for its writer at lower initial cost than the cost $C_0^{(x,y)}$ of replication, by means of the portfolio $(x'',y'') := (\frac{D(\mathrm{u})}{B_1},0) \approx (2.863,0)$ that consists of an investment in the bond only. The cost of creating this portfolio at time 0 is

$$C_0^{(x'',y'')} = x''B_0 \approx 2.863 < C_0^{(x,y)}.$$

At time 1, the liquidation value of this portfolio is

$$L_1^{(x'',y'')} = x''B_1 = 3.15 \geq D.$$

Thus (x'',y'') super-replicates D for its writer, but it does not replicate D. Moreover it is less expensive to create this portfolio at time 0 than the replicating portfolio (x,y) for D.

Example 8.29 illustrates the surprising fact that the Law of One Price does not hold in the presence of transaction costs. It played a central role in the pricing theory of earlier chapters because, in the absence of friction, a replicating strategy, if one exists, is always the least expensive super-replicating strategy (equivalently the most expensive sub-replicating strategy). Where there are transaction costs, however, it may be possible to super-replicate more cheaply by means of a strategy that incurs less transaction costs than the replicating strategy, for example by trading only in bonds as in the above example. For this reason replication (and by extension completeness) does not play as important a role as super-replication in the pricing and hedging of derivatives under transaction costs.

We conclude this brief discussion of models with transaction costs by citing the main pricing result for such models. The *bid price* and the *ask price* of a derivative D is defined as

$$\pi_D^a := \inf\{C_0^{\Phi} : \Phi \text{ super-replicates } D \text{ for its writer}\},$$
$$\pi_D^b := -\inf\{C_0^{\Phi} : \Phi \text{ super-replicates } D \text{ for its owner}\}.$$

Remark 8.30

In a friction-free model, it follows from Remark 8.26(2) and $C_0^{-\Phi} = -V_0^{\Phi}$ that

$$\pi_D^b = -\inf\{-V_0^{\Phi} : \Phi \text{ sub-replicates } D\}$$
$$= \sup\{V_0^{\Phi} : \Phi \text{ sub-replicates } D\},$$

and so the above definition of π_D^b coincides with its counterpart in Chapter 5.

Theorem 8.31

Let D be a European derivative in a viable model with proportional transaction costs, write $\bar{D} := (1+r)^{-T}D$, and let F_D be its set of fair prices at time 0. Then

$$F_D = \big\{\mathbb{E}_{\mathbb{Q}}(\bar{D}) : \mathbb{Q} \text{ an equivalent martingale measure}\big\}, \qquad (8.2)$$

$$\pi_D^a = \sup\big\{\mathbb{E}_{\mathbb{Q}}(\bar{D}) : \mathbb{Q} \text{ an equivalent martingale measure}\big\}, \qquad (8.3)$$

$$\pi_D^b = \inf\big\{\mathbb{E}_{\mathbb{Q}}(\bar{D}) : \mathbb{Q} \text{ an equivalent martingale measure}\big\} \qquad (8.4)$$

and

$$\big(\pi_D^b, \pi_D^a\big) \leq F_D \leq \big[\pi_D^b, \pi_D^a\big].$$

Moreover, F_D can be any of the sets (π_D^b, π_D^a), $(\pi_D^b, \pi_D^a]$, $[\pi_D^b, \pi_D^a)$ or $[\pi_D^b, \pi_D^a]$.

The proof of this result uses arguments similar to those in Section 5.6 and is omitted. Here is an example illustrating the fact that π_D^a and π_D^b can both be fair prices for a derivative D under proportional transaction costs.

Example 8.32

Consider a Cox-Ross-Rubinstein model with parameters $T = 1$, $r = 0$, $S_0 = 100$, $u = 1.2$, $d = 0.8$ that is subject to transaction costs at the rate $\gamma = 0.05$. The stock price evolution appears in Model 8.2.

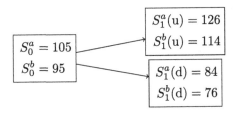

Model 8.2 One-period model with transaction costs in Example 8.32

Consider a put option with exercise date 1 and strike $K = 97$ in this model which offers a cash payoff $D = [S_1^b - K]_-$ at time 1. This means that $D(u) = 0$ and $D(d) = 21$, so if $\mathbb{Q} = (q, 1-q)$ is any equivalent martingale measure, then

$$\mathbb{E}_{\mathbb{Q}}(\bar{D}) = qD(u) + (1-q)D(d) = 21(1-q).$$

Now define

$$q^b := \tfrac{29}{38}, \quad S_0' := S_0^a = 105, \quad S_1'(u) := S_1^b(u) = 114, \quad S_1'(d) := S_1^b(d) = 76,$$

$$q^a := \tfrac{11}{42}, \quad S_0'' := S_0^b = 95, \quad S_1''(u) := S_1^a(u) = 126, \quad S_1''(d) := S_1^a(d) = 84.$$

It is straightforward to check that $\mathbb{Q}^b := (q^b, 1 - q^b)$ is an equivalent martingale measure with associated process $S' = (S'_t)_{t=0}^1$ and $\mathbb{Q}^a := (q^a, 1 - q^a)$ is an equivalent martingale measure with associated process $S'' = (S''_t)_{t=0}^1$. Moreover,

$$\mathbb{E}_{\mathbb{Q}^b}(\bar{D}) = 21(1 - q^b) = 4\tfrac{37}{38} \approx 4.974, \qquad \mathbb{E}_{\mathbb{Q}^a}(\bar{D}) = 21(1 - q^a) = 15.5,$$

after which (8.2) gives

$$F_D \supseteq \left[\mathbb{E}_{\mathbb{Q}^b}(\bar{D}), \mathbb{E}_{\mathbb{Q}^a}(\bar{D})\right] \approx [4.974, 15.5].$$

Consider first π_D^a. The definition (8.3) shows that

$$\pi_D^a \geq \mathbb{E}_{\mathbb{Q}^a}(\bar{D}) = 15.5. \tag{8.5}$$

Let us now find a replicating portfolio for the put option; note that a replicating portfolio also super-replicates for the writer, so its initial cost is an upper bound for π_D^a. A replicating portfolio (x, y) for the put must satisfy the system

$$L_1^{(x,y)}(\mathrm{u}) = D(\mathrm{u}), \qquad L_1^{(x,y)}(\mathrm{d}) = D(\mathrm{d}),$$

which is

$$x + 126y_+ - y_- = 0, \qquad x + 84y_+ - 76y_- = 21.$$

It is not hard to check that there is a unique solution $(x, y) = (63, -\tfrac{1}{2})$. The cost of creating (x, y) at time 0 is

$$C_0^{(x,y)} = xB_0 + y_+ S_0^a - y_- S_0^b = 63 - \tfrac{1}{2} \times 95 = 15.5.$$

Thus $\pi_D^a \leq C_0^{(x,y)} = 15.5$. It follows from (8.5) that $\pi_D^a = 15.5$, and so π_D^a is a fair price for D.

Consider now π_D^b. The definition (8.4) shows that

$$\pi_D^b \leq \mathbb{E}_{\mathbb{Q}^b}(\bar{D}) \approx 4.974.$$

Any portfolio (x, y) that satisfies

$$L_1^{(x,y)}(\mathrm{u}) = -D(\mathrm{u}), \qquad L_1^{(x,y)}(\mathrm{d}) = -D(\mathrm{d}) \tag{8.6}$$

is by definition a super-replicating portfolio for the owner of the put. The system (8.6) can be written as

$$x + 126y_+ - 114y_- = 0, \qquad x + 97y_+ - 103y_- = -21,$$

and it has a unique solution $(x, y) = (-63, \frac{21}{38}) \approx (-63, 0.553)$. The cash required to create (x, y) at time 0 is

$$C_0^{(x,y)} = -63 + \frac{21}{38} \times 105 = -4\frac{37}{38} \approx -4.974.$$

Thus $\pi_D^b \geq -C_0^{(x,y)} \approx 4.974$, and we conclude that $\pi_D^b = 4.974 \in F_D$.

In summary, both π_D^b and π_D^a are fair prices for D, and

$$F_D = [\pi_D^b, \pi_D^a] = [\mathbb{E}_{\mathbb{Q}^b}(\bar{D}), \mathbb{E}_{\mathbb{Q}^a}(\bar{D})] \approx [4.974, 15.5].$$

8.4 Topics for Further Study

The following are natural directions for further study, for which the material of this book could provide a stepping stone. This includes further study of the topics discussed in the previous sections in this chapter, as well as aspects of mathematical finance other than that of the arbitrage-free pricing of derivatives (the topic of this book).

Complete discrete-time models. Complete discrete-time models with a finite number of scenarios, such as the models in this book, are well-studied in the literature. The best known such model is the *Cox-Ross-Rubinstein model*, introduced by Cox et al. (1979), which we discussed in Chapter 6. The books by Shreve (2004a) and Elliott and Van der Hoek (2005) give a more extensive treatment, and because of its link to the continuous-time Black-Scholes model, many books on continuous-time models (see below) take the Cox-Ross-Rubinstein model as a starting point.

General discrete-time models. To simplify the theory all the models we have discussed have a finite number of scenarios. The key notions and results studied here (such as the Fundamental Theorems, the Law of One Price) carry over to discrete-time models with *infinitely many scenarios* Ω. The only difference is the sophistication of the proofs, which involve measure and integration theory, where finite probability and combinatorics suffice for finite Ω. Föllmer and Schied (2011) and Pliska (1997) are natural starting texts that treat this more general discrete-time setting, and also the books of Elliott and Kopp (2005) and Bingham and Kiesel (2004), which go on to cover continuous-time models too (see below).

Continuous-time models. The classic continuous-time model is the *Black-Scholes model* introduced by Black and Scholes (1973), which was briefly mentioned in Chapter 6. Etheridge (2002), Steele (2003), Shreve (2004b) and Platen and Heath (2009) offer accessible yet rigorous introductions to continuous-time mathematical models, including the Black-Scholes model. The enthusiastic reader may wish to consult the seminal text by Delbaen and Schachermayer (2006) on the Fundamental Theorem of Asset Pricing in continuous time. Any discussion of continuous-time models involves the sophisticated but powerful theory of stochastic analysis, a subject in its own right involving integration with random processes and its associated calculus.

More general financial models. Classic models such as the ones studied in this book involve many simplifying assumptions in order to make the mathematics tractable as well as to highlight the key ideas. Wilmott (2006), Joshi (2008) and Hull (2011) offer general introductions to models where the assumptions are relaxed to take into account transaction costs, dividends, liquidity risk, default (or credit) risk and exotic derivatives.

Other types of financial model. Much of the methodology developed in this book can be adapted to model interest rates and the associated fixed income securities such as (risky) bonds, swaps, caps and floors; for an introduction see Jarrow (2002), Brigo and Mercurio (2006) and Björk (2009).

Capinski and Zastawniak (2011) and Elton et al. (2003) offer introductions to the area of portfolio theory, which concerns itself with the maximization of an agent's wealth through the selection of suitable portfolios and trading strategies. The former text provides also an overview of many other aspects of mathematical finance.

Solutions to Exercises

Chapter 2

2.8.
(d) The existence of R^φ implies that $V_0^\varphi \neq 0$, hence not an arbitrage opportunity.

2.34.
Suppose that the model has a risk-neutral probability $\mathbb{Q} = (q, 1-q)$. Noting that $\bar{S}_1(\mathrm{d}) < \bar{S}_1(\mathrm{u})$ by (2.1) and $q \in (0,1)$, equation (2.16) gives

$$S_0 = \bar{S}_1(\mathrm{d}) + q\big(\bar{S}_1(\mathrm{u}) - \bar{S}_1(\mathrm{d})\big) > \bar{S}_1(\mathrm{d}),$$

$$S_0 = \bar{S}_1(\mathrm{u}) - (1-q)\big(\bar{S}_1(\mathrm{u}) - \bar{S}_1(\mathrm{d})\big) < \bar{S}_1(\mathrm{u}).$$

This leads to (2.4), and the model is viable by Theorem 2.9.

2.37.
$C_0 = \frac{50}{7} \approx 7.14286$ and $C_0' = \frac{25}{7} \approx 3.57143$.

Chapter 3

3.1.
Find x and x' such that the portfolio consisting of x of B and x' of B' has initial value 0.

N.J. Cutland, A. Roux, *Derivative Pricing in Discrete Time*,
Springer Undergraduate Mathematics Series,
DOI 10.1007/978-1-4471-4408-3, © Springer-Verlag London 2012

3.2.
Suppose that $q_1 := \lambda$ is given and then solve the system (3.5).

3.7.
Suppose that (3.15) holds true. Let (x, y) be any portfolio with zero initial value, that is

$$xB_0 + yS_0 = V_0^{(x,y)} = 0. \tag{S.1}$$

We show that (x, y) cannot be an arbitrage opportunity. There are three possibilities according to the sign of y:

- If $y = 0$, then (S.1) gives $x = 0$. This means that $V_1^{(x,y)} = 0$, so (x, y) is not an arbitrage opportunity.

- If $y > 0$, then it follows from (3.15) that

$$xB_1 + yS_1(\omega_n) < xB_1 + y(1+r)S_0 < xB_1 + yS_1(\omega_1),$$

which is equivalent to

$$V_1^{(x,y)}(\omega_n) < (1+r)V_0^{(x,y)} < V_1^{(x,y)}(\omega_1).$$

Since $V_0^{(x,y)} = 0$ by assumption, we deduce that $V_1^{(x,y)}(\omega_n) < 0$ so (x, y) is not an arbitrage opportunity.

- If $y < 0$, then a similar argument to the previous case leads to the inequality

$$V_1^{(x,y)}(\omega_1) < 0 < V_1^{(x,y)}(\omega_n),$$

which means that (x, y) is not an arbitrage opportunity.

Suppose now that (3.15) does not hold true. By (3.1) there are two possibilities for this, namely

$$S_1(\omega_1) < \cdots < S_1(\omega_1) \leq S_0(1+r) \tag{S.2}$$

or

$$S_0(1+r) \leq S_1(\omega_n) < \cdots < S_1(\omega_1). \tag{S.3}$$

Suppose that (S.2) holds. Then $(\frac{S_0}{B_0}, -1)$ is an arbitrage since $V_0^{(S_0/B_0,-1)} = 0$,

$$V_1^{(S_0/B_0,-1)} = |y|(S_0(1+r) - S_1) \geq 0$$

and

$$V_1^{(S_0/B_0,-1)}(\omega_1) = (S_0(1+r) - S_1(\omega_1)) > 0.$$

If (S.3) holds, then a similar argument shows that $(-\frac{S_0}{B_0}, 1)$ is an arbitrage opportunity. Thus if (3.15) is not satisfied, then the model is not viable.

3.13.
(a) Any risk-neutral probability for Model 3.2 takes the form

$$(q_1, q_2, q_3) = \left(\lambda, \tfrac{13}{15} - \tfrac{7}{3}\lambda, \tfrac{4}{3}\lambda + \tfrac{2}{15}\right)$$

where $\lambda \in (0, \tfrac{13}{35})$.

3.16.

(a) For any risk-neutral probability $\mathbb{Q}$, Theorem 3.5 gives

$$V_0^{\varphi'} = \mathbb{E}_\mathbb{Q}\big(\bar{V}_1^{\varphi'}\big) \leq \mathbb{E}_\mathbb{Q}(\bar{D}) \leq \mathbb{E}_\mathbb{Q}\big(\bar{V}_1^{\varphi}\big) = V_0^{\varphi}.$$

(b) The fact that $\pi_D^b \leq \pi_D^a$ is clear from Exercise 3.16(a) and (3.20)–(3.21). If D is attainable and φ replicates it, then φ is both super- and sub-replicating for D so $V_0^{\varphi} \leq \pi_D^b$ and $V_0^{\varphi} \geq \pi_D^a$, which gives $\pi_D^a = \pi_D^b = V_0^{\varphi}$. Equation (3.22) follows from Theorem 3.10.

3.17.

(a) Suppose that $\varphi = (x, y)$ super-replicates D and that D can be sold for $\pi \geq V_0^{\varphi}$ at time 0. The extended portfolio $\psi = (x + \frac{1}{B_0}(\pi - V_0^{\varphi}), y, -1)$ is an extended arbitrage since $V_0^{\psi} = 0$,

$$V_1^{\psi} = V_1^{\varphi} + (1+r)\big(\pi - V_0^{\varphi}\big) - D \geq (1+r)\big(\pi - V_0^{\varphi}\big) \geq 0,$$

and since φ does not replicate D there exists a scenario ω such that

$$V_1^{\psi}(\omega) = V_1^{\varphi}(\omega) + (1+r)\big(\pi - V_0^{\varphi}\big) - D(\omega) > (1+r)\big(\pi - V_0^{\varphi}\big) \geq 0.$$

Thus π is not a fair price for D.

3.22.

(b) $\pi_{C_1}^a = \frac{78}{7.7} \approx 10.1299.$

(c) $\pi_{C_1}^b = \frac{26}{3.3} \approx 7.8788.$

3.23.

(a) In view of (2.18), inequality (3.29) is equivalent to

$$xB_0 + y\bar{S}_1(\omega) \geq \bar{D}(\omega) \quad \text{for } \omega \in \Omega,$$

which is equivalent to $\varphi = (x, y)$ super-replicating D.

(b) The sequence $(V_0^{\varphi_k})_{k\in\mathbb{N}}$ is convergent, so there exists a number $L > 0$ such that $|V_0^{\varphi_k}| \leq L$ for $k \in \mathbb{N}$. By (3.29) this means that

$$y_k \Delta \bar{S}_1(\omega) \geq \bar{D}(\omega) - L \quad \text{for } k \in \mathbb{N}, \omega \in \Omega.$$

This implies (3.30) after defining

$$M := \max\big\{|\bar{D}(\omega)| : \omega \in \Omega\big\} + L.$$

(c) If the model is viable, then 3.31 follows directly from (3.16). For any $k \in \mathbb{N}$, if $y_k \geq 0$, then it follows from (3.30) that

$$0 \leq y_k \leq \frac{M}{|\Delta \bar{S}_1(\omega_n)|},$$

and if $y_k \leq 0$, then

$$-\frac{M}{\Delta \bar{S}_1(\omega_1)} \leq y_k \leq 0.$$

Thus the sequence $(y_k)_{k\in\mathbb{N}}$ is bounded. It is well known from analysis that any bounded sequence has a convergent subsequence.

(d) We restrict our attention to any subsequence of $(\varphi_k)_{k \in \mathbb{N}}$ for which the share holdings converge—such a subsequence exists by Exercise 3.23(c), and there is no ambiguity in denoting this again by $(\varphi_k)_{k \in \mathbb{N}} = (x_k, y_k)_{k \in \mathbb{N}}$. Since $(y_k)_{k \in \mathbb{N}}$ and $(V_0^{\varphi_k})_{k \in \mathbb{N}}$ converges, it follows that

$$\lim_{k \to \infty} x_k = \lim_{k \to \infty} \frac{1}{B_0}\left(V_0^{\varphi_k} - y_k S_0\right) = \frac{1}{B_0}\left(\pi_D^a - y S_0\right) = x.$$

Rearrangement of the last inequality gives $V_0^{(x,y)} = \pi_D^a$.

(e) The convergence of $(x_k, y_k)_{k \in \mathbb{N}}$ to (x, y) means that

$$\lim_{k \to \infty} V_1^{(x_k, y_k)}(\omega) = \lim_{k \to \infty}\left(x_k B_1 + y_k S_1(\omega)\right) = x B_1 + y S_1(\omega) = V_1^{(x,y)}(\omega)$$

for all $\omega \in \Omega$. Since $V_1^{(x_k, y_k)}(\omega) \geq D(\omega)$ for all $\omega \in \Omega$ and $k \in \mathbb{N}$, it follows that $V_1^{(x,y)}(\omega) \geq D(\omega)$ for all ω, so (x, y) super-replicates D.

3.34.

(a) Suppose that $\mathbb{Q}' = (q_1', \ldots, q_n')$ and $\mathbb{Q}'' = (q_1'', \ldots, q_n'')$ are risk-neutral probabilities, that $\lambda \in [0, 1]$, and define $\mathbb{Q} = (q_1, \ldots, q_n)$ by

$$q_j := \lambda q_i' + (1 - \lambda)\lambda q_j''$$

for $j = 1, \ldots, n$. Then $\mathbb{Q}$ is a probability measure since

$$0 < \min\{q_j', q_j''\} \leq q_j \leq \max\{q_j', q_j''\} < 1 \quad \text{for } j = 1, \ldots, n$$

and

$$\sum_{j=1}^{n} q_j = \lambda \sum_{j=1}^{n} q_j' + (1 - \lambda) \sum_{j=1}^{n} q_j'' = 1.$$

Moreover,

$$\mathbb{E}_{\mathbb{Q}}(\bar{S}_1) = \lambda \sum_{j=1}^{n} q_i' \bar{S}_1(\omega_j) + (1 - \lambda) \sum_{j=1}^{n} q_j'' \bar{S}_1(\omega_j) = S_0,$$

so $\mathbb{Q}$ is a risk-neutral probability. Thus the set of risk-neutral probabilities is convex.

(b) Suppose that $\mathbb{Q}$ and $\mathbb{Q}'$ are risk-neutral probabilities, and let

$$x := \lambda \mathbb{E}_{\mathbb{Q}}(D) + (1 - \lambda)\mathbb{E}_{\mathbb{Q}'}(D)$$

for any $\lambda \in [0, 1]$. Following the hint, the probability $\mathbb{Q}_\lambda$ of (3.37) is risk-neutral by Exercise 3.34(a). Moreover,

$$\mathbb{E}_{\mathbb{Q}_\lambda}(D) = \lambda \mathbb{E}_{\mathbb{Q}}(D) + (1 - \lambda)\mathbb{E}_{\mathbb{Q}}(D) = x,$$

which completes the proof.

3.38.

Construct an implicit arbitrage opportunity using Lemma 3.31(1).

3.41.

The convex hull appears in Figure S.1. It does not contain the origin.

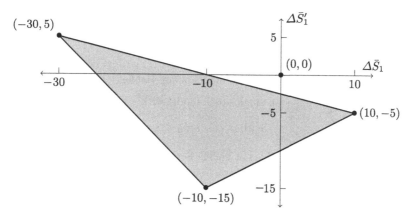

Figure S.1 Convex hull of stock prices and increments in Exercise 3.41

3.45.

Any hyperplane $\Delta \bar{S}_1' = m\Delta \bar{S}_1 + c$ with $m \in (-\frac{1}{6}, -\frac{1}{2})$ and $c < 0$ separates the origin from the convex hull of discounted stock price increments. Figure S.2 shows that the hyperplanes with equations $\Delta \bar{S}_1' = -\frac{1}{6}\Delta \bar{S}_1$ and $\Delta \bar{S}_1' = -\frac{1}{2}\Delta \bar{S}_1$ connect the origin with the convex hull, but both these hyperplanes are above the convex hull on the $(\Delta \bar{S}_1, \Delta \bar{S}_1')$ plane. This means that any non-zero vector (y, y'), for example $(y, y') = (-1, -4)$ pointing in the direction of the dark shaded area of Figure S.2 is an explicit arbitrage opportunity in this model.

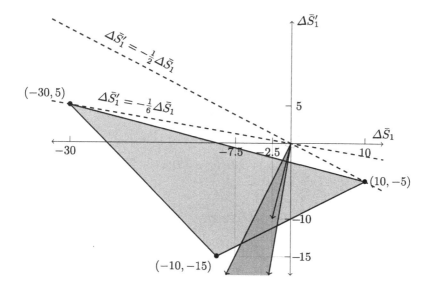

Figure S.2 Arbitrage opportunity in Exercise 3.45

3.50.

(a) Suppose that (x, y) is a strong arbitrage opportunity and define x' as in (3.40). Then $V_0^{(x', y)} = 0$ and

$$V_1^{(x', y)} = V_1^{(x, y)} - (1 + r)V_0^{(x, y)} > 0$$

since $-V_0^{(x, y)} > 0$ and $V_1^{(x, y)} \geq 0$.

(b) From (a) there is arbitrage, so the model is not viable.

(c) Consider a binary one-step single-stock model with

$$S_1(\text{d}) = (1 + r)S_0 < S_1(\text{u}).$$

By Theorem 2.9 this model is not viable. For any portfolio (x, y) with $V_0^{(x, y)} < 0$, we have

$$V_1^{(x, y)}(\text{d}) = xB_1 + yS_1(\text{d}) = (1 + r)V_0^{(x, y)} < 0,$$

and therefore the model does not contain strong arbitrage.

3.52.
$D_0 = \frac{20}{3} \approx 6.6667.$

3.62.
$D_0 = 8\frac{1}{3} \approx 8.333.$

3.63.

(a) Setting $q_1 := \lambda$ the general form of a risk-neutral probability is

$$\mathbb{Q}_\lambda := \left(\lambda, \frac{5}{3} - 5\lambda, \frac{2}{3} - 2\lambda, -\frac{4}{3} + 6\lambda\right)$$

for $\lambda \in \left(\frac{2}{9}, \frac{1}{3}\right)$. The general form can be expressed in other ways, for example by setting $q_2 := \mu$.

(b) A derivative D in this model is attainable if and only if

$$D(\omega_4) = -\frac{1}{6}D(\omega_1) + \frac{5}{6}D(\omega_2) + \frac{1}{3}D(\omega_3),$$

in which case its fair price is

$$D_0 = \frac{5}{3}D(\omega_2) + \frac{2}{3}D(\omega_3) - \frac{4}{3}D(\omega_4) = \frac{2}{9}D(\omega_1) + \frac{5}{9}D(\omega_2) + \frac{2}{9}D(\omega_3).$$

3.71.

(b) $\pi_D^b = \frac{40}{3} \approx 13.3333$ and $\pi_D^a = \frac{200}{9} \approx 22.2222.$

3.72.
Suppose that π is a fair price for D. We show by contradiction that $\pi' := \pi + c$ is a fair price for $D' := D + c(1 + r)$. To this end, suppose that π' is not a fair price for

D', that is, there exists an extended arbitrage opportunity $\psi_{D'} = (x, y, z)$ in D'. Let $\psi_D := (x + \frac{zc}{B_0}, y, z)$ be an extended portfolio in D. Then

$$V_0^{\psi_D} = xB_0 + y \cdot S_0 + z(\pi + c) = V_0^{\psi_{D'}},$$

$$V_1^{\psi_D} = xB_1 + y \cdot S_0 + z(D + c(1+r)) = V_1^{\psi_{D'}},$$

so ψ_D is an extended arbitrage opportunity in D. This contradicts the fact that π is a fair price for D; thus π' must be a fair price for D'.

Conversely, suppose that π' is a fair price for D'. Applying the above argument to D' and $-c$ shows that $\pi = \pi' - c$ is a fair price for $D = D' - c(1 + r)$.

3.73.
Following the hint, observe that

$$P_1 - C_1 = [S_1 - K]_- - [S_1 - K]_+ = K - S_1.$$

This payoff is attainable with replicating portfolio $(x, y) = (\frac{K}{B_1}, -1)$, so its unique fair price at time 0 is

$$V_0^{(K/B_1, -1)} = \frac{K}{1+r} - S_0.$$

Since the put option is not attainable, Theorem 3.70 gives

$$\{\pi + \pi' : \pi \text{ a fair price for } P_1, \pi' \text{ a fair price for } - C_1\}$$

$$= \{\pi + \pi' : \pi \in (\pi_{P_1}^b, \pi_{P_1}^a), \pi' \text{ a fair price for } - C_1\}.$$

This set has infinitely many elements, and so there exists a fair price π for P_1 and a fair price π' for $-C_1$ such that $\pi + \pi' \neq \frac{K}{1+r} - S_0$.

3.74.
(a) It is clear that any q satisfying (3.47) will also satisfy (3.46). Suppose now that $q = (q_1, \ldots, q_n)$ with $\sum_i q_i = 1$ and $q_i > 0$ for all i satisfies (3.46). This means that there exists a risk-neutral probability measure $\mathbb{Q} = (q_1, \ldots, q_n)$ on $\Omega = \{\omega_1, \ldots, \omega_n\}$. As D_0 is the unique fair price for D, Theorem 3.68 gives $D_0 = \mathbb{E}_{\mathbb{Q}}(\bar{D})$, that is,

$$D \cdot q = \mathbb{E}_{\mathbb{Q}}(D) = (1 + r)D_0.$$

Thus q satisfies (3.47).

(b) According to the theory of linear equations, the dimensions of the solution sets of the systems (3.46) and (3.47) are $n - \text{rank} \, A^T$ and $n - \text{rank} \begin{bmatrix} A^T \\ D \end{bmatrix}$ respectively, which gives (3.48). Thus D is a linear combination of the rows of A^T. That is, there exists a vector $(x, y^1, \ldots, y^m) \in \mathbb{R}^{m+1}$ such that

$$xB_1 + \sum_{j=1}^m y^j S_1^j(\omega_i) = D(\omega_i) \quad \text{for } i = 1, \ldots, n,$$

which means that D is attainable.

Chapter 4

4.24.
Show first that $V_t^{c\Phi} = cV_t^\Phi$ and $V_t^{\Phi+\Phi'} = V_t^\Phi + V_t^{\Phi'}$.

4.26.
Define a trading strategy $\Phi' = (x_t', y_t')_{t=0}^T$ by

$$(x_t', y_t') := \begin{cases} 0 & \text{if } s \le t, \\ \varphi_t & \text{if } t < s \le u, \\ \left(\frac{1}{B_u}V_u^\Phi, 0\right) & \text{if } t > u. \end{cases}$$

Then Φ' is self-financing and $(x_t', y_t') = \varphi_t$ for $t = s+1, \dots, u$. Moreover, $V_0^{\Phi'} = 0$ and

$$V_T^{\Phi'} = (1+r)^{T-u}V_u^{\Phi'} = (1+r)^{T-u}V_u^\Phi,$$

so Φ' is an arbitrage opportunity, and the model is not viable.

4.31.
It is sufficient to show for $t = 0, 1$, at every time-t node λ, the zero vector $0 \in \mathbb{R}^m$ is in the relative interior of the convex hull of the set of increments of the discounted prices $\{\Delta\bar{S}_{t+1}(\mu) : \mu \in \text{succ}\,\lambda\}$, where $\Delta\bar{S}_{t+1}(\mu) = \bar{S}_{t+1}(\mu) - \bar{S}_t(\mu)$. By Lemma 3.42 and Proposition 3.37 it follows that all single-step submodels of this model are viable, so that the model is viable by Theorem 4.29.

4.44.
The payoff of the Asian option is the random variable D with

$$D(\text{uu}) = 57, \qquad D(\text{ud}) = 21, \qquad D(\text{du}) = 0, \qquad D(\text{dd}) = 0.$$

Its fair price at time 0 is $D_0 = 24\frac{96}{121} \approx 24.7934$.

4.53.
(b) $C_0 = 16\frac{16}{49} \approx 16.3265$.

(c) $P_0 = 3\frac{59}{149} \approx 3.4014$.

4.54.
Rearrangement of (4.13) immediately gives (4.14). A trader could compute the right hand side of (4.14) using quoted stock, call and put prices, and compare it with the bank rate (LIBOR or other, representing the riskless rate r). If the trader finds that

$$r < \sqrt[T-t]{\frac{K}{S_t + P_t - C_t}} - 1,$$

then he may conclude that either the put or the call option is mis-priced (although he doesn't know which). This inequality implies that

$$C_t + \frac{K}{(1+r)^{T-t}} > P_t + S_t, \tag{S.4}$$

which means that either the call is too expensive or the put is too cheap. Consider the extended portfolio

$$(x, y, z, z') := \left(\tfrac{1}{B_t}(C_t - P_t - S_t), 1, -1, 1 \right)$$

in the bond, stock, call and put. This portfolio is created by writing and selling a call option, using the proceeds to buy a put and a share, and taking a bond position so that the value of this portfolio is 0 at time t. Note that (S.4) implies $x > -\frac{K}{B_T}$. Suppose that the trader holds the portfolio (x, y, z, a) unchanged from time t to time T. At time T, observe that

$$z[S_T - K]_+ + z'[S_T - K]_- = K - S_T,$$

so the value of the extended portfolio (x, y, z, z') is

$$xB_T + yS_T + K - S_T = xB_T + K > 0.$$

Thus the trader has found an extended arbitrage opportunity.
    If the trader finds that

$$r > \sqrt[T-t]{\frac{K}{S_t + P_t - C_t}} - 1,$$

then a similar argument shows that the extended portfolio $(-x, -y, -z, -z')$ gives rise to an extended arbitrage opportunity.

4.56.
At time t the trader is in possession of a derivative with payoff $D = S_T - F_0$ at time T.

Chapter 5

5.6.
By the definition of $\mathcal{F}_t$, if $\omega \in A$ for some $A \in \mathcal{F}_t$, then $\Omega_{\omega \uparrow t} \subseteq A$, so

$$\Omega_{\omega \uparrow t} \subseteq \bigcap \{ A \in \mathcal{F}_t \mid \omega \in A \}.$$

For the opposite inclusion we have $\omega \in \Omega_{\omega \uparrow t}$ and $\Omega_{\omega \uparrow t} \in \mathcal{F}_t$, so that

$$\bigcap \{ A \in \mathcal{F}_t \mid \omega \in A \} \subseteq \Omega_{\omega \uparrow t}.$$

5.22.
For the nodes at time 1, we have $(p_u, p_d) = (\tfrac{3}{4}, \tfrac{1}{4})$. The conditional probabilities of the nodes at time 2 are

$$(p_{uu}, p_{ud}) = \left(\tfrac{1}{3}, \tfrac{2}{3} \right), \qquad (p_{du}, p_{dd}) = \left(\tfrac{1}{2}, \tfrac{1}{2} \right).$$

The conditional probabilities of the nodes at time 3 are

$$(p_{\omega_1}, p_{\omega_2}) = \left(\tfrac{2}{5}, \tfrac{3}{5} \right), \qquad (p_{\omega_3}, p_{\omega_4}) = \left(\tfrac{1}{2}, \tfrac{1}{2} \right),$$

$$(p_{\omega_5}, p_{\omega_6}) = \left(\tfrac{2}{3}, \tfrac{1}{3} \right), \qquad (p_{\omega_7}, p_{\omega_8}) = \left(\tfrac{4}{5}, \tfrac{1}{5} \right).$$

5.28.
For any node λ at any time t, the random variable $\mathbb{E}_{\mathbb{Q}}(X|\mathcal{F}_t)$ is constant on Ω_λ, so

$$\sum_{\omega \in \Omega_\lambda} \mathbb{Q}(\omega)\mathbb{E}_{\mathbb{Q}}(X|\mathcal{F}_t)(\omega) = \mathbb{Q}(\Omega_\lambda)\mathbb{E}_{\mathbb{Q}}(X|\mathcal{F}_t)(\lambda) = \sum_{\omega \in \Omega_\lambda} \mathbb{Q}(\omega)X(\omega)$$

by Definition 5.27. This gives (5.7) in the special case $A = \Omega_\lambda$.

Proposition 5.8(2) says that every $A \in \mathcal{F}_t$ can be written as a disjoint union $A = \bigcup_{k=1}^{l} \Omega_{\lambda_k}$, where $\lambda_1, \ldots, \lambda_l$ are nodes at time t. Then (5.7) follows since

$$\sum_{\omega \in A} \mathbb{Q}(\omega)\mathbb{E}_{\mathbb{Q}}(X|\mathcal{F}_t)(\omega) = \sum_{k=1}^{l}\sum_{\omega \in \Omega_{\lambda_k}} \mathbb{Q}(\omega)\mathbb{E}_{\mathbb{Q}}(X|\mathcal{F}_t)(\omega)$$

$$= \sum_{k=1}^{l}\sum_{\omega \in \Omega_{\lambda_k}} \mathbb{Q}(\omega)X(\omega) = \sum_{\omega \in A} \mathbb{Q}(\omega)X(\omega).$$

5.29.
(a) The root node $\emptyset$ is the only node at time 0, and $\Omega_\emptyset = \Omega$, so for all ω,

$$\mathbb{E}_{\mathbb{Q}}(X|\mathcal{F}_0)(\omega) = \frac{1}{\mathbb{Q}(\Omega)}\sum_{\omega \in \Omega} \mathbb{Q}(\omega)X(\omega) = \mathbb{E}_{\mathbb{Q}}(X).$$

(b) The random variable X is constant on Ω_λ for each node λ at time t, so

$$\mathbb{E}_{\mathbb{Q}}(X|\mathcal{F}_t)(\lambda) = \frac{1}{\mathbb{Q}(\Omega_\lambda)}\sum_{\omega \in \Omega_\lambda} \mathbb{Q}(\omega)X(\omega) = \frac{1}{\mathbb{Q}(\Omega_\lambda)}X(\lambda)\sum_{\omega \in \Omega_\lambda} \mathbb{Q}(\omega) = X(\lambda).$$

(c) The random variable X is constant on Ω_μ for every node μ at time $t+1$. Thus for any time-t node λ,

$$\mathbb{E}_{\mathbb{Q}}(X|\mathcal{F}_t)(\lambda) = \frac{1}{\mathbb{Q}(\Omega_\lambda)}\sum_{\omega \in \Omega_\lambda} \mathbb{Q}(\omega)X(\omega)$$

$$= \frac{1}{\mathbb{Q}(\Omega_\lambda)}\sum_{\mu \in \text{succ }\lambda} \mathbb{Q}(\Omega_\mu)X(\mu) = \sum_{\mu \in \text{succ }\lambda} q_\mu X(\mu).$$

5.32.
(a) For any node λ at time t,

$$\mathbb{E}_{\mathbb{Q}}(X+Y|\mathcal{F}_t)(\lambda) = \frac{1}{\mathbb{Q}(\Omega_\lambda)}\sum_{\omega \in \Omega_\lambda} \mathbb{Q}(\omega)\big(X(\omega)+Y(\omega)\big)$$

$$= \mathbb{E}_{\mathbb{Q}}(X|\mathcal{F}_t)(\lambda) + \mathbb{E}_{\mathbb{Q}}(Y|\mathcal{F}_t)(\lambda).$$

(b) The random variable Y is constant on Ω_λ for each node λ at time t, so

$$\mathbb{E}_{\mathbb{Q}}(XY|\mathcal{F}_t)(\lambda) = \frac{1}{\mathbb{Q}(\Omega_\lambda)}\sum_{\omega \in \Omega_\lambda} \mathbb{Q}(\omega)X(\omega)Y(\lambda) = Y(\lambda)\mathbb{E}_{\mathbb{Q}}(X|\mathcal{F}_t)(\lambda).$$

5.35.

The conditional probabilities are

$$q_{\lambda_1} = \tfrac{4}{5}, \qquad q_{\lambda_2} = \tfrac{1}{5}, \qquad q_{\omega_1} = \tfrac{1}{2}, \qquad q_{\omega_2} = \tfrac{1}{2}, \qquad q_{\omega_3} = \tfrac{1}{4}, \qquad q_{\omega_4} = \tfrac{3}{4}.$$

Use these the check whether the martingale property holds at each node.

5.38.

At the root and d nodes there is a unique conditional risk-neutral probability. At the mode u it is

$$(q_{\omega_1}, q_{\omega_2}, q_{\omega_3}) = \left(\lambda, \tfrac{6}{5} - 2\lambda, -\tfrac{1}{5} + \lambda\right)$$

for any $\lambda \in (\tfrac{1}{5}, \tfrac{3}{5})$. For each λ this gives the equivalent martingale measure $\mathbb{Q}_\lambda$ given by

$$\mathbb{Q}_\lambda(\omega_1) = \tfrac{2}{3}\lambda, \qquad\qquad \mathbb{Q}_\lambda(\omega_4) = \tfrac{19}{60},$$

$$\mathbb{Q}_\lambda(\omega_2) = \tfrac{4}{5} - \tfrac{4}{3}\lambda, \qquad\qquad \mathbb{Q}_\lambda(\omega_5) = \tfrac{1}{60},$$

$$\mathbb{Q}_\lambda(\omega_3) = -\tfrac{2}{15} + \tfrac{2}{3}\lambda.$$

5.39.

(a) Suppose that $\mathbb{Q}$ and $\mathbb{Q}'$ are equivalent martingale measures, and for any $\alpha \in [0, 1]$ define

$$\mathbb{Q}^\alpha(\omega) := \alpha\mathbb{Q}(\omega) + (1 - \alpha)\mathbb{Q}'(\omega) \tag{S.5}$$

for all ω. Then $\mathbb{Q}$ is an equivalent probability since $\mathbb{Q}^\alpha(\omega) > 0$ for all ω, and

$$\sum_{\omega \in \Omega} \mathbb{Q}^\alpha(\omega) = \alpha \sum_{\omega \in \Omega} \mathbb{Q}(\omega) + (1 - \alpha) \sum_{\omega \in \Omega} \mathbb{Q}'(\omega) = 1.$$

At every node λ at time $t < T$ we have

$$\mathbb{E}_{\mathbb{Q}^\alpha}(\bar{S}_{t+1}|\mathcal{F}_t)(\lambda)$$

$$= \frac{1}{\mathbb{Q}^\alpha(\Omega_\lambda)}\left(\alpha \sum_{\omega \in \Omega_\lambda} \mathbb{Q}(\omega)\bar{S}_{t+1}(\omega) + (1 - \alpha) \sum_{\omega \in \Omega_\lambda} \mathbb{Q}'(\omega)\bar{S}_{t+1}(\omega)\right)$$

$$= \frac{1}{\mathbb{Q}^\alpha(\Omega_\lambda)}\left(\alpha\mathbb{Q}(\Omega_\lambda)\mathbb{E}_{\mathbb{Q}}(\bar{S}_{t+1}|\mathcal{F}_t)(\lambda) + (1 - \alpha)\mathbb{Q}'(\Omega_\lambda)\mathbb{E}_{\mathbb{Q}'}(\bar{S}_{t+1}|\mathcal{F}_t)(\lambda)\right)$$

$$= \frac{\alpha\mathbb{Q}(\Omega_\lambda) + (1 - \alpha)\mathbb{Q}'(\Omega_\lambda)}{\mathbb{Q}^\alpha(\Omega_\lambda)}\bar{S}_t(\lambda) = \bar{S}_t(\lambda),$$

so $\mathbb{Q}^\alpha$ is an equivalent martingale measure.

(b) Suppose that $\mathbb{Q}'$ and $\mathbb{Q}''$ are risk-neutral probabilities, and let

$$x := \alpha\mathbb{E}_{\mathbb{Q}}(D) + (1 - \alpha)\mathbb{E}_{\mathbb{Q}}(D)$$

for any $\alpha \in [0, 1]$. Equation (S.5) defines an equivalent martingale measure $\mathbb{Q}^\alpha$ with the property that

$$\mathbb{E}_{\mathbb{Q}^\alpha}(D) = \alpha\mathbb{E}_{\mathbb{Q}}(D) + (1 - \alpha)\mathbb{E}_{\mathbb{Q}}(D) = x.$$

5.46.

(a) Suppose that $z, z' \in \mathcal{A}$, and $\alpha, \beta \in \mathbb{R}$. There exist predictable $y = (y_t)_{t=1}^T$ and $y' = (y_t')_{t=1}^T$ such that

$$z = \left(\sum_{t=1}^T y_t(\omega_1) \cdot \Delta \bar{S}_t(\omega_1), \ldots, \sum_{t=1}^T y_t(\omega_n) \cdot \Delta \bar{S}_t(\omega_n) \right),$$

$$z' = \left(\sum_{t=1}^T y_t'(\omega_1) \cdot \Delta \bar{S}_t(\omega_1), \ldots, \sum_{t=1}^T y_t'(\omega_n) \cdot \Delta \bar{S}_t(\omega_n) \right).$$

Writing $\alpha y + \beta y'$ for the predictable process $(\alpha y_t + \beta y_t')_{t=1}^T$, it follows that

$$\alpha z + \beta z' = \left(\sum_{t=1}^T (\alpha y + \beta y')_t (\omega_1) \cdot \Delta \bar{S}_t(\omega_1), \ldots, \sum_{t=1}^T (\alpha y + \beta y')_t (\omega_n) \cdot \Delta \bar{S}_t(\omega_n) \right).$$

Thus $\alpha z + \beta z' \in \mathcal{A}$, and $\mathcal{A}$ is a subspace of $\mathbb{R}^n$.

(b) The model is viable, so there is no predictable process $y = (y_t)_{t=1}^T$ such that

$$\sum_{t=1}^T y_t \cdot \Delta \bar{S}_t \geq 0, \qquad \sum_{t=1}^T y_t(\omega) \cdot \Delta \bar{S}_t(\omega) > 0 \quad \text{for some } \omega \in \Omega;$$

that is, there is no predictable process y such that

$$\left(\sum_{t=1}^T y_t(\omega_1) \cdot \Delta \bar{S}_t(\omega_1), \ldots, \sum_{t=1}^T y_t(\omega_n) \cdot \Delta \bar{S}_t(\omega_n) \right) \in \mathcal{R}.$$

Thus $\mathcal{A} \cap \mathcal{R} = \emptyset$.

(c) By the Separating Hyperplane Theorem (Lemma 3.46) there exists a vector $\theta = (\theta_1, \ldots, \theta_n) \in \mathbb{R}^n$ such that

$$\theta \cdot z = 0 \quad \text{for all } z \in \mathcal{A},$$

$$\theta \cdot z > 0 \quad \text{for all } z \in \mathcal{R}.$$

For each j the vector $e_j = (0, \ldots, 1, \ldots 0)$ with 1 in the jth position is in $\mathcal{R}$ and so $\theta_j = \theta \cdot e_j > 0$. Defining $q = (q_1, \ldots, q_n)$ by

$$q_j := \theta_j / \sum_{k=1}^n \theta_k$$

for all j gives (5.12)–(5.13), while $\sum_{j=1}^n q_j = 1$ and $q_j > 0$ for all j.

(d) The definition of $\mathcal{A}$ gives

$$\mathbb{E}_{\mathbb{Q}} \left(\sum_{t=1}^T y_t \cdot \Delta \bar{S}_t \right) = 0$$

for all predictable $y = (y_t)_{t=1}^T$. Since $\mathbb{Q}$ is equivalent, it follows from Theorem 5.41 that $\mathbb{Q}$ is an equivalent martingale measure.

5.51.
$P_0 = 2\frac{457}{519} \approx 2.8805$.

5.54.
Suppose that Φ^D replicates D, and that Φ^E replicates E. By Exercise 4.24 the trading strategy $\Phi := \alpha\Phi^D + \beta\Phi^E$ is self-financing and $V_T^\Phi = \alpha D + \beta E$, so Φ replicates D. By Theorem 5.47 and Exercise 5.32(a), the unique fair price of $\alpha D + \beta E$ at any time t is

$$\frac{1}{(1+r)^{T-t}}\mathbb{E}_\mathbb{Q}(\alpha D + \beta E) = \alpha D_t + \beta E_t.$$

5.55.
D and G can be expressed as linear combinations of the call options.

5.61.
We have

$$D_1(\mathrm{u}) = 1018\tfrac{2}{11} \approx 1018.1818,$$

$$D_1(\mathrm{d}) = 77\tfrac{3}{11} \approx 77.2727,$$

$$D_0 = 640\tfrac{60}{121} \approx 640.4959.$$

5.65.
The definition of super- and sub-replication gives $V_T^\Phi \leq D \leq V_T^{\Phi'}$, from which it follows that

$$\mathbb{E}_\mathbb{Q}\big(\bar{V}_T^\Phi\big) \leq (1+r)^{-T}\mathbb{E}_\mathbb{Q}(D) \leq \mathbb{E}_\mathbb{Q}\big(\bar{V}_T^{\Phi'}\big). \qquad (\text{S.6})$$

Theorem 5.40 then gives (5.16).

If D is not attainable, then there exist ω, ω' such that $V_T^\Phi(\omega) < D(\omega)$ and $V_T^{\Phi'}(\omega') > D(\omega')$, so both inequalities in (S.6) are strict. This in turn implies that both inequalities in (5.16) are strict.

5.70.
(c) The set of fair prices for the put option is $(0, 2.84)$.

5.71.
Assuming (1), suppose that π is not a simple fair price for D. Thus there exists a simple extended arbitrage, which by definition is also an extended arbitrage opportunity for D. Thus π is not a fair price for D either, which is (3).

We now show that (2) implies (1). Suppose that $\pi = V_0^\Phi$, where $\Phi = (\varphi_t)_{t=0}^T$ super-replicates D but it is not a replicating strategy for D. The simple extended trading strategy $\Psi = (\varphi_t, -1)_{t=0}^T$ satisfies $V_0^\Psi = 0$, $V_T^\Psi \geq 0$ and since Φ does not replicate D, there is some ω such that $V_T^\Psi(\omega) > 0$. Thus Ψ is a simple extended arbitrage opportunity, and π is not a fair price for D, which is (1).

Similarly, if $\pi = V_0^\Phi$, where $\Phi = (x_t, y_t)_{t=0}^T$ sub-replicates D but is not a replicating strategy for it, then the simple extended trading strategy $\Psi = ((-x_t, -y_t), 1)_{t=0}^T$ is a simple extended arbitrage opportunity, implying (1).

Suppose that (3) holds, so π is not a fair price for D. We show that this implies (2), thus completing the proof. In the case of attainable D, this means that $\pi \neq D_0 = \pi_D^+ = \pi_D^-$. By Proposition 5.67 there exists a trading strategy $\Phi = (x_t, y_t)_{t=0}^T$ sub- or super-replicating D and such that $\pi = V_0^\Phi$. Since D is attainable, it has a replicating strategy $\Phi^D = (x_t', y_t')_{t=0}^T$ with $D_0 = V_0^{\Phi'}$. Since $V_0^\Phi \neq V_0^{\Phi'}$, Theorem 4.38 shows that Φ does not replicate D, and we conclude (2).

If D is not attainable, (3) implies by Theorem 5.63 that $\pi \geq \pi_D^+$ or $\pi \leq \pi_D^-$. Proposition 5.67 then shows that there exists a sub- or super-replicating strategy Φ for D such that $\pi = V_0^\Phi$; the strategy Φ does not replicate D due to the non-attainability of D. This is (2), and completes the proof.

Chapter 6

6.4.

(a) Model S.3 gives a tree representation of this model, which could be condensed into a recombinant tree.

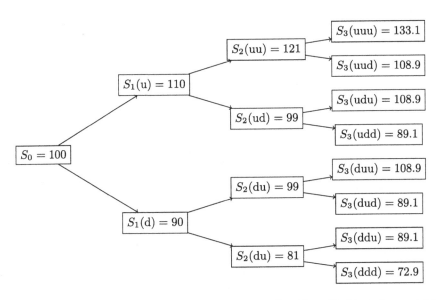

Model S.3 Stock price evolution in three-step Cox-Ross-Rubinstein model in Exercises 6.4 and 6.9

6.9.

(b) $D_0 = 6.3625$.

6.13.

$D_0 \approx 42.1296$.

6.15.
(a) $P_0^B = (1+r)^{-T} B(A-1; T, q)$.

6.19.
$P_0 \approx 23.2896$.

6.20.
(a) $C_0^{AN} = S_0 B'(A; T, q')$.

6.24.
(b) $\hat{P}_t(z) = K(1+r)^{t-T} B(A_t(z) - 1; T - t, q) - z B(A_t(z) - 1; T - t, q')$.

6.26.
Let f be any function such that

$$\int_{-\infty}^{\infty} \frac{1}{\sqrt{2\pi}} e^{-\frac{1}{2}y^2} f(y) \, dy$$

is well defined. Then

$$\mathbb{E}_{\mathbb{P}}\big(f(Y)\big) = \mathbb{E}_{\mathbb{P}}\left(f\left(\frac{X-\mu}{\sqrt{\nu}}\right)\right)$$

$$= \int_{-\infty}^{\infty} \frac{1}{\sqrt{2\pi\nu}} e^{-\frac{(z-\mu)^2}{2\nu}} f\left(\frac{z-\mu}{\sqrt{\nu}}\right) dz$$

$$= \int_{-\infty}^{\infty} \frac{1}{\sqrt{2\pi}} e^{-\frac{1}{2}y^2} f(y) \, dy$$

after changing variables from z to $y = \frac{z-\mu}{\sqrt{\nu}}$ in the integral. This means that Y is a Gaussian random variable with mean 0 and variance 1.

6.30.
At any node λ at time $t - 1$ in the nth Cox-Ross-Rubinstein model we have

$$\mathbb{Q}_n\big(C_t^{(n)} = \sqrt{\Delta_n}|\Omega_\lambda\big) = \mathbb{Q}_n(\Omega_{\lambda u}|\Omega_\lambda) = q_n,$$

$$\mathbb{Q}_n\big(C_t^{(n)} = -\sqrt{\Delta_n}|\Omega_\lambda\big) = \mathbb{Q}_n(\Omega_{\lambda d}|\Omega_\lambda) = 1 - q_n.$$

This is independent of λ and so

$$\mathbb{Q}_n\big(C_t^{(n)} = \sqrt{\Delta_n}\big) = q_n, \qquad \mathbb{Q}_n\big(C_t^{(n)} = -\sqrt{\Delta_n}\big) = 1 - q_n.$$

The definitions of mean and variance immediately give (6.25)–(6.26).

Suppose that $c_1, \ldots, c_n$ are numbers in the set $\{-\sqrt{\Delta_n}, \sqrt{\Delta_n}\}$, and let $\omega \in \Omega_n$ be the unique scenario $\omega = \omega_1 \omega_2 \cdots \omega_n$ where

$$\omega_t := \begin{cases} u & \text{if } c_t = \sqrt{\Delta_n}, \\ d & \text{if } c_t = -\sqrt{\Delta_n}, \end{cases}$$

and let k be the number of u's among the ω_t's. Then

$$Q_n\left(\bigcap_{t=1}^{T}\{C_t^{(n)} = c_t\}\right) = Q_n\left(\Omega_{\omega_1\cdots\omega_n}\right)$$

$$= Q_n(\omega) = q_n^k(1-q_n)^{n-k} = \prod_{t=1}^{n}Q_n\left(C_t^{(n)} = c_t\right).$$

Thus for any $x_1,\ldots,x_n$

$$Q_n\left(\bigcap_{t=1}^{n}\{C_t^{(n)} \leq x_t\}\right) = \prod_{t=1}^{n}Q_n\left(C_t^{(n)} \leq x_t\right),$$

and so $C_1^{(n)},\ldots,C_n^{(n)}$ are independent.

6.35.

(a) The formula for q_n in (6.22) gives

$$\lim_{n\to\infty}q_n = \lim_{n\to\infty}\frac{e^{\frac{1}{2}\sigma^2\Delta_n} - e^{-\sigma\sqrt{\Delta_n}}}{e^{\sigma\sqrt{\Delta_n}} - e^{-\sigma\sqrt{\Delta_n}}} = \lim_{n\to\infty}\frac{e^{\frac{1}{2}\sigma^2\Delta_n} - e^{-\sigma\sqrt{\Delta_n}}}{2\sinh(\sigma\sqrt{\Delta_n})}.$$

Since $x = \sigma\sqrt{\Delta_n} = \sigma\sqrt{\frac{T}{n}} \downarrow 0$ as $n\to\infty$ and all the functions in the right hand limit are continuous, we may write

$$\lim_{n\to\infty}q_n = \lim_{x\downarrow 0}\frac{e^{\frac{1}{2}x^2} - e^{-x}}{2\sinh x}.$$

Equation (6.42) holds true since the functions in the limit are all continuous at 0.

Equation (6.43) follows in similar fashion from

$$\lim_{n\to\infty}2\sqrt{nT}\left(q_n - \tfrac{1}{2}\right) = \lim_{n\to\infty}\frac{T}{\sqrt{\Delta_n}}\frac{e^{\frac{1}{2}\sigma^2\Delta_n} - \cosh(\sigma\sqrt{\Delta_n})}{\sinh(\sigma\sqrt{\Delta_n})}$$

$$= \lim_{x\downarrow 0}\frac{\sigma T}{x}\frac{e^{\frac{1}{2}x^2} - \cosh x}{\sinh x}.$$

(b) Application of L'Hôpital's rule to the right hand limit in (6.42) gives

$$\lim_{n\to\infty}q_n = \lim_{x\to 0}\frac{xe^{\frac{1}{2}x^2} + e^{-x}}{2\cosh x} = \frac{1}{2},$$

and therefore

$$\lim_{n\to\infty}\text{var}_{Q_n}\left(H^{(n)}\right) = \lim_{n\to\infty}4Tq_n(1-q_n) = T.$$

(c) The Taylor expansions of $e^{\frac{1}{2}x^2}$ and $\cosh x$ around $x=0$ give

$$e^{\frac{1}{2}x^2} - \cosh x = \sum_{k=0}^{\infty}\frac{(\frac{1}{2}x^2)^k}{k!} - \sum_{k=0}^{\infty}\frac{x^{2k}}{(2k)!} = \sum_{k=0}^{\infty}a_k x^{2k},$$

and (6.44) follows because $a_0 = a_1 = 0$. Substitution into (6.43) then gives

$$\lim_{n\to\infty} 2\sqrt{nT}\left(q_n - \tfrac{1}{2}\right) = \lim_{x\to 0} \frac{\sigma T}{\sinh x} \sum_{k=2}^{\infty} a_k x^{2k-1}$$

and by L'Hôpital's rule

$$\lim_{n\to\infty} 2\sqrt{nT}\left(q_n - \tfrac{1}{2}\right) = \lim_{x\to 0} \frac{\sigma T}{\cosh x} \sum_{k=2}^{\infty} (2k-1)a_k x^{2k-2} = 0.$$

Chapter 7

7.10.

Suppose that $P_0^E < P_0^A$. Write and sell an American put for P_0^A, buy a European put for P_0^E, and invest the difference $P_0^A - P_0^E > 0$ in bonds; denote this extended portfolio by ψ. There are now two possibilities.

(1) The American option is not exercised until time T, when $P_T^E = P_T^A$. Then the final value of the portfolio ψ is

$$P_T^E - P_T^A + P_0^A - P_0^E = P_0^A - P_0^E > 0.$$

(2) In some scenario ω the American option is exercised at some time $t < T$. To fulfil his obligation at that time, the writer should borrow K and use it to buy one share. This amounts to modifying the portfolio ψ to ψ', say, consisting of 1 European put, 1 share, and bonds worth $-K$ in addition to the original investment in bonds, worth $P_0^A - P_0^E$. Make no further changes, so at expiry time T, the value $V_T^{\psi'}$ of the extended portfolio ψ' for any scenario ω' with $\omega' \uparrow t = \omega \uparrow t$ is

$$V_T^{\psi'} = [K - S_T]_+ - (K - S_T) + C_0^A - C_0^E \geq C_0^A - C_0^E > 0$$

because $a_+ \geq a$ for any a and $C_0^A > C_0^E$.

7.17.

(a) The risk neutral probability at each node is $(0.8, 0.2)$. Routine calculation gives (writing $Y_t = \mathbb{E}(\hat{Z}_{t+1}^X | \mathcal{F}_t)$ for $t < 3$) the tree representation in Figure S.4 of X, Y and Z^X. It is advantageous to exercise the option whenever $X > Y$, which is at the nodes d, ud, du, dd.

(b) If the interest rate is changed to 0 then the risk neutral probability at each node changes to $(\tfrac{2}{3}, \tfrac{1}{3})$. A similar calculation gives $Z_0^X = 29.648$ which is the same as $\mathbb{E}(\bar{X}) = C_0^E$ using the new equivalent martingale measure.

7.18.

(b) $C_0^A = C_0^E = 18.34$.

(c) $P_0^A = 13.703$ and $P_0^E = 12.04$.

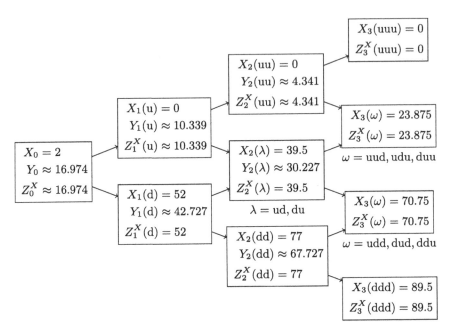

Figure S.4 Payoff, continuation value and price of American option in Exercise 7.17

7.22.
For $t > 0$, $\{\tau \geq t\} = \Omega \setminus \{\tau \leq t - 1\}$. If τ is a stopping time then $\{\tau \leq t - 1\} \in \mathcal{F}_{t-1}$, hence $\{\tau \geq t\} \in \mathcal{F}_{t-1}$.

Conversely, suppose that $\{\tau \geq t\} \in \mathcal{F}_{t-1}$ for each t. Then if $t < T$,

$$\{\tau = t\} = \{\tau \geq t\} \setminus \{\tau \geq t + 1\} \in \mathcal{F}_t.$$

Moreover, $\{\tau = T\} = \{\tau \geq T\} \in \mathcal{F}_{T-1} \subset \mathcal{F}_T$, so τ is a stopping time.

7.24.
τ_1 is a stopping time but τ_2 is not.

7.27.
(a) We know that $\bar{V}^\Phi$ is a $\mathbb{Q}$-martingale for any equivalent martingale measure $\mathbb{Q}$, so Doob's Optional Sampling Theorem gives that $\mathbb{E}_\mathbb{Q}(\bar{V}_\tau^\Phi) = V_0^\Phi$. Since $V_\tau^\Phi \geq 0$ then $\bar{V}_\tau^\Phi \geq 0$ also and so $V_0^\Phi \geq 0$.

(b) If $V_\tau^\Phi(\omega) > 0$ for some ω then we must have $\mathbb{E}_\mathbb{Q}(\bar{V}_\tau^\Phi) > 0$, so $V_0^\Phi > 0$.

7.31.
If π is unfair to the advantage of the writer, the definition gives a strategy Φ such that $V_0^\Phi = \pi$ and for every stopping time (or exercise time) τ, $V_\tau^\Phi \geq X_\tau$ and $V_\tau^\Phi(\omega) > X_\tau(\omega)$ for at least one ω. This is a strictly super-replicating strategy. Similarly if π is unfair to the advantage of the owner by means of a strategy Φ and a stopping time τ, then

$V_0^\Phi = \pi$ and $V_\tau^\Phi \leq X_\tau$ and $V_\tau^\Phi(\omega) < X_\tau(\omega)$ for at least one ω. Thus Φ, τ is a strictly sub-replicating pair for X.

Conversely, a strategy Φ with $\pi = V_0^\Phi$ that strictly super-replicates X is precisely what is required by the definition to show that π is unfair to the advantage of the writer. Similarly a strictly sub-replicating pair Φ, τ is precisely what is required by the definition to show that π is unfair to the advantage of the owner.

7.47.

Suppose that $\tau(\omega) = t$ and $\omega \uparrow t = \omega' \uparrow t$. By definition $V_t^\Phi(\omega) = X_t(\omega)$ for every minimal super-replicating strategy Φ, and, since V^Φ and X are both adapted, $V_s^\Phi(\omega') = V_s^\Phi(\omega)$ and $X_s(\omega) = X_s(\omega')$ for all $s \leq t$. Thus $\tau(\omega') = t$.

7.48.

Suppose that $\sigma(\omega) = \max\{\tau_1(\omega), \tau_2(\omega)\} = t$ and $\omega' \uparrow t = \omega \uparrow t$. Then for $i = 1, 2$ we have $\tau_i(\omega) \leq t$ so $\tau_i(\omega) = \tau_i(\omega')$; thus $\sigma(\omega) = \sigma(\omega')$, which shows that σ is a stopping time.

Since each τ_i is optimal then taking the characterization (3) from Theorem 7.44 we have that $X_{\tau_i} = V_{\tau_i}^\Phi$ for every minimal super-replicating strategy Φ for X. For any ω either $\sigma(\omega) = \tau_1(\omega)$ or $\sigma(\omega) = \tau_2(\omega)$, and it follows that $X_\sigma = V_\sigma^\Phi$ for every minimal super-replicating strategy Φ for X. Hence σ is optimal for X.

The existence of a greatest optimal stopping time for X follows since there are only finitely many stopping times altogether.

7.49.

(a) Suppose that D is attainable with replicating strategy Φ. This means that $V_T^\Phi = D$, and so X^D is attainable by means of the hedging pair (Φ, τ) where $\tau \equiv T$. Conversely, if X^D is attainable by means of a hedging pair (Ψ, σ) this means that $V_T^\Psi \geq D$ and $V_\sigma^\Psi = X_\sigma^D$.

For any ω with $\sigma(\omega) = T$, clearly $V_T^\Psi(\omega) = D(\omega)$. We will show that this is true also if $\sigma(\omega) = t < T$. In this case, then $V_t^\Psi(\omega) = X_t^D(\omega) = 0$. Since $V_T^\Psi \geq D \geq 0$ this means that $V_T^\Psi(\omega') = D(\omega') = 0$ for all $\omega' \in \Omega_{\omega \uparrow t}$. Thus $V_T^\Psi = D$ and D is attainable.

(b) If X_D is attainable, then from (a) any hedging strategy Φ for X_D is a replicating strategy for D and the fair price for X_D is $\pi_{X_D} = V_0^\Phi = \mathbb{E}_\mathbb{Q}(\bar{V}_T^\Phi) = \mathbb{E}_\mathbb{Q}(\bar{D}) = D_0$.

(c) From (a), $\tau \equiv T$ is an optimal stopping time, so it is clearly the greatest. The discussion in (b) shows that if σ is an optimal stopping time with $\sigma(\omega) = t < T$ then $D(\omega') = 0$ for all $\omega' \in \Omega_{\omega \uparrow t}$. This suggests that $\tau_{\min}$ is defined by

$$\tau_{\min}(\omega) = \begin{cases} T & \text{if } D(\omega) > 0, \\ \min\{t : D(\omega') = 0 \quad \text{for all } \omega' \in \Omega_{\omega \uparrow t}\} & \text{if } D(\omega) = 0. \end{cases}$$

This is a stopping time; and it is optimal since for any equivalent martingale measure $\mathbb{Q}$ we have $\mathbb{E}_\mathbb{Q}(\bar{X}_{\tau_{\min}}) = \mathbb{E}_\mathbb{Q}(\bar{D}) = D_0 = \pi_{X_D}$. Thus it is the smallest optimal stopping time for X_D.

From this it is clear that a stopping time σ is an optimal stopping time for X_D if and only if $\sigma \geq \tau_{\min}$.

7.54.

If $V^\Phi \geq Z^X$ then $V^\Phi \geq X$ because $Z^X \geq X$. Conversely, if Φ super-replicates X then $V^\Phi \geq X$ and so $\bar{V}^\Phi \geq \bar{X}$. But $\bar{V}^\Phi$ is a supermartingale, so by Theorem 7.51 $\bar{V}^\Phi \geq \bar{Z}^X$ and so $V^\Phi \geq Z^X$.

7.56.

Let $Y = M' - A'$ be any decomposition of a supermartingale Y such that $A'_0 = 0$ and A' is predictable and increasing, and M' is a martingale. Then $A'_0 = A_0$ and for $t > 0$

$$A'_t - A'_{t-1} = (M'_t - Y_t) - (M'_{t-1} - Y_{t-1}).$$

Taking conditional expectations gives

$$A'_t - A'_{t-1} = \mathbb{E}_{\mathbb{Q}}(A'_t - A'_{t-1}|\mathcal{F}_{t-1}) = \mathbb{E}_{\mathbb{Q}}((M'_t - Y_t) - (M'_{t-1} - Y_{t-1})|\mathcal{F}_{t-1})$$

$$= M'_{t-1} - \mathbb{E}_{\mathbb{Q}}(Y_t|\mathcal{F}_{t-1}) - M'_{t-1} + Y_{t-1}$$

$$= Y_{t-1} - \mathbb{E}_{\mathbb{Q}}(Y_t|\mathcal{F}_{t-1}) = A_t - A_{t-1}$$

which shows that the decomposition constructed in the above proof of Doob's Theorem is the only possible one.

7.59.

The process Z^X for this option was found in Exercise 7.17. Using the formula in the proof of Theorem 7.55 we obtain the Doob decomposition of $\bar{Z}^X$ which appears in Figure S.5. The hedging strategy $\Phi = (x_t, y_t)_{t=0}^3$ is a replicating strategy for the European derivative with payoff $(1+r)^3 M_3$ at time 3, and is obtained using the techniques of Chapter 4. It appears in Figure S.6.

The solution to Exercise 7.17 shows that τ_* is given by the following table:

ω	uuu	uud	udu	udd	duu	dud	ddu	ddd
$\tau_*(\omega)$	3	3	2	2	1	1	1	1

7.62.

Suppose that $\tau_{\max}(\omega) = t < T$. Then if $\omega' \uparrow t = \omega \uparrow t$ we have $\mathbb{E}_{\mathbb{Q}}(\bar{Z}^X_{s+1}|\mathcal{F}_s)(\omega') = \mathbb{E}_{\mathbb{Q}}(\bar{Z}^X_{s+1}|\mathcal{F}_s)(\omega)$ and $\bar{Z}^X_s(\omega') = \bar{Z}^X_s(\omega)$ for all $s \le t$ and so $\tau_{\max}(\omega') = t$.

7.64.

Calculation using the formula for X gives $X_0 = X_1(\mathrm{d}) = X_2(\mathrm{ud}) = X_2(\mathrm{dd}) = 0$; $X_1(\mathrm{u}) = 7.5$, $X_2(\mathrm{uu}) = 7\frac{11}{12}$, $X_2(\mathrm{du}) = 8\frac{7}{12}$. The process Z^X is given by backwards recursion using the defining equations (7.9)–(7.10). Then find the Doob decomposition $\bar{Z}^X = M - A$ using the formulae in the proof of Theorem 7.55. Writing $\hat{M}_t = (1+r)^t M_t$ and $\hat{A}_t = (1+r)^t A_t$ and, for $t < T$ only, $Y_t = (1+r)^t \mathbb{E}_{\mathbb{Q}}(\bar{Z}^X_{t+1}|\mathcal{F}_t) = \frac{1}{1+r}\mathbb{E}_{\mathbb{Q}}(Z^X_{t+1}|\mathcal{F}_t)$ this gives the tree of values in Figure S.7. From this we have the fair price $Z^X_0 = 5\frac{35}{72} \approx 5.486$.

The stopping time $\tau_{\max}$ is found using (7.18). Reading from the tree diagram gives $\tau_{\max}(\mathrm{u}) = 1$ and $\tau_{\max}(\mathrm{du}) = \tau_{\max}(\mathrm{dd}) = 2$. A hedging strategy for X is simply a replicating strategy for the derivative $D = \hat{M}_2$ that is found using the techniques of Chapter 4, and is given in Figure S.8.

7.65.

First construct the process Z^X by backwards recursion using the defining equations (7.9)–(7.10). Then find the Doob decomposition $\bar{Z} = M - A$ using the formulae in the proof of Theorem 7.55. Writing $\hat{M}_t = (1+r)^t M_t$ and $\hat{A}_t = (1+r)^t A_t$ and, for $t < T$ only, $Y_t = (1+r)^t \mathbb{E}_{\mathbb{Q}}(\bar{Z}^X_{t+1}|\mathcal{F}_t) = \frac{1}{1+r}\mathbb{E}_{\mathbb{Q}}(Z^X_{t+1}|\mathcal{F}_t)$ this gives the tree of values in Figure S.9.

From this we can read off $\tau_{\min}$ using the formula $\tau_{\min}(\omega) = \min\{t : Z^X_t(\omega) = X_t(\omega)\}$, giving $\tau_{\min}(\mathrm{u}) = 1$, $\tau_{\min}(\mathrm{du}) = 2$ and $\tau_{\min}(\mathrm{ddu}) = \tau_{\min}(\mathrm{ddd}) = 3$. Similarly

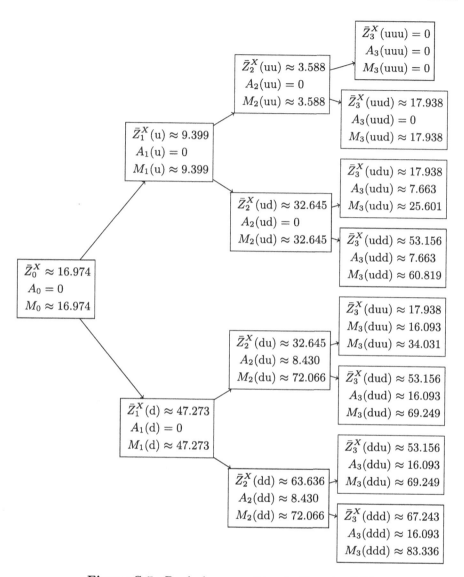

Figure S.5 Doob decomposition in Exercise 7.59

$\tau_{\max}$ is found using the formula (7.18), which is equivalent in the above notation to

$$\tau_{\max} = \begin{cases} T & \text{if } Y_t = Z_t^X \text{ for all } t < T, \\ \min\{t : Y_t < Z_t^X\} & \text{otherwise.} \end{cases}$$

Reading from the tree diagram gives $\tau_{\max}(\mathrm{ud}) = \tau_{\max}(\mathrm{du}) = 2$, $\tau_{\max}(\mathrm{uuu}) = \tau_{\max}(\mathrm{uud}) = \tau_{\max}(\mathrm{ddu}) = \tau_{\max}(\mathrm{ddd}) = 3$. A hedging strategy for X is simply a replicating strategy for the derivative $D = \hat{M}_3$ that is found using the techniques of Chapter 4, and is given in Figure S.10 below.

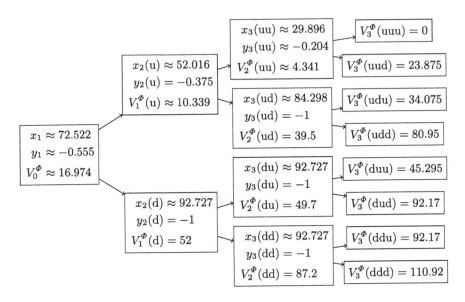

Figure S.6 Hedging strategy in Exercise 7.59

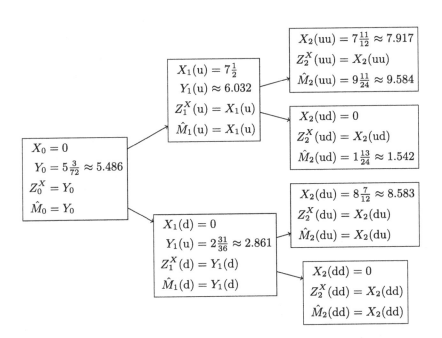

Figure S.7 Pricing the American option in Exercise 7.64

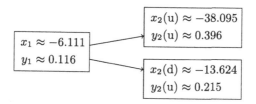

Figure S.8 Hedging strategy in Exercise 7.64

7.68.

Clearly $\bar{Z}_T^X = \bar{X}_T$. For $t < T$ the definition of Z_t^X is equivalent to

$$\bar{Z}_t^X = \max\Big\{\bar{X}_t, \sup_{\mathbb{Q}\in\mathcal{Q}} \mathbb{E}_\mathbb{Q}\big(\bar{Z}_{t+1}^X | \mathcal{F}_t\big)\Big\}. \qquad (S.7)$$

Thus $\bar{Z}_t^X \geq \mathbb{E}_\mathbb{Q}(\bar{Z}_{t+1}^X | \mathcal{F}_t)$ for all $\mathbb{Q}\in\mathcal{Q}$ so that $\bar{Z}^X$ is a $\mathcal{Q}$-supermartingale, and $\bar{Z}^X \geq \bar{X}$. Now show now by backwards induction that if Y is any other $\mathcal{Q}$-supermartingale that dominates $\bar{X}$ then $Y_t \geq \bar{Z}_t^X$ for all t. At time T

$$Y_T \geq \bar{X}_T = \bar{Z}_T^X.$$

For any $t < T$, the inequality $Y_{t+1} \geq \bar{Z}_{t+1}^X$ means that for all $\mathbb{Q}\in\mathcal{Q}$

$$Y_t \geq \mathbb{E}_\mathbb{Q}(Y_{t+1} | \mathcal{F}_t) \geq \mathbb{E}_\mathbb{Q}\big(\bar{Z}_{t+1}^X | \mathcal{F}_t\big).$$

In addition, since $Y_t \geq \bar{X}_t$ it follows that

$$Y_t \geq \max\Big\{\bar{X}_t, \sup_{\mathbb{Q}\in\mathcal{Q}} \mathbb{E}_\mathbb{Q}\big(\bar{Z}_{t+1}^X | \mathcal{F}_t\big)\Big\} = \bar{Z}_t^X$$

which concludes the inductive step.

7.75.

If Φ super-replicates X then $V_\sigma^\Phi \geq X_\sigma$ for every stopping time σ and if Φ' sub-replicates X then there is a stopping time τ such that $V_\tau^{\Phi'} \leq X_\tau$. For this particular τ we thus have $V_\tau^\Phi \geq V_\tau^{\Phi'}$ and so, taking any equivalent martingale measure $\mathbb{Q}$

$$V_0^\Phi = \mathbb{E}_\mathbb{Q}\big(\bar{V}_\tau^\Phi\big) \geq \mathbb{E}_\mathbb{Q}\big(\bar{V}_\tau^{\Phi'}\big) = V_0^{\Phi'}.$$

From this and the definition of π_X^a and π_X^b we see that $\pi_X^a \geq \pi_X^b$.

7.77.

If Φ super-replicates X^D then in particular $V_T^\Phi \geq X_T^D = D$ so Φ super-replicates D. Conversely, if Φ super-replicates D then $V_T^\Phi \geq D \geq 0$ and so for any $t < T$ we have, for any equivalent martingale measure $\mathbb{Q}$

$$V_t^\Phi = \mathbb{E}_\mathbb{Q}\big((1+r)^{t-T} V_T^\Phi | \mathcal{F}_t\big) \geq 0 = X_t^D$$

and so Φ super-replicates X^D. Consequently $\pi_{X^D}^a = \pi_D^a$.

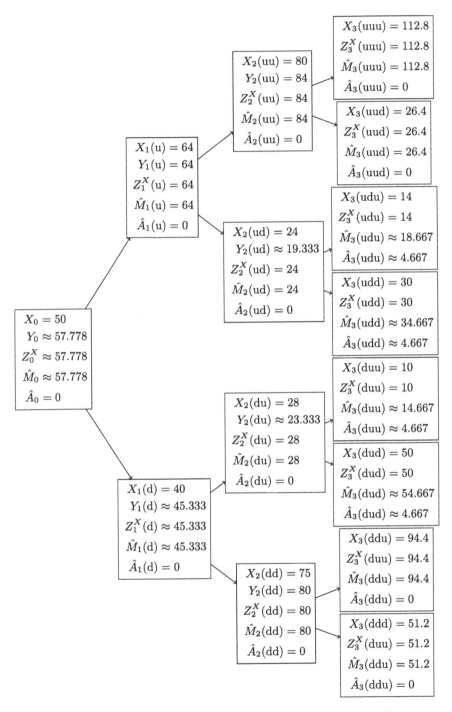

Figure S.9 Pricing the American option in Exercise 7.65

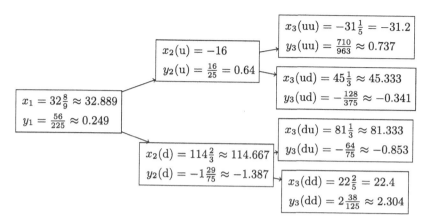

Figure S.10 Hedging strategy in Exercise 7.65

For the bid prices, first we see that if Φ sub-replicates D then it also sub-replicates X^D by means of the stopping time $\tau = T$. Thus $\pi_D^b \leq \pi_{X^D}^b$.

Conversely, if Φ sub-replicates X^D by means of a stopping time τ then $V_\tau^\Phi \leq X_\tau$. For any ω with $\tau(\omega) = T$ this means that $V_T^\Phi(\omega) \leq X_T^D(\omega) = D(\omega)$. For ω with $\tau(\omega) = t < T$ it means that $V_t^\Phi(\omega) \leq X_t^D(\omega) = 0$. The strategy Φ can be modified on $\Omega_{\omega\uparrow t}$ for $s > t$ by making the stock holding zero, giving a strategy Φ' such that $V_s^{\Phi'}(\omega') = (1 + r)^{s-t} V_t^\Phi(\omega) \leq 0 \leq D$ for $\omega' \in \Omega_{\omega\uparrow t}$ and $s \geq t$. Thus Φ' sub-replicates D and $V_0^{\Phi'} = V_0^\Phi$. Consequently $\pi_{X^D}^b \leq \pi_D^b$.

7.86.

Of course this follows immediately from Exercise 7.77 if we look ahead to Corollary 7.88 and Theorem 7.90 below.

However, working directly from the definitions, note that if τ is a stopping time then $\bar{X}_\tau^D \leq \bar{X}_T^D = \bar{D}$ so

$$\sup_{Q \in \mathcal{Q}} \mathbb{E}_Q\big(\bar{X}_\tau^D\big) \leq \sup_{Q \in \mathcal{Q}} \mathbb{E}_Q\big(\bar{X}_T^D\big) = \sup_{Q \in \mathcal{Q}} \mathbb{E}_Q(\bar{D}) = \pi_D^+.$$

But T is a stopping time and so

$$\pi_{X^D}^+ := \sup_{\tau}\sup_{Q \in \mathcal{Q}} \mathbb{E}_Q\big(\bar{X}_\tau^D\big) = \sup_{Q \in \mathcal{Q}} \mathbb{E}_Q\big(\bar{X}_T^D\big) = \pi_D^+.$$

Similarly

$$\inf_{Q \in \mathcal{Q}} \mathbb{E}_Q\big(\bar{X}_\tau^D\big) \leq \inf_{Q \in \mathcal{Q}} \mathbb{E}_Q\big(\bar{X}_T^D\big) = \pi_D^-$$

and so

$$\pi_{X^D}^- := \sup_{\tau}\inf_{Q \in \mathcal{Q}} \mathbb{E}_Q\big(\bar{X}_\tau^D\big) = \inf_{Q \in \mathcal{Q}} \mathbb{E}_Q\big(\bar{X}_T^D\big) = \pi_D^-.$$

7.98.

$F_X = (10\frac{3}{5}, 22\frac{3}{5})$.

Bibliography

Bingham, N.H., Kiesel, R.: Risk Neutral Valuation: Pricing and Hedging of Financial Derivatives, 2nd edn. Springer, London (2004)

Björk, T.: Arbitrage Theory in Continuous Time, 3rd edn. Oxford Finance. Oxford University Press, London (2009)

Black, F., Scholes, M.: The pricing of options and corporate liabilities. J. Polit. Econ. **81**(3), 637–654 (1973)

Brigo, D., Mercurio, F.: Interest Rate Models—Theory and Practice: With Smile, Inflation and Credit, 2nd edn. Springer, Berlin (2006)

Capinski, M., Zastawniak, T.: Mathematics for Finance: An Introduction to Financial Engineering, 2nd edn. Springer Undergraduate Mathematics Series. Springer, London (2011)

Cox, J.C., Ross, S.A., Rubinstein, M.: Option pricing: A simplified approach. Math. Finance **7**, 229–263 (1979)

Delbaen, F., Schachermayer, W.: The Mathematics of Arbitrage. Springer Finance. Springer, Berlin (2006)

Elliott, R.J., Kopp, P.E.: Mathematics of Financial Markets, 2nd edn. Springer Finance. Springer, New York (2005)

Elliott, R.J., Van der Hoek, J.: Binomial Models in Finance. Springer, New York (2005)

Elton, E.J., Gruber, M.J., Brown, S.J., Goetzmann, W.N.: Modern Portfolio Theory and Investment Analysis, 6th edn. Wiley, New York (2003)

Etheridge, A.: A Course in Financial Calculus. Cambridge University Press, Cambridge (2002)

N.J. Cutland, A. Roux, *Derivative Pricing in Discrete Time*,
Springer Undergraduate Mathematics Series,
DOI 10.1007/978-1-4471-4408-3, © Springer-Verlag London 2012

Föllmer, H., Schied, A.: Stochastic Finance: An Introduction in Discrete Time, 3rd edn. de Gruyter, Berlin (2011)

Hull, J.C.: Options, Futures, and Other Derivatives, 8th edn. Pearson Education, Upper Saddle River (2011)

Jarrow, R.A.: Modelling Fixed-Income Securities and Interest Rate Options, 2nd edn. Stanford University Press, Stanford (2002)

Joshi, M.: The Concepts and Practice of Mathematical Finance, 2nd edn. Cambridge University Press, Cambridge (2008) chapter 3

Platen, E., Heath, D.: A Benchmark Approach to Quantitative Finance. Springer Finance, Springer, Berlin (2009)

Pliska, S.R.: Introduction to Mathematical Finance: Discrete Time Models. Blackwell Sci., Oxford (1997)

Shreve, S.E.: Stochastic Calculus for Finance I: The Binomial Asset Pricing Model. Springer, New York (2004)

Shreve, S.E.: Stochastic Calculus for Finance II: Continuous-Time Models. Springer, New York (2004)

Steele, J.M.: Stochastic Calculus and Financial Applications. Stochastic Modelling and Applied Probability, vol. 45, Springer, Berlin (2003)

Wilmott, P.: Paul Wilmott on Quantitative Finance, 2nd edn. Wiley, New York (2006)

Index

N.J. Cutland, A. Roux, *Derivative Pricing in Discrete Time*,
Springer Undergraduate Mathematics Series,
DOI 10.1007/978-1-4471-4408-3, © Springer-Verlag London 2012